## PATENTS ON LIFE

This volume brings a unique collection of legal, religious, ethical, and political perspectives to bear on debates concerning biotechnology patents, or 'patents on life'. The ever-increasing importance of biotechnologies has generated continual questions about how intellectual property law should treat such technologies, especially those raising ethical or social-justice concerns. Even after many years and court decisions, important contested issues remain concerning ownership of and rewards from biotechnology – from human genetic material to genetically engineered plants – and regarding the scope of moral or social-justice limitations on patents or licensing practices. This book explores a range of related issues, including questions concerning morality and patentability, biotechnology and human dignity, and what constitutes fair rewards from genetic resources. It features high-level international, interfaith, and cross-disciplinary contributions from experts in law, religion, and ethics, including academics and practitioners, placing religious and secular perspectives into dialogue to examine the full implications of patenting life.

Thomas C. Berg is James L. Oberstar Professor of Law and Public Policy at the University of St. Thomas School of Law. He is a leading scholar of Church–State relations and a leader in exploring the intersection of intellectual property, religion, and social ethics.

Roman Cholij is Research Associate at the Von Hügel Institute for Critical Catholic Inquiry, St Edmund's College, University of Cambridge and a Chartered Trade Mark Attorney. He has written two books, several book chapters, and numerous articles on various aspects of law and of theology.

Simon Ravenscroft is the Arts Outreach Fellow at Magdalene College, and a Research Associate of the Von Hügel Institute for Critical Catholic Inquiry at St Edmund's College, University of Cambridge. He has published on a range of topics across the fields of theology, philosophy, and social theory.

# Patents on Life

RELIGIOUS, MORAL, AND SOCIAL JUSTICE ASPECTS OF BIOTECHNOLOGY AND INTELLECTUAL PROPERTY

Edited by

THOMAS C. BERG

University of St. Thomas (Minnesota) School of Law

ROMAN CHOLIJ

St. Edmund's College, University of Cambridge

SIMON RAVENSCROFT

Magdalene College, University of Cambridge

CAMBRIDGE
UNIVERSITY PRESS

**CAMBRIDGE**
**UNIVERSITY PRESS**

University Printing House, Cambridge CB2 8BS, United Kingdom

One Liberty Plaza, 20th Floor, New York, NY 10006, USA

477 Williamstown Road, Port Melbourne, VIC 3207, Australia

314–321, 3rd Floor, Plot 3, Splendor Forum, Jasola District Centre, New Delhi – 110025, India

79 Anson Road, #06–04/06, Singapore 079906

Cambridge University Press is part of the University of Cambridge.

It furthers the University's mission by disseminating knowledge in the pursuit of education, learning, and research at the highest international levels of excellence.

www.cambridge.org
Information on this title: www.cambridge.org/9781108428682
DOI: 10.1017/9781108659802



First published 2020

Printed in the United Kingdom by TJ International Ltd. Padstow Cornwall

*A catalogue record for this publication is available from the British Library.*

ISBN 978-1-108-42868-2 Hardback

# Contents

# Contributors

**Margo A. Bagley** is Asa Griggs Candler Professor of Law, Emory University School of Law, and Hardy Cross Dillard Professor of Law, University of Virginia School of Law. Among her recent publications is the chapter 'Towering Wave or Tempest in a Teapot? Synthetic Biology, Access & Benefit Sharing, and Economic Development', in Susy Frankel and Daniel Gervais, eds., *Intellectual Property on the Internet and the Connection with Human and Economic Development* (Victoria University Press, 2017).

**Thomas C. Berg** is James Oberstar Professor of Law and Public Policy, University of St. Thomas School of Law (Minnesota). He is the author of, among other works, 'Intellectual Property and the Preferential Option for the Poor', *Journal of Catholic Social Thought* 5 (Winter 2008): 193–233.

**Michael J. Broyde** is Professor of Law at Emory University and Projects Director of its Center for the Study of Law and Religion. He and Steven Weiner have been studying and writing on matters of Jewish law together for forty years; their most recent article together was 'Understanding Rights in Context: Freedom of Contract or Freedom from Contract?', *Journal of the Beth Din of America* 1 (Spring 2012): 48–65.

**Roman Cholij** is a Research Associate of the Von Hügel Institute for Critical Catholic Inquiry at St Edmund's College, University of Cambridge, and a chartered trade mark attorney. Among his publications in the fields of theology and law is 'IP in Christian Law', *Intellectual Property Quarterly* 3 (2012): 137–148.

**Stephen M. Colecchi**, D.Min., served as the Director of the Office of International Justice and Peace of the United States Conference of Catholic Bishops from 2004 to 2018. Among his various publications, he is author of the book *In the Footsteps of Jesus: Resource Manual on Catholic Social Teaching* (United States Conference of Catholic Bishops, 2005).

**Paul J. Heald** is the Richard W. and Marie L. Corman Research Professor at the University of Illinois. His other publications on patenting life forms include 'Veggie Tales: Pernicious Myths about Patents, Innovation, and Crop Diversity in the Twentieth Century', *Illinois Law Review* (2012): 1051–1102 (with Susannah Chapman).

**Michael A. Kock** is founder of *dr. kock consulting*, an IP and strategy consulting service, and was from 2007 to 2017 Head of Intellectual Property for Syngenta Crop Protection AG. His recent publications on plant IP issues include (with Floris Ten Have), 'The "International Licensing Platform – Vegetables": A Prototype of a Patent Clearing House in the Life Science Industry', *Journal of Intellectual Property Law and Practice* (2016).

**Kathleen Liddell** is the Herschel Smith Lecturer of Intellectual Property Law and Director of the Centre for Law, Medicine and Life Sciences at the University of Cambridge. She is the author of *The Limits of Consent* (Oxford University Press, 2009).

**Monsignor Osvaldo Neves de Almeida**, an expert in international public law, was an official of the Secretariat of State of the Holy See, with wide-ranging responsibilities from 1992 until his retirement in 2017. He is currently based in Rio de Janeiro, Brazil, where he continues to be an advisor to the Secretariat of State. He is a priest of the Prelature of Opus Dei.

**Simon Ravenscroft** is the Arts Outreach Fellow at Magdalene College and a Research Associate of the Von Hügel Institute for Critical Catholic Inquiry at St Edmund's College, University of Cambridge. His publications include 'Charité, amour et don dans l'oeuvre d'Ivan Illich', in *Ivan Illich, l'alchimiste des possibles*, Martin Fortier and Thierry Paquot, eds. (Lemieux Editeur, 2016): 165–211.

**Christopher Rennie-Smith** is a European patent consultant, and former chairman of a Board of Appeal and member of the Enlarged Board of Appeal at the European Patent Office.

**Mohammed El Said**, PhD, is Professor in International Trade and Intellectual Property Law, Lancashire Law School, University of Central Lancashire. His recent publications include 'TRIPS-Plus, Public Health and Performance-Based Rewards Schemes Options and Supplements for Policy Formation in Developing and Least Developed Countries', *American University International Law Review* 31 (2016): 373–444.

**Joshua D. Sarnoff** is a professor of law at DePaul University College of Law and a faculty member in and former director of the Center for Intellectual Property Law and Information Technology (CIPLIT). He is the author of, among other works, 'Patent Eligible Inventions after *Bilski*: History and Theory', *Hastings Law Journal* 63 (December 2011): 53–125.

**Ingrid Schneider** is Professor of Political Science in the Centre for Ethics in Information Technologies at the Department of Informatics, and in the Centre for Bio-Governance, at the Universität Hamburg, Germany. Her most recent publication is *The Politics of Big Data: Big Data – Big Brother?*, edited with Ann Rudinow Saetnan and Nicola Green (Routledge, 2018).

**Katerina Sideri** is Marie Curie Individual Fellow based at the University of Ghent. Her most recent book is *Bioproperty, Biomedicine and Deliberative Governance: Patents as Discourse on Life* (Routledge, 2014).

**Steven S. Weiner** is Deputy General Counsel, Intellectual Property for SRI International, an independent not-for-profit research institute, and a part-time lecturer at the University of Pennsylvania's Wharton School. He and Michael Broyde have been studying and writing on matters of Jewish law together for forty years; their most recent article together was 'Understanding Rights in Context: Freedom of Contract or Freedom from Contract?', *Journal of the Beth Din of America* 1 (Spring 2012): 48–65.

**Paul J. Wojda** is Associate Professor of Theology, University of St. Thomas (Minnesota), and author of, among other things, 'Adoration, Veneration, Plastination: Theo-Liturgical Reflections', in T. Christine Jespersen, Alicita Rodriquez, and Joseph Starr, eds., *The Anatomy of Body Worlds: Critical Essays on the Plastinated Cadavers of Gunther von Hagens* (McFarland & Company, 2009): 211–27.

# Prefatory Note and Acknowledgements

The information in this book is generally current through early 2019, except for short discussions of selected subsequent developments at the conclusions of Chapter 1 and Chapter 16.

* * *

The editors collectively wish to thank the institutions that sponsored or funded this book project and the Patents on Life conference (Cambridge, November 2015) that gave birth to it:

- The Von Hügel Institute for Critical Catholic Inquiry (VHI), at St Edmund's College, Cambridge, and especially John Loughlin, Philip McCosker, Lidia Ripamonti, and Alban McCoy, for hosting the conference and supporting the project in numerous other ways.
- The Terrence J. Murphy Institute for Catholic Thought, Law, and Public Policy at the University of St. Thomas, and especially Elizabeth Schiltz, Seanne Harris, and William Junker, for generous financial and logistical support.
- Porticus UK, with special thanks to Jane Leek, who made Porticus's funding possible.

Thomas Berg adds his thanks to Roman and Simon; to his St. Thomas Law colleagues Megan McNevin, research librarian, and Henry Bishop, administrative assistant, for manuscript preparation; to Deans Robert Vischer and Joel Nichols for institutional support; to Audrey Chapman, Ruth Okediji, David Opderbeck, and Frank Pasquale, who provided inspiration for several of the ideas in the project; to St. Thomas law students Gina Bugliosi, Nick Davis, and Thomas Wheeler, for research assistance; and to his wife, Maureen Kane Berg, for her many varieties of support.

Roman Cholij would like to thank in particular Professor John Loughlin, former director of the VHI, for his support, encouragement, and good judgement that led to the success of the Patents on Life conference. To the speakers and contributors of the conference and, of course, to all the authors of this work, a special debt is owed for sharing the fruits of their research, knowledge, and wisdom. Special thanks also to his colleague Simon Ravenscroft, for his excellent professionalism and dedication to the VHI Patents on Life project from its earliest days; and finally, but not least, to his wife, Louise, for her contributions to the patents conference and constant support throughout the project.

For their vision and good humour, Simon Ravenscroft would like to thank Tom and Roman; for the Murphy Research Fellowship that funded his involvement in a good chunk of this project, personal thanks to the Terrence J. Murphy Institute; and for conviviality, collegiality, and other more profound forms of support during the period in which this book was compiled, he is grateful to James Pearson; Christian Coppa; Ragnar Misje Bergem; Deborah, Massimo, Luca, and Isabella Di Martino; his parents, Tim and Karen Ravenscroft; and above all to Ruth Jackson.

Roman and Simon would like to offer special thanks to Tom for his considerable efforts in compiling the manuscript, but also for his exceptional dedication, professionalism, and heroic patience in many matters besides that have led to this book seeing the light of day.

# Opening

# 1

# An Introduction to Religious and Political Discourse on Life Patents

*Roman Cholij*

## I. BACKGROUND TO THIS BOOK

The inspiration for the collection of essays in this volume, from contributors of widely differing backgrounds, came from an international conference convened at the University of Cambridge in 2015, on the theme 'Patents on Life: Through the Lenses of Law, Religious Faith and Social Justice'. The conference was organised jointly by the Von Hügel Institute (VHI) for Critical Catholic Inquiry at St Edmund's College and the Murphy Institute for Catholic Thought, Law, and Public Policy at the University of St. Thomas, Minnesota. For the first time on British soil, an interdisciplinary group of experts met to discuss intellectual property (IP), representing the Vatican and the Roman Catholic Church; other faith groups; judges, lawyers, and officials from the IP world; academics; and representatives from industry and NGOs. A prime purpose of this extraordinary meeting was to allow different groups in society and stakeholders in the complex world of IP to communicate and to hear and learn from each other's points of view and concerns on diverse but related matters falling under the conference theme. The conference was inspired in turn by a direct invitation to the VHI from the Papal Representative, the Apostolic Nuncio, to the UN in Geneva to produce a report that might assist the Holy See in its interventions as Permanent Observer at the Intergovernmental Committee on Intellectual Property, Genetic Resources, Traditional Knowledge and Folklore (IGC) of the World Intellectual Property Organization (WIPO). The objective was to provide a context for how the patent system could be used in a new, constructive, and positive way to achieve the objectives of a social good – in this case, the fair distribution of benefits from patented inventions reliant on indigenous genetic resources and traditional knowledge.[1] The Vatican's engagement with

[1] *Patents on Genetic Resources? A Catholic Perspective for the World Intellectual Property Organization*, Caritas in Veritate Foundation Working Papers (Chambésy: FCIV, 2013): http://www.fciv.org/downloads/FCIV%20WP3%20Patents%20on%20Genetic%20Resources.pdf. Portions of the report serve as the basis for Chapter 12 in this volume.

WIPO on this theme dates back to 2001 and illustrates, perhaps in a somewhat unexpected way, the relevance of religious discourse in IP debates.

## II. RELIGIOUS DISCOURSE IN THE PUBLIC FORUM: DIALOGUE, NOT EXCLUSION

It is estimated that on 19 May 2018, more than two billion people around the world tuned in to watch the television broadcast of the religious ceremony at which Prince Harry and Meghan Markle were made man and wife (and Duke and Duchess of Sussex) in St George's Chapel at Windsor Castle in the United Kingdom. That same number was exposed to the blockbuster performance of Bishop Michael Curry, head of the Episcopal Church in the United States, who gave the sermon of his life. The Bishop proclaimed to the listening world, and to the bemused royal family and guests and amused bride and groom, that love, with its source in God as revealed by Jesus Christ in his teaching, life, and death, has the transformative power to change governments, to change businesses, and to change the world. Reflecting on the principle that 'love is not selfish and self-centred', Curry proclaimed in his own inimitable style: 'He didn't die for anything he could get out of it. Jesus did not get an honorary doctorate for dying. ... He gave up his life; he sacrificed his life, for the good of others ... for the wellbeing of the world.'[2]

Contrast this with the famous intervention of the atheist Alastair Campbell, Prime Minister Tony Blair's Director of Strategy and Communications, that 'we don't do God',[3] which reflected a tension in British society regarding the relevance of religion and the use of religious language in government. This comment, however, was by no means an authoritative declaration that religion has no place in the public square.[4] Campbell's comment is illustrative of modern liberal political thought's claim that there is a neutral mode of public reason that operates over and above religious differences, as articulated pre-eminently in the twentieth century by John Rawls.[5] However, these views have been seriously challenged by a number of recent philosophers and theologians who argue that public dialogue with religious authorities is necessary in the search for mutual understanding and is a better option for the public good than trying to relegate such views to the private domain.[6] In any case, the

[2] The full text of the sermon can be found on the Archbishop of Canterbury's website: https://www.archbishopofcanterbury.org/speaking-and-writing/sermons/power-love-bishop-michael-currys-sermon-wedding-prince-harry-and.

[3] Colin Brown, 'Campbell Interrupted Blair as He Spoke of His Faith: "We Don't Do God"', *The Telegraph*, 4 May 2003: https://www.telegraph.co.uk/news/uknews/1429109/Campbell-interrupted-Blair-as-he-spoke-of-his-faith-We-dont-do-God.html.

[4] Despite the affirmation by a British Judge of the High Court that the law 'has no place for Christianity'. *R. (on the application of Johns)* v. *Derby City Council* [2011] EWHC 375 (admin); [2011] F.L.R. 2094. This was a case where homosexual rights took precedence over religious beliefs.

[5] E.g. John Rawls, *Political Liberalism* (New York: Columbia University Press, 1993).

[6] See, for example, John Milbank, *Theology and Social Theory*, 2nd ed. (Oxford: Blackwell, 2006); Jeffrey Stout, *Democracy and Tradition* (Princeton: Princeton University Press, 2004).

empirical reality is that the majority of the world's population, including within Western Europe, subscribes to religious or metaphysical views and codes of conduct, nor can this reality be ascribed simply to a pre-scientific and pre-enlightenment mindset.[7] Mahatma Gandhi, the Reverend Martin Luther King Jr, Archbishop Oscar Romero of San Salvador, Archbishop Desmond Tutu of Cape Town, and Pope John Paul II are just some of the high-profile examples of religious figures who have influenced public life in recent times in a significant way for the common good. Scholars have likewise shown that the Roman Catholic Church in particular has been one of the most visible religious forces in the domain of world politics and that this involvement has been an effective force for the global advancement of democracy over the past several decades.[8]

Using a different style from Bishop Curry, the Vatican,[9] with its current head Pope Francis, engages actively with people and institutions from around the world, regardless of religious affiliation, on matters of social and global importance (including patents and biotechnology), referencing its own body of collective experience and social teachings.[10] It is involved at national and international levels, including variously addressing the UN and national governments, participating in the work of UN agencies, addressing the diplomatic corps accredited to the Holy See periodically on social issues of international concern, issuing encyclical letters for the attention of 'all men and women of good will', and much else besides. For example, in the encyclical letter *Laudato Si'*, in which biotechnology is mentioned, Pope Francis wrote, 'Now faced as we are with global environmental deterioration, I wish to address every person living on this planet'.[11] And speaking to a meeting of Judges in the Vatican, he emphasised that 'the Enlightenment slogan that the Church must not be involved in politics has no application here, for the Church must be involved in the great political issues of our day . . .. [P]olitical life is one of the highest forms of charity.'[12]

7 According to Pew Research, in 2015 just Christians and Muslims together made up 4.1 billion of a total global population of 7 billion. The largest religious group is Christianity: http://www.pewresearch.org/fact-tank/2017/04/05/christians-remain-worlds-largest-religious-group-but-they-are-declining-in-europe/.

8 David Hollenbach, S.J., *The Common Good & Christian Ethics* (Cambridge: Cambridge University Press, 2002), 98.

9 The Vatican City, strictly speaking, is an independent state established by the Lateran Treaty of 1929 between the Kingdom of Italy and the Holy See, the latter being the central governing body of the entire Roman Catholic Church. The Pope is head of both.

10 Convenient summaries of the Catholic Church's Social Teaching are found in Pontifical Council for Justice and Peace, *Compendium of the Social Doctrine of the Church* (Rome: PCJP, 2004).

11 Pope Francis, *Laudato si'*, encyclical letter, Vatican website, 24 May 2015, §3, http://w2.vatican.va/content/francesco/en/encyclicals/documents/papa-francesco_20150524_enciclica-laudato-si.html.

12 Pope Francis, 'Statement to the "Judges' Summit on Human Trafficking and Organized Crime"', Vatican City, 3 June 2016, https://w2.vatican.va/content/francesco/en/speeches/2016/june/documents/papa-francesco_20160603_summit-giudici.html.

## III. ENGAGEMENT WITH PATENT ISSUES

A relatively new area of engagement for the Roman Catholic Church at official levels, and indeed for all religious bodies, is the world of IP. The various essays in Part II of this volume attest to the newness of that engagement. This is not because this area of law is particularly new (the first British patent, granted to a Dutchman John of Utynam for special techniques in making stained-glass windows, was granted in 1449). What is new are concerns about the broader impacts on members of society, at different levels, and the implications for social justice on whole sections of society in developing nations, of recent developments in the scope of subject matter protection for holders of IP rights, concentrated in the hands of a few. There is concern about IP's new geographical spread across the world based on legal models from developed countries that are not necessarily suitable everywhere. More particularly, and as discussed throughout this book, a need for attentiveness, if not concern, among churches, civic groups, and other relevant parties arises from, first, the use or misuse of the legal instrument known as TRIPS, attached to the World Trade Organization (WTO) free trade agreement (brought to being in 1994), in which, inter alia, patent protection for imported technologies from developed countries into developing countries is a prerequisite for free trade. Such concerns arise, second, from the implications of patents in the field of biotechnology, the development of which has exploded since the 1970s, which challenge moral and ethical principles and sensitivities. These developments have contributed substantially to (a) the development of a body of literature that questions the assumptions of the traditional rationale of the patenting system – namely, reward and incentive to innovate – and also to (b) political and organisational initiatives and actions that seek to address the real and perceived imbalances of the system when implemented according to narrow legal interpretations.

## IV. NATURE OF INTELLECTUAL PROPERTY AND OF PATENTS: REFLECTIONS ON TRIPS

### A. *IP Basics*

IP protects applications of ideas and information that are of commercial value. IP does protect works of creativity and self-expression of an artist, writer, researcher, or musician that might be non-commercial in nature (so that they are not unfairly and without permission copied and commercialised by others). But the system is adapted to allow holders of IP to reap the fruits of their labour and investment, usually through monetisation, in a proportionate way under conditions and time periods set in law. Although there are many types of IP, the four main pillars of the system are patents (our concern here), which give temporary protection to technological inventions; copyright, which covers a broad category of works protecting inter alia literary, artistic, and musical works, as well as software; trademarks, which protect brands; and design rights, protecting the appearance of things.

## B. TRIPS

TRIPS provides minimum requisites for patent law that each WTO member is in principle required to implement in its territory if not already present. Patents are to be available, for a term of not less than twenty years (Art 33), for both products and processes in all fields of technology (including biotechnology dealing with living matter and pharmaceuticals), providing that the product or process (a) is new, (b) involves an 'inventive step', and (c) is capable of industrial application (Art 27.1). Sufficient information must be disclosed to allow the invention 'to be carried out by a person skilled in the art' (hence facilitating knowledge and technology transfer). Members are barred from disallowing patents just because they originate from outside their territory or because patented products are imported rather than being locally produced. Some exceptions and limitations are permitted, although narrowly construed in practice, one being inventions 'the prevention within their territory of the commercial exploitation of which is necessary to protect *ordre* public or morality' (Art 27.2). All members are required to provide protection for plant varieties, by either patent or an 'effective sui generis system' or a combination of both. The rights conferred on a patent holder are essentially the negative right of excluding an unauthorised party from making, using, offering for sale, selling, or importing a product (or one using a patented process). Holders can sell their rights, assign, transfer by succession or license them.[13]

However, as the various contributors in this volume argue, patent law is just as much a social, moral, and political issue as a legal and technical one.[14] Liddell and Ravenscroft, for example, explain that ethical issues implicitly permeate all aspects of patent doctrine, including definitions of invention, novelty, inventive step, utility, and disclosure. The explicit morality exclusion, optional in TRIPS but incorporated as Article 53(a) of the European Patent Convention (EPC) – although not incorporated in US patent law – in particular should be interpreted as a 'policy lever' which tailors patent law to its overarching objective of promoting socially beneficial inventions in a manner compatible with fair and just social organisation.[15]

Broyde and Weiner and El Said, respectively, introduce perspectives from the Jewish and Islamic traditions on the broader aspects of IP and as applied to

[13] *Trade Related Aspects of Intellectual Property Rights (TRIPS) Agreement*, Annex 1C of the Marrakesh Agreement Establishing the World Trade Organization, Articles 27–29, https://www.wto.org/english/docs_e/legal_e/27-trips_01_e.htm.

[14] See also Shobita Parthasarathy, *Patent Politics: Life Forms, Markets, and the Public Interest in the United States and Europe* (Chicago: University of Chicago Press, 2017).

[15] See also Kathleen Liddell, 'Immorality and Patents: The Exclusion of Inventions Contrary to Ordre Public and Morality', in *New Frontiers in the Philosophy of Intellectual Property*, ed. Annabelle Lever (Cambridge: Cambridge University Press, 2012), 140–71.

contemporary issues in life patenting. Broyde and Weiner present the case that a distinctly *halakhic* theory of intellectual property can be derived from the application of Talmudic unfair competition principles to IP questions, these being equitable in nature and operating to protect broad social interests. This provides a contrast to the personal property basis undergirding much of secular, contemporary patent law. El Said, on the other hand, looks to the principles of *Maslaha* ('public interest') and the objectives of *Sharia*, to suggest how the conflict which arises between accessibility (for example, in relation to genetic innovation) and restrictions on private ownership would be resolved.

Bagley, representing a directly Biblical-based perspective, looks at a particular issue of huge social importance in relation to developing countries: access to medicines. She un-packages the social and ultimately biblical underpinning of compulsory licencing exceptions, which are provided for in TRIPS under very precise conditions (reducing considerably premium prices) but are vigorously opposed in practice by major pharmaceutical companies as a form of disguised IP infringement and IP 'theft'. Bagley provocatively turns the theft concept on its head, proposing that in fact it is the IP holders that should be considered guilty of 'theft' when they effectively act against the interests of those in dire need of medicines and are losing little actual reward because of the compulsory licence arrangement in place.

The power politics dimension of TRIPS has also been decisive in the Vatican's own interventions at the international level on behalf of developing nations and in the cause of social justice, especially in the areas of healthcare and food security (see the Neves and Colecchi chapters). In 2009 this led to what was for some a remarkable, even unwelcome, public statement by Pope Benedict XVI: 'On the part of rich countries there is excessive zeal for protecting knowledge through an unduly rigid assertion of the right to intellectual property, especially in the field of health care.'[16] Understood in context, however, the statement is actually thoroughly unremarkable.[17]

Catholic social doctrine understands intellectual property to have a social function requiring that the interests of patent holders be balanced against the legitimate needs and rights of the rest of society. These rights ultimately rest on the foundation of human dignity and invoke other higher principles, such as human global solidarity and the universal destination of the earth's goods. Thomas Aquinas defined law as 'an ordinance of reason for the common good' (*Summa Theologica*, I-II, 90.4). Patent law therefore should likewise be oriented to the common good, not skewed towards the private good of the patent holder whose interests might at times conflict

[16] Pope Benedict XVI, *Caritas in Veritate*, encyclical letter, Vatican website, 29 June 2009, §22, http://w2.vatican.va/content/benedict-xvi/en/encyclicals/documents/hf_ben-xvi_enc_20090629_caritas-in-veritate.html.

[17] The background to this statement is analysed in Roman Cholij, 'IP in Christian Law', *Intellectual Property Quarterly* 17 (2012), 137–48.

with the common good in his pursuit of maximum monetisation of his market monopoly. These tensions play out most obviously in the field of access to essential medicines in healthcare, as well as in the field of food security. This social function of patents and all IP translates in papal teaching as the 'social encumbrance' or 'social mortgage' on IP ownership – a theme developed in the contributions here of Neves and Wojda.

Critics note that TRIPS came about through the powerful lobbying activities of US corporations to protect their business interests. In this, the corporations proved to be successful. The business models of the pharmaceutical, seed and pesticides, music, film, and software businesses were dependent on powerful IP protection. However, little regard was paid to the actual developmental needs of poorer nations in need of access to the free market, who were effectively bullied into opening up their countries to IP protection for foreign companies, even if this did not benefit their country but did the opposite.[18] These corporations also appeared to pay no heed to the social history of IP globally and in their own country. The United Nations Development Programme (UNDP) noted in a report that

> [m]any of today's advanced economies refused to grant patents throughout the 19th and early 20th centuries, or found legal and illegal ways of circumventing them – as illustrated by the many strategies used by European countries during the industrial revolution .... They formalized and enforced intellectual property rights gradually as they shifted from being net users of intellectual property to being net producers; several European countries ... completed what is now standard protection only in the 1960s and 1970s.[19]

The Lord Justice Jacob, a judge in the Court of Appeal of England and Wales, has commented about TRIPS: '[T]o require nations such as Bangladesh, upon pain of trade sanctions, to have substantial intellectual property laws and a system of enforcement of such laws is to ask for the impossible. Although TRIPS provided some leeway for compliance, that leeway is nowhere near enough. It is hardly surprising that it causes much resentment.' He notes also that TRIPS is a major departure from the driving force of earlier patents (such as the first one granted in the UK). This patent 'was created so as to encourage exploitation within the realm [the teaching of the art of making stained glass by a Dutchman to English apprentices]. TRIPS and the modern worldwide patent system do not do that. They provide the monopoly but not the local industry to go with it. So it does not create an incentive to invest in third world countries. Again that is resented and again that is hardly surprising.'[20] In fact, according to Jagdish Bhagwati, noted Indian economist, the

18 *The TRIPS Agreement and the Global Politics of Intellectual Property Reform in Developing Countries* (Oxford: Oxford University Press, 2009), 8–9.

19 UNDP, *Human Development Report 2001: Making New Technologies Work for Human Development* (Oxford: Oxford University Press, 2001), 102.

20 'Is Intellectual Property the Grit in the Wheels of Industry?', in Robin Jacob, *IP and Other Things: A Collection of Essays and Speeches* (London: Bloomsbury, 2015), 120–130, at 127–8.

WTO has become 'primarily a collector of intellectual property-related rents on behalf of multinational corporations'.[21]

TRIPS actually does include some flexibilities such as delayed implementation for least developed economies (although even then separate 'TRIPS-Plus' trade agreements might override these) and reference to 'Objectives' (Article 7) and 'Principles' (Article 8). Article 7 states explicitly that protection and enforcement of IPR should be 'conducive to social and economic welfare, and to a balance of rights and obligations'. Article 8 in theory allows countries to adopt measures 'necessary to protect public health and nutrition, and to promote the public interest in sectors of vital importance to their socio-economic and technological development'. The practice, however, has proven to be different. At the Doha Round of trade negotiations among the WTO membership (which took place in Doha, Qatar, in 2001, with special focus on fairer treatment of developing countries), the Doha Declaration on TRIPS and Public Health was negotiated against a backdrop of disagreements regarding the extent to which the TRIPS Agreement allows WTO member countries to facilitate access to essential medicines to address pandemics such as HIV/AIDS. The Declaration, which referenced Articles 7 and 8, included matters such as the right to grant compulsory licences and the freedom to determine the grounds upon which licences are granted. Although by 2003 some progress had been made in implementing the spirit of the Declaration (an interim waiver allowing restricted compulsory licencing), overall the Doha Round was unsuccessful in remedying the imbalances that had been addressed.[22]

## v. GENETIC RESOURCES AND PATENTS

The power dynamics at the WTO, essentially dividing the Northern Hemisphere from the Southern, have also been playing out at other international fora, including at WIPO. WIPO itself has, since 2007, followed more closely a trajectory of paying more than lip service to issues of social justice in relation to the global IP regime through adoption of the Development Agenda, reflecting the Human Development policies and programmes of other UN agencies.[23] However, there is currently a stalemate because of differences between representatives of developed and developing countries at the WIPO IGC on Intellectual Property and Genetic Resources in the drafting of a new international legal instrument on the patenting of innovations using genetic resources

21 IPR Commission, *Integrating Intellectual Property Rights and Development Policy* (London: UK Commission on Intellectual Property Rights, 2002), 177, http://www.iprcommission.org/papers/pdfs/final_report/CIPRfullfinal.pdf.

22 The Doha Texts are found on the WHO website at https://www.wto.org/english/tratop_e/dda_e/dda_e.htm.

23 See, for example, Neil Weinstock Netanel (ed.), *The Development Agenda. Global Intellectual Property and Developing Countries* (Oxford: Oxford University Press, 2009), and the WIPO website: http://www.wipo.int/ip-development/en/agenda/.

(GR) and/or Traditional Knowledge (TK).[24] The text would also incorporate within the IP system the objectives of the Convention on Biological Diversity (CBD), an international legal instrument for the conservation and sustainable use of biological diversity, presented for signature at the Rio Earth Summit in 1992. The CBD required that access to genetic resources be on the basis of prior informed consent, and on mutually agreed terms that provide a fair and equitable sharing of the benefits arising out of the utilization of such resources.[25] The Conference of Parties to the CBD, in the 'Bonn Guidelines', further recommended disclosure of the origin of GR and TK in the patent applications of those set to benefit commercially from this IP, thereby facilitating better monitoring of effective arrangements to benefit the holders of GR and TK – often the poor and marginalised. This task was eventually passed on to WIPO.[26]

## VI. THE CONTRIBUTION OF GENETIC RESOURCES AND TRADITIONAL KNOWLEDGE TO MEDICINE

It is uncontroversial that historically the value of GR and TK to medicine and healthcare has been significant. It has been estimated that 33 per cent to 50 per cent of modern pharmaceutical preparations were derived originally from plants.[27] A recent testimony of the ongoing importance of plant-based medicines is the award in 2015 of the Nobel Prize for Medicine to Tu Youyou, the first Chinese woman ever to receive such an award, for the discovery of the most successful anti-malarial drug on the market. Tu's team found information in a 1,600-year-old text indicating that people in 400 CE used sweet wormwood (*Artemisia annua L.*), which is known in Chinese as *qinghaosu*, to treat malaria with some success. The team used this information to create the anti-malarial drug.[28]

Many genetic resources that possess significant economic and social value are located in territories (such as Brazil or Peru in the case of Amazonian forest-derived products) that have been inhabited since time immemorial by native communities

24 The work and background of the IGC is available online: http://www.wipo.int/tk/en/igc/.

25 UN, *Convention on Biological Diversity*, 5 June 1992, Article 1, https://www.cbd.int/convention/text/default.shtml.

26 For a fuller discussion, see Chapter 12 and the report, *Patents on Genetic Resources?* (cited in note 1). A supplementary agreement to the CBD, and deriving from it, the Nagoya Protocol on Access to Genetic Resources and the Fair and Equitable Sharing of Benefits Arising from their Utilization (ABS), came into effect on 12 October 2014. This aims to facilitate ABS but is independent of the work of the WIPO IGC.

27 An example is digitalis, a popular cardiac medication, identified as the active component of the foxglove leaf. For other examples, see Ryan Abbott, 'Documenting Traditional Medical Knowledge', WIPO, March 2014, http://www.wipo.int/export/sites/www/tk/en/resources/pdf/medical_tk.pdf.

28 See Rhodi Lee, 'Malaria Cure Based on Traditional Chinese Medicine Wins Tu Youyou Nobel Prize', *Tech Times*, 12 October 2015, http://www.techtimes.com/articles/93754/20151012/malaria-cure-based-on-traditional-chinese-medicine-wins-tu-youyou-nobel-prize.htm.

who possess extensive knowledge of their traditional medicinal use. Such knowledge often provides 'clues' to researchers within the pharmaceutical industry for the development elsewhere of prospective new therapeutic drugs.

Genetic resources linked to traditional knowledge held by native communities – the indispensable 'clue' to the potential proven medicinal uses of a herb, for example – can potentially save private companies large amounts of money in research and development costs. The genetic resource could be a plant, fungus, microorganism, or animal poison/toxin (such as from snakes, frogs, and toads) from which substances could be extracted and isolated. The search for new drugs based on GR and TK is known as 'bioprospecting' by its protagonists. However, the unilateral economic exploitation of these resources and knowledge, without benefit sharing with those providing the original GR and TK, is also considered by many non-commercial groups as a misappropriation and termed 'biopiracy'.[29]

Chapter 12 of this book, 'Genetic Resources and Patents: In Search of Ethical Solutions to Global IP Discord' (Cholij), develops these themes and that of mandatory disclosure of GR in patent applications as sought by countries of the South, but mostly resisted by countries of the North, in the text-based negotiations of a new WIPO IGC Treaty. It explores the reasons for the resistance, but also the more compelling reasons for change in light of the 'new thinking' on the social functions of IP, and the religious social ethics doctrines of the Roman Catholic Church incorporated in its body of Catholic Social Teaching.

## VII. BIOTECHNOLOGY

Beyond the distributive justice issues of the exploitation of biological material, ethical concerns have grown most prominently in recent times in the rapidly expanding research fields of bioscience and biotechnology and the patenting thereof. These developments involve living matter in the form of whole or parts of organisms, genes, genomic sequencing, genetic engineering, and synthetic biology, and the responsible use of the new precision gene editing tools of CRISPR-Cas9. These concerns revolve around the meaning and implications of patent ownership over living matter, the social and moral consequences of pushing back the boundaries of what can be patented, and the disjuncture between moral and legal acceptability for certain types of technology. Critics question in particular the path to government approbation of reward for controversial biotechnologies through patents, restriction of access

[29] See ETC Group, Patents and Biopiracy, http://www.etcgroup.org/issues/patents-biopiracy: 'Biopiracy, a term originally coined by ETC Group, refers to the appropriation of the knowledge and genetic resources of farming and indigenous communities by individuals or institutions that seek exclusive monopoly control (patents or intellectual property) over these resources and knowledge.' The literature on biopiracy is very extensive. See, e.g., Daniel F. Robinson, *Confronting Biopiracy: Challenges, Cases and International Debates* (London: Earthscan, 2010); Cynthia M. Ho, 'Biopiracy and Beyond: A Consideration of Socio-Cultural Conflicts with Global Patent Policies', *University of Michigan Journal of Law Reform* 39 (2006), 433–542.

to essential genetic information that cannot be 'worked around' (with the restriction on research and diagnostic tools dependent on such information), and the social and environmental implications of monopolies over self-replicating living material, especially in agriculture.

## VIII. BACKGROUND TO THE BIOTECHNOLOGY INDUSTRY

Although the biotechnology industry has grown exponentially since the 1970s and 1980s, it is not a modern phenomenon. Selective breeding of plants and animals, for example, has been used by farmers and breeders since time immemorial. This, along with techniques for grafting and creating hybrids, can be considered a proto-form of biotechnology, showcased in the Green Revolution of the 1930s to 1960s, which saw massive increases in agricultural production worldwide. Through these traditional techniques, the genetic makeup of living organisms were being changed by man, although in a 'natural' manner that did not involve interventions at a molecular level. And this is not to mention 'natural' genetic mutations that constantly occur which can be harnessed.

The watershed moment leading into the modern era of the industry occurred in 1972 with the advent of DNA recombinant technology, more popularly known as 'genetic engineering', from the research of two Californian scientists, Herbert Boyer and Stanley Cohen. DNA recombinant technology is a process in which a gene from one organism is spliced into the chromosome of another organism, such as a bacterium like E. coli, so that the recipient organism will produce the protein coded for by the spliced gene. Genetic sequences are produced that would otherwise not be found in the genome. Hallam Stevens, who has documented the development of the biotech industry, writes:

> Cohen and Boyer's discovery formed the basis for what became the biotech industry. Globally the combined revenue of this industry was over $250 billion in 2013. . . . But the industry continues to grow rapidly: it includes not only the sale of drugs and therapeutics, but also genetic tests, personalized medicine, the sale of technologies such as DNA chips and sequencing machines, trade in patents, the running of global clinical trials, and the use of biotechnologies in agriculture.[30]

## IX. LIVING THINGS TREATED AS CHEMICALS FOR PATENTING PURPOSES

Inventions traditionally are thought of in terms of mechanical or engineered objects, or perhaps new chemical or pharmacological combinations that lead to new

[30] Hallam Stevens, *Biotechnology and Society. An Introduction* (Chicago: University of Chicago Press, 2016), 47. The first licensed drug generated using recombinant DNA technology was recombinant human insulin that replaces animal-derived substitutes.

therapies. The extension of the concept to living organisms, or their parts, has challenged the IP system, but has also shown that the system has developed in response to industry pressures. The biotech industry argues that it needs to secure adequate protection through patents against unauthorised copying or reverse engineering, given the large high-risk investments involved in research and development and seeing products through the regulatory stages before going to market. It is estimated, for example, that bringing a new plant biotechnology-derived trait to market costs around $136 million and takes on average thirteen years.[31] Critics, however, argue that industry has bullied legislators and the interpreters of legislation to accommodate the IP system to industry's full demands, often without adequate consultations and considerations of the overall effects on society or the true need for patent incentives to enable innovation in the field.

The defining moment in the United States that helped to significantly boost investments in biotechnology – but at the same time drew the world of religion for the first time into the technical arena of patents – was the landmark 1980 decision of the US Supreme Court *Diamond v. Chakrabarty*.[32] The case, which involved a human-made genetically engineered bacterium capable of breaking down crude oil, is treated in this volume by Wojda and Heald. The Court ruled that a modified, man-made microorganism could be classed as a 'composition of matter' or 'an article of manufacture' for patenting purposes and therefore be treated as a patentable invention. Living things could be treated as if they were just chemicals.

The scope of patent protection in the United States had now been officially extended by the courts, rather than explicitly by the legislature, not only to microorganism *products* but to microorganisms themselves.[33] The *Chakrabarty* decision emboldened the biotechnology industry to argue that patents should issue on seeds, whole plants, animals, tissues, cells, genes, and other materials that had commercial value. In congressional hearings on the Biotechnology Competitiveness Act (1988), expert witnesses testified that the United States was locked in a 'global race against time to assure our eminence in biotechnology', a race in which the USPTO had an important role to play.[34]

31 Philips McDougall, 'The Cost and Time Involved in the Discovery, Development and Authorisation of a New Plant Biotechnology Derived Trait', Consultancy Study for Crop Life International, September 2011, https://croplife.org/wp-content/uploads/2014/04/Getting-a-Biotech-Crop-to-Market-Phillips-McDougall-Study.pdf.

32 447 U.S. 303 (1980).

33 Europe, however, had in fact for a number of years already accepted patents for microorganisms. For example, in 1975, the German Federal Supreme Court ruled that microorganisms were patentable. Also, the UK patent office had already granted a patent on Chakrabarty's oil-eating bacterium during the 1970s without publicity. But it was only in 1988 that the EPO granted the first patent on plants.

34 Mark Sagoff, 'Patented Genes: An Ethical Appraisal', *Issues in Science and Technology Online: Perspectives* (1998), http://online.sfsu.edu/rone/GEessays/PatentedGeneEthics.htm.

## X. REACTIONS OF RELIGIOUS GROUPS TO THE NEW SCOPE OF PATENT PROTECTION

Wojda's and Heald's chapters describe the objections of a moral nature, from civic and religious groups, to the effects of the *Chakrabarty* decision. Essentially, they argued that allowing patents on life forms effectively commodified life or living things, demeaning their significance. Living matter, in religious thinking, is a creation of God and not of man. The riposte from industry and the law profession, on the other hand, is that these objectors, lay and religious, have (1) confused the notions of legally patentable matter with naturally occurring matter and (2) confused patenting with owning (since patents do not confer legal ownership of the invention but just give the right to prevent others from using or possessing it).

Although it is undoubtedly true that some of the interventions of religious leaders post-*Chakrabarty* have been more 'prophetic' than sophisticated, this is not to say that valid messages have not been communicated.[35] Indeed, as Sarnoff explains in his essay on a related issue, exclusions from the subject matter of patents for scientific and natural discoveries and for fundamental, abstract ideas were justified historically on religious and deontological as well as utilitarian moral grounds. Crossing the line between discovery and invention, as traditionally understood, therefore has religious and moral overtones. To claim the 'given' of living matter in any form as a human invention has moral relevance.

Legally, of course, property ownership is distinct from patent ownership. Yet in fairness to those with intuitive reactions against patenting of life forms, one of the important rights of real property ownership is to exclude others from using or possessing, just as with patents. Therefore, the two concepts are not totally different, even if they are distinct.[36] But ultimately and more importantly, serious matters of public policy are being raised, even if implicitly, by these interventions. Questions raised include that of the appropriateness of the judiciary becoming instruments of societal change without proper accountability. An example that moral critics present is the giving of rights to patent owners over self-replicating patented seed, enabling them in law to sue farmers for replanting or selling new seed from their own legitimately grown plants.

## XI. A CASE STUDY: MONSANTO

Monsanto (now a wholly owned subsidiary of Bayer) is the largest biotechnology seed company in the world; among other things, it has dominated the market in genetically modified (GM) seeds. One of its best-selling products is 'Roundup'

[35] Cf. Annabelle Lever, 'Is It Ethical to Patent Human Genes?', in *Intellectual Property and Theories of Justice*, ed. Axel Gosseries, Alain Marciano, and Alain Strowell (New York: Palgrave Macmillan, 2008), 246–64; Audrey R. Chapman, 'Religious Contributions to the Debate on the Patenting of Human Genes', *University of St. Thomas Law Journal* 10 (2013), 650–81.

[36] Lever, 'Is it Ethical?'

herbicide containing glyphosate, which when used on its GM seeds (such as for soybeans, canola, corn, alfalfa, sugar beets, and cotton) selectively only kills the weeds. The seeds are bioengineered using recombinant DNA technology (called 'Roundup Ready' technology). The herbicide and the technology have been hugely successful in the market, but along with the success, the company has filed some 147 patent infringement suits against farmers in the United States, based on the alleged IP 'theft' of replanting saved, naturally reproduced seed. Saving seed from one's crop for replanting, exchange, or selling has always been a traditional right of farmers before the advent of patented seed.[37] It now arguably constitutes infringement (and breach of contract), as well as arguably fostering a climate of suspicion amongst neighbouring farmers. On its own website, in defence of its legal actions against farmers, Monsanto approvingly notes that 'when one farmer sees another farmer saving patented seed, they will often report them'.[38] Even pollen drift that leads to contamination of a non-GM neighbouring field could be the basis for legal action to force royalty payments.[39] Unsurprisingly, Monsanto has been the subject of much critical commentary.[40] The corporation in turn defends its position in public media campaigns, including the media pages on its website, attempting to demonstrate that there is more than one side to the complex issue of GM seed protection.

## XII. DIALOGUE, NOT CONFRONTATION

In this book, Michael Kock, the former global head of IP for another giant biotech seed company, Syngenta, also presents the industry case for patenting seeds. However, expressing sensitivity to the legitimate concerns of other players and stakeholders, he writes: '[T]he current confrontational debate needs to evolve to a collaborative dialogue based on trust. Recognizing that both small and large players, seed companies, farmers and politicians have a role to play and can only answer the huge challenges in a collaborative approach is key.' Interestingly, and unusually for an industry representative and patent lawyer, Kock refers to the approach of Pope Francis approvingly (thereby also implicitly recognising the value of religious interventions in IP debates). In his encyclical *Laudato Si'* of 24 May 2015, the Pope discusses GM seeds and then has this to say (§135):

37 A fuller discussion of farmer privileges and plant breeder rights requires consideration of the international sui generis right of UPOV for the protection of plant varieties and the International Treaty on Plant Genetic Resources for Food and Agriculture (ITPGRFA), both of which are outside the scope of our chapter.

38 'Saved Seeds and Farmer Lawsuits', Statement of 11 April 2017, https://monsanto.com/company/media/statements/lawsuits-against-farmers/.

39 A much cited case is that of Percy Schmeiser, a Canadian canola farmer sued for infringement in Canada but claiming his Roundup Ready plants were the result of contamination. *Monsanto Canada Inc.* v. *Schmeiser* [2004] 1 S.C.R. 902 (can).

40 For a biblically based critique, see Margo Bagley, 'The Wheat and the (GM) Tares: Lessons for Plant Patent Litigation from the Parables', *University of St. Thomas Law Journal* 10 (2013), 683–713.

> Certainly, these issues require constant attention and a concern for their ethical implications. A broad, responsible scientific and social debate needs to take place, one capable of considering all the available information and of calling things by their name. It sometimes happens that complete information is not put on the table; a selection is made on the basis of particular interests, be they politico-economic or ideological. This makes it difficult to reach a balanced and prudent judgement on different questions, one which takes into account all the pertinent variables. Discussions are needed in which all those directly or indirectly affected (farmers, consumers, civil authorities, scientists, seed producers, people living near fumigated fields, and others) can make known their problems and concerns, and have access to adequate and reliable information in order to make decisions for the common good, present and future. This is a complex environmental issue; it calls for a comprehensive approach which would require, at the very least, greater efforts to finance various lines of independent, interdisciplinary research capable of shedding new light on the problem.

Such a collaborative approach, as articulated by Pope Francis, is also a principle shared by many of the contributors to this book. The premise is that patents on advanced biotechnologies and bioscience present complex issues, having significant impacts on society over and above matters of economics and profits.

## XIII. MORALITY AND PUBLIC ORDER

Social justice issues, such as those affecting farmers (in both developing and developed countries), can be argued to fall under the general umbrella of the patent law heading of 'morality and public order'. The patent system, without adequate legislative checks and balances appropriate to the subject matter of the patent and the social context of its application, can unwittingly become an instrument of social harm in its consequences, where the only normative rules are the rules of the market. In the European system of patenting there is an explicit, if underused, exception to patenting in the 'morality and public order' exclusion enshrined in Article 53 of the European Patent Convention, examined closely here by Liddell and Ravenscroft and Rennie-Smith. As the former argue, potentially this could be a tool to be used expansively to address imbalances against the public interest. Rennie-Smith in his review of the jurisprudence of the Boards of Appeal at the European Patent Office (EPO) shows how the exclusion has worked in practice concerning controversial biotechnology inventions. He points out a distinction between exploitation of the invention and exploitation of the patent, which effectively, he says, gives the exclusion narrower possibilities. He nonetheless argues that its application can be enhanced by greater engagement on the part of religious organisations and other non-commercial entities, following procedures already available.

## XIV. CONTROLLING THE SCOPE OF PATENTABILITY: DNA SEQUENCING AND GENE PATENTS

Ethical objections can be raised as much toward undue restrictions of technical knowledge through lack of inventiveness as toward the manner by which an invention might be exploited.

Some religious authorities have made public their views in this regard too. An example concerns the patenting of gene sequences during the unfolding of the international Human Genome Project (HGP; 1990–2003), an initiative that is now leading to very important advances in medicine, agriculture, animal welfare, and other areas.[41] Speaking on 28 October 1998 to the Pontifical Academy of Sciences, a body of leading scientists of international renown (including the recently deceased Cambridge scientist Stephen Hawking),[42] Pope John Paul II pointedly expressed his disapproval of patenting discoveries of the human genome: '[W]e rejoice that numerous researchers have refused to allow discoveries made about the genome to be patented. Since the human body is not an object that can be disposed of at will, the results of research should be made available to the whole scientific community and cannot be the property of a small group.'[43] We can interpret this statement, in the context of the Pope's whole speech, as expressing the conviction that the human genome in its natural state is part of the common heritage of mankind, as well as the conviction that patenting human body parts, however small, is a form of commodification that attacks human dignity.

The Pope's views in fact echoed many other critics of biotech patents, who base their arguments on fundamental principles of patent law, in particular those of discovery and non-obviousness, and its traditional objectives. The US National Human Genome Research Institute, for example, currently endorses patents on new inventive methods of DNA sequencing but draws the line at the restriction on basic genetic information, claiming this should fall outside patenting. Such restrictions threaten to inhibit or unduly constrain biomedical research and the translation of research discoveries to clinical applications. The Pope's statement in fact quickly followed on the so-called Bermuda Principles of the leaders of the HGP, which set forth the expectation that all DNA sequence information should be released into publicly available databases within twenty-four hours of being generated. Despite this, it is estimated that about a fifth of the human genome is subject to patent claims. Once a gene sequence is claimed, it is impossible to 'invent around' the sequence, as it occurs in nature; that result arguably undercuts one of the very reasons for patent law, namely to stimulate further innovation.[44] The UK Nuffield

41 Leroy Hood and Lee Rowen, 'The Human Genome Project: Big Science Transforms Biology and Medicine', *Genome Medicine* 5 (2013), 79; Stevens, *Biotechnology and Society*, 174–91.

42 On the Academy, see the official website, http://www.casinapioiv.va.

43 'Address to the Plenary Session "Human Genome"', 28 October 1998: http://www.casinapioiv.va /content/accademia/en/magisterium/johnpaulii/28october1994.html.

44 National Human Genome Research Institute, 'Intellectual Property and Genomics', https://www .genome.gov/19016590/intellectual-property/.

Council on Bioethics, funded by the Wellcome and Medical Research Council, stated in 2002 that patents on genes should be the exception rather than the rule. It criticised patent office practice for being too lax in applying the criteria of novelty and inventiveness and for considering genetic information in DNA sequences as if it were equivalent to chemical compounds.[45]

These lines of thinking were given a dramatic boost in the United States by the Supreme Court's ruling in *Association for Molecular Pathology v. Myriad Genetics*, where it was explicitly affirmed that an isolated naturally occurring DNA segment is a product of nature and as such cannot be patented (thereby also invalidating several thousand human gene patents that were already issued by the USPTO). In contrast, so-called complementary DNA (cDNA) – an artificial product designed to mirror the coding parts of genes – is eligible for protection, as it is not naturally occurring. Scholars have nonetheless noted problem areas in this judgment, and the decision applies only within the United States. In Australia, for example, the High Court has ruled, on purely technical grounds, that not even cDNA is patentable, this still being regarded as genetic information per se rather than a 'manner of manufacture' as required in Australian patent law.[46] Interestingly, in an earlier instance to the ruling in 2014, the Full Federal Court, emphasised that '[t]his case is not about the wisdom of the patent system' and that it is not the role of the courts to decide 'whether, for policy or moral or social reasons, patents for gene sequences should be excluded from patentability'.[47] At the time, even the media had campaigned against the Myriad patents with newspapers running titles such as 'Claim for gene an affront to nature' and 'Genetic inheritance belongs to all humanity'.[48]

In addition to objections relating to restriction of access to genetic information, where most biotechnology patent jurisprudence has occurred (at least in Europe) is in regard to the intrinsic morality of the biological 'invention' itself. Rennie-Smith's chapter walks us through some major cases at the EPO, where he was formerly President of the Enlarged Board of Appeal. These cases involved plants, animals, and human embryos, such as, respectively, the *Plant Cells* case, concerning genetically engineered plant cells which were resistant to a class of herbicide; the Harvard *Oncomouse* case, concerning a genetically modified mouse for use in cancer research; and *Use of Embryos/WARF*, which forbade the patenting of a claim to human embryonic stem cell cultures, which could only be prepared by a method

45 Nuffield Council on Bioethics, *The Ethics of Patenting DNA: A Discussion Paper* (London: NCOB, 2002).

46 For an overview of the US case, see Emma Barraclough, 'What Myriad Means for Biotech', *WIPO Magazine*, August 2013, http://www.wipo.int/wipo_magazine/en/2013/04/article_0007.html. For Australia, see *D'Arcy* v. *Myriad Genetics Inc* [2015] HCA 35. Jurisprudence since that time has interpreted the ruling narrowly in favour of industry. For a legal practitioner's perspective, see http://www.spruson.com/the-evolving-patentability-genetic-material-australia-post-myriad/.

47 *D'Arcy* v. *Myriad Genetics Inc* [2014] FCAFC 115.

48 Cf. Adam Denley and James Cherry, 'Decoding Gene Patents in Australia', Cold Spring Harb Perspect Med, 5(1): a020909 (2015 Jan): https://www.ncbi.nlm.nih.gov/pmc/articles/PMC4292076/.

that involved the destruction of the human embryos from which the cultures were derived.

Schneider's chapter, however, questions the examination practice at the EPO in cases involving controversial technologies around stem cells, gametes, and genome editing (using CRISPR). She argues that there should be more transparency and accountability, with a need for a better balance to secure the proper interpretation of the *ordre public* and morality exemption in European patent law in accordance with the purposes and intentions of the European legislator and with the EU's Charter of Fundamental Rights. This is to ensure that patents do not serve as incentives to pursue controversial research of a kind, which is contrary to European values, *ordre public*, and public policy.

Sideri's chapter takes up the theme of the CRISPR technology, focusing in particular on its potential application to germ line modification. She argues that, assuming that for certain medical conditions this technology might actually become ethically permissible, such technology ought to be available outside the confines of the patent system, and that 'delinkage' mechanisms could be established where funding for R&D leading to innovations in the field is not dependent on patents being issued. In reaching this conclusion, Sideri examines closely the various justifications for the patent system.

## XV. CONCLUSION

The rapid pace of developments in biotechnology and bioscience requires a corresponding alertness to these developments by bioethicists, including religious bioethicists. To this end, the Vatican in 1994 established the Pontifical Academy for Life to study and advise on important scientific developments relating to human life. One of its current projects is 'Human Genome Editing: Scientific Perspectives and Ethical and Social Implications'.[49]

In the spirit of the background to this book, we end this introductory chapter with words taken from a recent speech by Pope Francis:

> While the Church applauds every effort in research and application directed to the care of our suffering brothers and sisters, she is also mindful of the basic principle that 'not everything technically possible or doable is thereby ethically acceptable'. Science like all other human activities is conscious that certain limits must be

[49] Pontifical Academy for Life, 'Human Genome Editing: Scientific Perspectives and Ethical and Social Implications', http://www.academyforlife.va/content/pav/en/projects/human-genome-editing.html. Other foundational Vatican documents on bioethics include Pope John Paul II, *Evangelium Vitae*, encyclical letter, 25 March 1995, https://w2.vatican.va/content/john-paul-ii/en/encyclicals/documents/hf_jp-ii_enc_25031995_evangelium-vitae.html; and the document from the Congregation for the Doctrine of the Faith, *Instruction Dignitas Personae on Certain Bioethical Questions*, 8 September 2008, http://www.vatican.va/roman_curia/congregations/cfaith/documents/rc_con_cfaith_doc_20081208_dignitas-personae_en.html.

respected for the good of humanity itself, and that a sense of ethical responsibility is needed.[50]

The editors of this book do not necessarily endorse or agree with all the views or opinions of the individual authors as expressed here, and the authors do not necessarily agree on all points amongst themselves. Nevertheless, all of the writers back the essential proposition that ethically unacceptable technologies or socially harmful practices arising from the use of technology should not be encouraged by awards from the patent system.[51]

50 Pope Francis, 'Address to Participants at the International Conference Organized by the Pontifical Council for Culture on Regenerative Medicine', Sala Regia, 28 April 2018, https://w2.vatican.va/content/francesco/en/speeches/2018/april/documents/papa-francesco_20180428_conferenza-pcc.html. See also the website of the conference to which the Pope spoke: http://vaticanconference2018.com/.

51 An example that has been widely reported in the world press is the case of the Chinese scientist He Jianku, who announced to colleagues at the International Human Genome Editing Summit at the University of Hong Kong, 27–29 November 2018, that he had successfully edited the genes of twin babies, ostensibly to make them HIV-resistant. Following this announcement, the organisers in a statement denounced his 'unexpected and deeply disturbing' claims, adding, 'Even if the modifications are verified, the procedure was irresponsible and failed to conform with international norms'. 'Hong Kong Conference Slams "Deeply Disturbing" Baby Gene-Editing Claim by Chinese Scientist', *Hong Kong Free Press*, 29 November 2018, https://www.hongkongfp.com/2018/11/29/hong-kong-conference-slams-deeply-disturbing-baby-gene-editing-claim-chinese-scientist/.
As this book entered final production, new developments in the US Senate illustrated both the power of the biotech industry and the timeliness of this set of essays. Draft legislation proposed by bipartisan senators would remove from § 101 of the US Patent Act any judicially-created exceptions to patent eligibility including "laws of nature" or "natural phenomena", thereby overturning, among other things, the Supreme Court's *Myriad* decision. At a Senate subcommittee hearing in early June 2019, witnesses included Joshua Sarnoff, a contributor to this volume who testified in opposition. For the hearings and his (and others') testimony, see https://blog.juristat.com/101-reform-senate-hearing. See also Chapter 16, p. 305, in this volume.

# 1

# Life Patents, Law, and Morality

# 2

# Morality, Religion, and Patents

*Kathleen Liddell and Simon Ravenscroft*[*]

## INTRODUCTION

Patent law offers inventors the attractive possibility of up to twenty years' exclusive rights in the commercial marketplace. The invention must be new, non-obvious, explained to other skilled scientists, and have a clear use. Often inventions concern intricate industrial machinery, chemical compounds, textiles, drill bits, electrical circuits, and other dry, technical paraphernalia – but not always. Imagine that an inventor has realised how scientists can add and delete traits in the human genome in a way that can be inherited by future generations (without the consent of these future generations or their parents); or how to increase the fizz in alcopop drinks, with the side effect that drinkers become intoxicated more quickly; or a method to test a fetus more swiftly and inexpensively for disorders, with the option of abortion. Should nation states grant patents to these inventions? In what way should religious and ethical arguments be able to influence this question?

This chapter advances the argument that religious and ethical reasoning have a role to play in policy debates about patent law and also in some patent law cases. In some quarters, this is a controversial suggestion, since patent law is often seen as a highly technical area of law, scientific invention, and economic policy. It is also controversial because ethical reasoning can be fraught and difficult in pluralist societies, given the wide variety of beliefs adhered to, especially when religion enters the equation.

We begin at the most general level, by arguing that in democratic, pluralist societies, moral and religious arguments have a legitimate contribution to make to public discourse *tout court*, including on matters of law and policy. We then make a case for the relevance of religious and moral deliberation for patent law in particular, given that some of the inventions and new technologies that seek patent protection have significant repercussions for wider society, and patent protection is a way of encouraging and supporting their development.

We also consider ways in which religion and ethics might be said to count as relevant evidence not only in patent policy debates, but also in patent proceedings.

* Authors listed alphabetically.

We address this against the background of the 'politics of knowledge' that has arisen in Europe and the United States, and we include reference to the explicit immorality exclusion found in European patent legal systems.

Given that the interpretation and application of the immorality exclusion has been controversial with lawyers, we propose an alternative, potentially more fruitful approach to the exclusion, treating it as a 'policy lever'. Such an approach clarifies the scope and purpose of the exclusion by contextualising it within the broader aims of patent law, to encourage socially beneficial inventions in a manner compatible with fair and just social organisation. Understood in this light, the exclusion offers judges and patent offices room for meaningful moral deliberation, which ought to include receiving non-standard evidence from moral philosophy or religious sources where relevant, yet without requiring them to define what is moral or immoral in an absolute or timeless way. This echoes the arguments raised at the beginning of the chapter about religious and ethical reasoning being incorporated in ways that are sensitive to the plurality of beliefs in liberal democratic societies.

## I. ETHICS AND RELIGION IN PUBLIC DISCOURSE

For philosophers, morality is a framework of norms that organises good life. For some, the fundamental norm they would put forward is to act so that one maximises good consequences, with the nature of 'good consequences' defined in various ways, but most often in terms of utility or pleasure. For others, morality is less concerned with good consequences and more with doing what is right – following one's duty – regardless of the outcome. Immanuel Kant, for example, argued that to be moral is to 'act only in accordance with that maxim through which you can at the same time will that it should become a universal law',[1] as well as to 'act that you use humanity, whether in your own person or in the person of any other, always at the same time as an end, never merely as a means'.[2] Debates between utilitarians or consequentialists (the first group) and deontologists (the second) are well-known and long-standing. To these two ethical frameworks we can add a third – virtue ethics – which lies closer to that of many religious traditions. This involves reasoning from a certain fundamental conception of the human good – for example, the idea that humans realise their nature in loving relationship with God – giving thought to how one's actions contribute to the realisation of that highest good in daily life. This is not the same as consequentialist reasoning, because where the latter thinks of action instrumentally, separating means from ends, the former thinks of action in terms of how it contributes to the development of a person's overall character: a form of life and set of habits that lead

[1] Immanuel Kant, *Groundwork of the Metaphysics of Morals*, trans. Mary Gregor (Cambridge: Cambridge University Press, 1997), 4:421.

[2] Kant, *Groundwork*, 4:429.

towards or away from this highest good. Good habits are defined as virtues, and bad habits as vices (as for Plato, Aristotle, Augustine, or Aquinas).[3]

Most modern societies are pluralist in the sense that these various ethical frameworks and their adherents co-exist and are free to enter into peaceful debate and dialogue with one another. (Note that by 'pluralism' here, we are describing what we take to be an *existing feature* of these societies, rather than advocating a particular philosophical doctrine to be imposed upon them.) With regard to law and public policy, unless we want those who at any particular point in time possess the most social and political power to impose their system of morality on others (a very authoritarian kind of liberal democracy), then we have to find a way of reaching decisions and conclusions that takes account of this fact of moral pluralism. Few, of course, would take issue with this general point; yet the question of how we get there – which is really a question of how moral and religious perspectives, sometimes taken as private matters, should contribute to public discourse – is notoriously tricky. For example, many political philosophers are suspicious of the idea of religious arguments being used in the public forum. One of the most prominent of these was the late, liberal theorist John Rawls. Rawls, and those like him, argued that public reasoning should be restricted to arguments that appeal to principles and ideals that no reasonable person could reasonably reject.[4] As such, forms of argument that rest on comprehensive religious or ethical premises, which others in a pluralistic society might reasonably reject, ought not to be considered legitimate in the public forum. The picture here is of a mode of public discourse in which everyone argues not necessarily from the same terms, but in an effort to find overlapping consensus recognising the inevitability of reasonable disagreement in their high order beliefs. In recent years, the Princeton political and religious philosopher Jeffrey Stout has built on established criticisms of Rawls to propose an alternative to this.[5] He acknowledges that one can be sympathetic to elements of Rawls's argument – the idea of reasoning about public matters in ways that are not biased towards one's race, gender, or economic status, for example – while realising that it is unrealistic and unnecessary to expect citizens to bracket their most deeply held commitments when reasoning in public.[6] For Stout, Rawls's chief failure was being unable to imagine public discourse as constructively *dialogical* – that is, proceeding through an

3 For a summary of these positions and others, see Michael Sandel, *Justice: What's The Right Thing to Do?* (London: Allen Lane, 2009).

4 John Rawls, *Political Liberalism* (New York: Columbia University Press, 1993).

5 Jeffrey Stout, *Democracy and Tradition* (Princeton: Princeton University Press, 2004).

6 Stout, *Democracy and Tradition*, 63–91. Liddell has written that this is not an accurate representation of Rawls's argument, and Stout himself acknowledges that Rawls softened his position in later work (though remains unconvinced regarding the significance of this shift; p. 69). Liddell cites, for example, the second edition of *Political Liberalism* (New York: Columbia University Press, 1996), li–liii, where Rawls updates the view expressed in the first edition on pp. 247–8. It is, nonetheless, certainly fair to say that Rawls thought that comprehensive moral and political reasons appealing to, for instance, religious convictions should be used sparingly in public debates. See further Kathleen Liddell, 'Biolaw and Deliberative Democracy' (unpublished DPhil diss., University of Oxford, 2003), 75–8, 121–7, 144, 247–9.

acknowledgement and elaboration of the differing beliefs of citizens. The restrictions he makes are unrealistic because, as many religious critics have noted, this expectation denies the public scope of many religious beliefs. As Nicholas Wolterstorff has argued, 'It belongs to the *religious convictions* of a good many religious people in our society that *they ought to base* their decisions concerning fundamental issues of justice *on* their religious convictions'.[7] Furthermore, there are many important issues that cannot be resolved 'solely on the basis of arguments from commonly held principles'.[8]

Stout's model rests on what he calls a 'widespread and steady commitment, on the part of citizens, to talk things through with citizens unlike themselves', leading to 'the kind of exchange where each person's deepest commitments can be recognised for what they are and assessed accordingly'.[9] This is helpful in considering the role religious and moral arguments have to play in deliberations over law and policy in democratic, pluralistic societies that wish to properly attend to the particularity of citizens' beliefs. Adherents of comprehensive religious and moral systems may reason openly from these premises, but they must also acknowledge the existence of alternative perspectives, and the fact of reasonable disagreement about what is good and right.

## II. ACCEPTING RELIGIOUS REASONING AS A FORM OF MORAL REASONING IN PUBLIC DEBATES ABOUT PATENTS

We have observed a marked reluctance to allow space for religious reasoning in public discourse about patents. The discipline of 'bioethics' has penetrated many life science policy debates, including the boundaries and contours of patenting ethically controversial inventions. However, it is still relatively uncommon to hear religious perspectives during mainstream public debates about patent protection.[10] This is due to resistance not only from the patent profession, which sees itself as a highly technical field far removed from theology, but also from the fields of bioethics and moral philosophy. Generally speaking, bioethicists ignore religious reasoning and are very swift to disengage attempts to bring it into the work of ethical advisory commissions.[11] One of us (Liddell) researched the way in which ethical advisory commissions do their work and asked during an interview, 'What relevance or

7 Nicholas Wolterstorff, 'The Role of Religion in Decision and Discussion of Political Issues', in *Religion in the Public Square: The Place of Religious Convictions in Political Debate*, by Nicholas Wolterstorff and Robert Audi (New York: Rowman and Littlefield, 1997), 105.

8 Stout, *Democracy and Tradition*, 10.

9 Stout, *Democracy and Tradition*, 10, 297.

10 Exceptions include the conference that initiated this book; the Conference on Bioethical Issues of Intellectual Property in Biotechnology, Tokyo, Japan, 6–7 September 2004; Mark J. Hanson, 'Religious Voices in Biotechnology: The Case of Gene Patenting', in *Claiming Power Over Life: Religion and Biotechnology Policy*, ed. Mark J. Hanson (Washington DC: Georgetown University Press, 1999), 72–111; and statements from religious institutions, such as those referred to by Osvaldo Neves de Almeida in this volume.

11 Exceptions tend to be bioethicists who regard themselves as religious; for example, consider the work of Audrey Chapman, Ronald Cole-Turner, or David Albert Jones.

importance do reasons that appeal to religious beliefs have to an advisory commission?' The response was:

> Zero! It is only important as clues to what people's prejudices are. In a society like ours that is multi-faith and multicultural you can never appeal to what society should do [on the basis] of a particular religion. It would be persuasive only to the numbers of people believing that faith. You have to provide something that's going to convince other people universally. If you produce an argument that's only going to convince Catholics – fine! you've got that market sewn up, but it doesn't say anything to society because nobody who isn't a Catholic is going to have a motive for what Catholic theology requires. So I think, as a point of principle, it should not be done.[12]

In our view, it is unfair to characterise non-religious, reflective bioethical reasoning as completely different from reflective theological reasoning. Bioethical and philosophical thinking rests on assumptions, some of which are tacit, and thus also rely to some extent on a type of faith or faiths and are not fully objective or 'rational'. Similarly, religious reasoning has its own rationales based on its own differing assumptions about the nature of the world and of what it means to be human, and it is reasonable for this to have a place in the design and implementation of public policy. It is often said that the difference between religion and bioethics is that the former vests ultimate authority in 'God' – which is perverse or untrustworthy for those who do not believe in God. The alternative offered in bioethics is that ultimate authority vests in human reason. What is overlooked in this critique is that to place ultimate faith in the powers of human reason is also perverse and untrustworthy for those who see the fallibility of humans. There are other parallels in the way that religious and bioethical moral reasoning occur. They both try to organise and characterise consistent, coherent, and important values, and prescribe how to address situations when these values are in tension; they both refer to key texts/scripts, wise authority figures, practice-based cases, reason, and established traditions of thought and doctrine.

Once one accepts the equity of including bioethical moral *and* religious moral reasoning in pluralist public debates, there is no need to separate religious and moral issues. In the remainder of this essay, we use 'morality' as encompassing religious and bioethical moral perspectives.

## III. MORALITY AND RELEVANT KNOWLEDGE IN PATENT LAW

Moving to the specific policy realm that is the subject of this essay, moral (including religious) argument is more relevant to the highly technical world of intellectual property and patents than some might assume. To begin with, ethical issues

[12] Personal interview, Kathleen Liddell with the commissioner appointed to the Human Genetics Commission, 7 January 2003: Liddell, 'Biolaw and Deliberative Democracy', 125.

permeate many, if not all, aspects of patent doctrine in an implicit way, including definitions of invention, novelty, inventive step, disclosure, and defences and their scope.[13] The interpretation of inventive step is not, for example, simply about economics and competitive markets; it is also about protecting people's freedoms appropriately, specifically their freedom to do and make things, and only limiting that freedom in morally justifiable and proportionate ways. Thus the exclusive rights that follow with a patent are only granted where the inventor realised and disclosed something non-obvious. The notion that ethical considerations are always and already bound up with patent law in this kind of way is often overlooked by those in the patent profession, who argue that the patent system should not mire itself in the subjectivity of morality, on the basis that the profession lacks the necessary moral training or expertise. Broad ethical and moral considerations are here to stay in patent law, whether we like it or not.

Interestingly, moreover, many patent systems also allow explicit ethical objections to what can be patented. The most notable of these appears in Article 53(a) of the European Patent Convention (EPC), stating, 'European patents shall not be granted in respect of inventions the commercial exploitation of which would be contrary to "*ordre public*" or morality'. An explicit provision like Article 53(a) means that patent protection is denied to 'immoral inventions', regardless of how novel or inventive they may be. Significantly, it places the requirement to define 'immorality' squarely on the shoulders of lawyers, including patent examiners and judges. The rule has a very similar cousin in the Biotechnology Directive of the European Union,[14] and that directive has been implemented in national patent laws following the same or similar phrasing. Indeed, ethical exclusions such as this are quite commonplace. In a study led by Lionel Bently for the World Intellectual Property Organization, Shamnad Basheer surveyed seventy-three member countries of the World Trade Organization and found that all but six of them had explicit morality exclusions in their patent regimes along similar lines (the exceptions are Australia, Bangladesh, Canada, Guyana, Uganda, and the United States, to which we return below).[15] EU countries also agreed in 1998, as part of the Biotechnology Directive, on a list of inventions that automatically trigger the morality exclusion. These include processes for cloning human beings; processes for modifying the germ line genetic identity of human beings; uses of human embryos for industrial and commercial purposes; and processes for modifying the genetic identity of animals, which are

[13] See Nuffield Council on Bioethics, *The Ethics of Patenting DNA: A Discussion Paper* (London: NCOB, 2002); Rainer Moufang, 'Patenting of Human Genes, Cells and Parts of the Body? The Ethical Dimensions of Patent Law', *International Review of Intellectual Property and Competition Law* 25 (1994), 487–515.

[14] Directive 98/44/EC of the European Parliament and of the Council of 6 July 1998 on the legal protection of biotechnological inventions, OJ l/213, Art 6(1).

[15] Shamnad Basheer (with Shashwat Purohit and Prashant Reddy), 'Patent Exclusions That Promote Public Health Objectives', Annex IV in Lionel Bently et al., *Exclusions from Patentability and Exceptions and Limitations to Patentees' Rights* (Geneva: WIPO, 2010), 56, 58, 62, 71, 73.

likely to cause them suffering without any substantial medical benefit to man or animals, and also animals resulting from such processes.[16] The list has subsequently been added to the implementing regulations of the EPC under Rule 28(a)–(d), and thus now binds all EPC member countries. It is not, however, meant to be exhaustive or replace the general immorality exclusion.[17] Rather, it provides a straightforward résumé of technologies and activities considered (at the time of the debates) too immoral to justify patent protection, which can be applied without needing to consider the meaning of 'morality' in greater detail. Where an invention falls outside the four categories, it may still be challenged for being contrary to *ordre public* and morality.

It is significant that the United States does not have a provision like Article 53(a), given that the country is quite dominant in policy discussions concerning patent law. This contrast between the US and European systems is linked to further important differences in the extent to which non-technical forms of knowledge, such as moral or religious reasoning, have been able to influence patent proceedings in these regions. Shobita Parthasarathy has explored these differences in detail as an issue of the politics of knowledge – that is, of 'how a policy domain defines relevant knowledge and also how it handles challenges to its definition of relevant knowledge'.[18] She explains how Europe's *ordre public* clause 'inadvertently created an opportunity for outsiders to introduce new forms of knowledge and expertise into the European patent policy domain', with the result that the expertise barrier[19] in Europe has proven notably more porous than in the United States.[20] For example, in the extended European proceedings relating to the Harvard Oncomouse – a type of laboratory mouse genetically modified to develop cancerous tumours so as to make them suitable for cancer research – a wide variety of non-standard evidence was submitted by third parties under Article 53(a). This included ethics articles, reports from medical ethics committees, religious documents and expertise, as well as petitions and surveys.[21] In the United States, by contrast, there is very little legitimate opportunity to raise explicit moral objections.[22]

The question of what is counted as relevant knowledge in patent proceedings is an important one in democratic, pluralistic societies, given the ethical controversies that are often ignited by new inventions, particularly in the biotechnological arena. This relates back to our preceding discussion regarding the forms of moral and

16 Directive 98/44/EC, Art. 6(2).

17 *T-315/03 HARVARD/Transgenic Animals* [2005] EPOR 31, Reasons 6.1.

18 Shobita Parthasarathy, 'Whose Knowledge? What Values? The Comparative Politics of Patenting Life Forms in the United States and Europe', *Policy Sciences* 44 (2011), 267–88 (the direct quote can be found on p. 268).

19 Defined as the rules of a policymaking domain that make it difficult for those without specialist or 'legitimate' knowledge to engage as equals.

20 Parthasarathy, 'Whose Knowledge', 271; cf. *id.*, 268, 270–6.

21 *Id.*, 277–9.

22 *Id.*, 284. Cf. Margo Bagley, 'Patent First, Ask Questions Later: Morality and Biotechnology in Patent Law', *William and Mary Law Review* 45 (2003): 469–547.

religious reasoning that are appropriate in public policy discussions generally. In a recent essay exploring how the Unitary Patent Package[23] will alter the institutional framework within which the morality provisions are adjudicated upon in Europe, Aisling McMahon argues:

> It is crucial that patent law is not viewed as solely a technical/scientific endeavour. The grant of a patent and patent law generally may give rise to fundamental questions of rights and/or important social questions particularly in relation to biotechnological inventions. Indeed, the Biotechnology Directive makes express reference to the protection of fundamental rights and for the need for patent law to be applied in a way which safeguards human dignity.[24]

Patent proceedings, then, should provide a forum for appropriate moral deliberation when it is relevant, alongside consideration of the technical and scientific matters that are their usual forte. Yet the morality provision in Europe has proven highly controversial, and the surrounding confusion has led some to argue that it should not be adopted in the United States, and that it should be applied narrowly or cautiously in Europe, or that the category of 'immorality' should be interpreted legalistically and in a way that is largely devoid of moral content.[25] In the remainder of this chapter, we will summarise these difficulties, before proposing how they might be ameliorated somewhat by an approach that better contextualises the morality provision within the overarching purposes of patent doctrine as a whole.[26]

## IV. INTERPRETING THE EXPLICIT IMMORALITY PROVISION

### A. *Challenges*

Much of the legal controversy around the immorality exclusion has rested on two questions: first, what aspect of the invention or its commercial exploitation is the focus of the moral inquiry; and second, what standard of immorality is required to

23 The UPP will introduce a unitary European patent that will be available across all participating EU States, governed by the newly created Unified Patent Court.

24 Aisling McMahon, 'An Institutional Examination of the Implications of the Unitary Patent Package for the Morality Provisions: A Fragmented Future Too Far?', *International Review of Intellectual Property and Competition Law* 48 (2017): 61–2.

25 Stephen Crespi, 'Biotechnology Patenting: The Wicked Animal Must Defend Itself', *European Intellectual Property Review* 17 (1995): 431–41; Bagley, 'Patent First'; A. M. Viens, 'Morality Provisions in Law Concerning the Commercialization of Human Embryos and Stem Cells', in *Embryonic Stem Cell Patents: European Law and Ethics*, eds. Aurora Plomer and Paul Torremans (Oxford: Oxford University Press, 2009), 85–115; G0002/06 *Wisconsin Alumni Research Foundation (WARF) / Stem Cells* (2009) EPOR 15; Edinburgh Patent EP 06953351 (Opposition Division), 21 March 2003.

26 What follows is a contraction and modification of an argument that can be found in full in Kathleen Liddell, 'Immorality and Patents: The Exclusion of Inventions Contrary to Ordre Public and Morality', in *New Frontiers in the Philosophy of Intellectual Property*, ed. Annabelle Lever (Cambridge: Cambridge University Press, 2012), 140–71.

trigger the exclusion. In other words, how, from a lawyer's perspective, does one assess whether an invention is immoral to the point that, unlike other inventions, it should not be granted patent protection? There are several possible answers to the first question, and this has to do in part with the way the exclusion has been drafted at the international, regional, and national level. The exclusion says that you cannot patent an invention *the commercial exploitation of which* is contrary to morality. It is the word 'exploitation' here that is key. For example, do we mean 'exploitation' in terms of the implications of using the invention? This might mean the immorality of using a letter bomb, or (for some) a drug for euthanasia, or of using genetically modified crops because of the harm they might do to the environment or to those who consume related foods. Or, do we mean exploitation in terms of the implications of *patenting* an invention – that is, where it would be immoral to hand over the commercial and financial benefits of a patent? When you do that, you grant a time-limited property right, you commodify the subject of the patent, and you grant the right to exclude others from using the invention, so as to allow the raising of monopoly profits as a way of rewarding and incentivising innovation. Interpreting the morality provision this way, it might be argued that it is unethical to grant patents over genetic tests for breast cancer, for example, since this would unfairly limit access to those tests, and preclude those in need of a diagnosis from seeking the necessary medical treatment in response. Or do we mean exploitation in the sense of taking advantage of preceding unethical scientific research (a 'poisoned fruit' argument)? This might, for example, be where an invention is based on research with non-consenting subjects, like prisoners or children, or research that has involved cruelty to animals, or research deemed contrary to human dignity because it involves the use of human embryos.

These foci, of course, are not mutually exclusive, and this adds to the difficulty. Arguably, all three could be considered relevant to a judge adjudicating on Article 53(a). Indeed, in the proceedings around *HARVARD/Transgenic Animals* concerning the Oncomouse patent, opponents raised arguments in all three categories: it was argued that the innovation was immoral because of the pain it caused the animals, because it was inappropriate to commodify the animals in such a way (and thus grant exclusive rights to buy, sell, keep, and even create an entire animal population), and because the preceding research was unethical because it was both unnatural and posed a risk to the wider environment if the creatures escaped. Thus far, the European Patent Office (EPO) has only been willing to consider objections of the first type, where the utilisation or performance of the invention would be considered immoral. They have not been willing to consider moral/socio-economic objections that patenting might lead to high prices or restricted access, or that it leads to an inappropriate commodification or propertisation of animals, plants, or embryos.[27] There have been a number of recent

[27] G0001/98 *NOVARTIS/Transgenic Plant*, Reasons 3.9; T0315/03 *HARVARD/Transgenic Animals*, Reasons 4.2.

cases concerning embryo stem-cell lines heard in the Court of Justice of European Union (CJEU), in which the CJEU has, somewhat unwittingly, engaged with the third, 'poisoned fruit' type of argument, but the EPO has thus far been unwilling to do that.[28]

The second question, that of the standard of immorality required to trigger the exclusion, raises a number of sub-issues, of which we will explore two of particular relevance to our main discussion of the role of moral arguments in patent proceedings. The first is how to define 'immorality' in the context of patents; the second immediately follows, and concerns what kinds of evidence may be used to establish that immorality. For example, should we use public opinion surveys, or arguments from moral philosophy or from religious teachings, or human rights declarations, or the findings of ethics committees? This brings us back to the politics of knowledge controversy mentioned previously, regarding what counts as permissible knowledge in this sort of technical legal forum. The EPO itself has swung between a number of different approaches. In 1990 in *T19/90 HARVARD/Oncomouse*, the Technical Board of Appeal took a typically utilitarian 'balancing' approach, weighing possible harms against possible benefits. It was held that the bar to patenting ought to depend 'mainly on a careful weighing up of the suffering of animals and possible risks to the environment on the one hand, and the invention's usefulness to mankind on the other'.[29] In 1994, the EPO took a different line in a case involving the patenting of an isolated human DNA sequence coding for the human protein H2-relaxin, developed from tissue samples obtained from pregnant women. Here it was held that the immorality test should be whether a large majority of the public would regard an invention as 'so abhorrent that the grant of patent rights would be inconceivable'.[30] This is a demanding test, given that the diverse character of European society makes finding anything so widely considered abhorrent rather difficult. Perhaps for this reason the EPO has since shown little enthusiasm for this test, and in 1995 a third approach was offered, as follows:

> The concept of morality is related to the belief that some behaviour is right and acceptable whereas other behaviour is wrong, this belief being founded on the totality of the accepted norms which are deeply rooted in a particular culture. For the purposes of the EPC, the culture in question is the culture inherent in European society and civilisation. Accordingly, under Article 53(a) EPC, inventions the exploitation of which is not in conformity with the conventionally-accepted standards of conduct pertaining to this culture are to be excluded from patentability as being contrary to morality.[31]

28 For more on these cases, see Sideri, this volume, and Schneider, this volume.

29 T19/90 *HARVARD/Oncomouse* (Examination), Reasons 5.

30 T0272/95 *HOWARD FLOREY/Relaxin* (Opposition), Reasons 6.2.1 and 6.3–6.

31 T0356/93 *PLANT GENETIC SYSTEMS*, Reasons 6. See further, Rennie-Smith, this volume.

The key shift here is from the demand for a clear public consensus regarding what is morally abhorrent, to the less demanding test of finding beliefs that are 'deeply rooted' and 'conventionally-accepted'. More recently, the EPO has suggested that of these three tests, the third supplied the working definition of morality, while the balancing approach taken in the first should be preferred in animal manipulation cases, in the absence of reliable evidence on public opinion.[32]

### B. *Article 53(a) as a Policy Lever*

Some of the difficulties and confusions in applying the morality provision can be mitigated by approaching Article 53(a) as a 'policy lever', as defined by Dan Burk and Mark Lemley.[33] In their article, Burk and Lemley argued that patent doctrine should be understood not as a set of rules and protocols to be applied legalistically, but rather as an intricate set of policy levers designed to achieve the fundamental goal of patent law, defined for them as the promotion of new inventions. Writing in a US context, they did not include a morality provision along the lines of Article 53(a) in their list of policy levers, yet it seems highly suited to their framework. Their approach requires first defining the goal of the policy domain to which these levers relate. This is easier in patent law than in other areas, since it is widely agreed that the purpose of patent law is utilitarian: to encourage inventions and new technologies. Yet what is sometimes forgotten is that, put more comprehensively, the goal of granting patents is to encourage *socially beneficial* inventions in a manner compatible with *fair and just* social organisation. The aim of patent law is not, in other words, to encourage any new invention at any cost, but only those that contribute to fair and just social organisation. With this in mind, it becomes easier to see the policy goal of the morality exclusion as to exclude inventions that are undesirable according to this measure.

Taking this line sheds much-needed light on some of the quandaries discussed above, such as the focus of the moral inquiry mandated by Article 53(a). As we noted, the EPO has thus far only considered moral objections to patenting of the first type mentioned, namely, objections regarding the implications of using an invention. However, with the aforementioned approach, this could reasonably be extended to the others mentioned as well. For example, there is no reason why it should not be appropriate for patent authorities to consider the implications of *patenting* an invention, since conferring property status on things that should not – for reasons linked to our common heritage, or to human rights, or human dignity – be regarded as tradable commodities can be considered contrary to fair and just social organisation. Similarly, where a patent might, for example, restrict access to a certain medical treatment, it would be consistent to consider objections arguing that such

[32] T0315/03 *HARVARD/Transgenic Animals* (2005) EPOR 31, Reasons 10.10.

[33] Dan L. Burk and Mark A. Lemley, 'Policy Levers in Patent Law', *Virginia Law Review* 89 (2003), 1575–1696.

a restriction is inconsistent with fairness and justice (though there may also be other instruments that are better suited to address this sort of issue).[34] Moreover, objections based on the 'poisoned fruit' argument, concerning the moral status of the prior research that contributed to the invention, should also be considered, since it is in no way socially beneficial for unethical or dangerous research to be rewarded by the patent system or encouraged by the State.

Crucially, patent authorities in this context would not need to define immorality in an absolute and timeless fashion, but could simply consider how an invention bears in a particular way upon some of the questions just mentioned, while taking into account the social fact of moral pluralism and the empirical information available. The fact that the EPO has, for the time being, approved two tests for immorality (the balancing approach of *HARVARD/Oncomouse* and the conventionally accepted norms approach of *PLANT GENETIC SYSTEMS*) need not be seen as incoherent or problematic. Either test may be appropriate when assessing whether an invention is socially harmful or contributes to unjust social organisation, depending on the nature of the invention in question and the broader moral and social issues implicated. Furthermore, in the making of such assessments, a wide range of evidence should be permissible given the wider social aim of the system (bringing us back to our opening discussion). This should include, where relevant, non-scientific and non-technical evidence, such as from moral philosophy or religious reasoning, or surveys of public opinion, or the findings of ethics committees. In a morally plural society, all of these forms of evidence can help build up a picture of conventionally accepted norms and standards, as well as of the many and various ethical issues that might be raised by a new technology according to the belief systems of different citizen groups. All of this can help judges and patent examiners in their role as skilled legal decision makers. That they might adjudicate in this manner is well within the ordinary duties and capacities of such decision makers, who recognise that law leaves discretion for those in such positions to develop and shape the law in ways that promote desirable goals and behaviour.

## CONCLUSION

We began this chapter arguing that it is unrealistic in pluralistic societies to think that public discourse should be conducted only on the basis of commonly held principles. As such, despite the suspicions of some political philosophers, it is right that substantive moral *and* religious viewpoints have a place in public policy deliberations of all kinds, as part of a broader public dialogue animating the democratic life of an engaged citizenry. Once this is acknowledged, it follows that moral and religious reasoning has an important contribution to make to the specific

[34] Such as Crown use exemptions, compulsory licensing, or research use exemptions (see Liddell, 'Immorality and Patents', 160–1).

policy area of patent law too, given that patent proceedings can involve issues of great moral significance for wider society. This should not be feared; it provides an opportunity for the patent system to fulfil its broader social role, which is to encourage inventions, but only where they are socially beneficial and contribute to just and fair social organisation. The explicit immorality exclusion found in European patent legal systems offers a valuable tool for optimising patent policy when understood in this light, as a policy lever, and is an advisable inclusion in all national patent systems. Countries without such a provision, most notably the United States, are disadvantaged by its omission. To insist, against this, on an approach to patenting that is narrowly scientific and technical, or which lays claim to a spurious scientific 'objectivity', denies the broader role that the patent system is intended to play in society as a whole.

# 3

# Religious and Moral Grounds for Patent-Eligible Subject Matter Exclusions[1]

*Joshua D. Sarnoff*[2]

## INTRODUCTION

Patents in general, and patent-eligible subject matter (i.e., what can be patented) in particular, have significant links to religion. This is especially true in regard to patents on "inventions" derived from discoveries of natural products and their properties and of other natural phenomena. Such inventions have been the subject of important judicial decisions and of public concerns about the propriety of owning nature. For just one recent and controversial example, the US Supreme Court held in *Association for Molecular Pathology v. Myriad Genetics*[3] that isolated DNA (in that case DNA that coded for breast cancer genes) is not a patent-eligible invention, as it is a naturally occurring product, although cDNA (complementary DNA), which is not naturally occurring but contains the same genetic coding sequences (with non-functional DNA sequences removed), is a patent-eligible invention.[4]

As Chris Dent has discussed, the English (Protestant) patent system arose in part in response to Venetian patent practices, the need to fund and develop technology for a religious war against (Catholic) Spain, and developing legal constraints on royal authority.[5] Exclusions from the subject matter of patents were developed for scientific

[1] Much of the material for this chapter is taken from an earlier publication, Joshua D. Sarnoff, *Patent-Eligible Inventions after Bilski: History and Theory*, 63 Hastings L.J. 53 (2011); from amici briefs filed in the US Supreme Court on behalf of law professors in *Brief of Fifteen Law Professors as Amici Curiae in Support of Petitioners*, Ass'n for Molecular Pathology v. Myriad, Inc., 569 U.S. 576 (2013), and *Brief of Nine Law Professors as Amici Curiae in Support of Petitioners*, Mayo Collaborative Servs. v. Prometheus Labs., Inc., 566 U.S. 66 (2012); and from preliminary drafts of Joshua D. Sarnoff, Patents and Morality: Religion, Science, Nature, and the Law (forthcoming).

[2] Too many people contributed inputs to the formation of this chapter to thank adequately. However, I would like to specifically name Professor Thomas Berg, the Terrence J. Murphy Institute, and the von Hügel Institute, and thank them for inviting me to the conference and for encouraging me to prepare this chapter.

[3] 569 U.S. 576 (2013).

[4] *See id.* at 589–96.

[5] *See* Chris Dent, *Religion and the Early Modern Patent System, 1560–1660*, 6 Oxford J. L. & Relig. 580, 582–90 (2017).

and natural discoveries and for fundamental, abstract ideas (science, nature, and ideas).[6] These exclusions were justified historically on Lockean religious, deontological, and utilitarian moral grounds.

This chapter briefly traces the history of such eligibility exclusions, as they developed from changing conceptions of the natural world and within patent law in the United Kingdom and were transplanted (with some differences) to the United States. It then explains why, under American patent law, such discoveries were to be treated as prior art against the applicant who disclosed them. Consequently, in order to validly claim any patent-eligible invention, an applicant had to do more than merely supply a practical (useful) application for his scientific, natural, or abstract discovery. For a claimed invention to be patent-eligible, the creative advance over the prior art constituting the invention had itself to be inventive – that is, the "inventive concept" had to be reflected *in* the application itself rather than in the scientific or natural discovery that the invention embodied and applied to achieve a practical effect.

The chapter concludes with a brief comparative discussion of the line between discovery and patent-eligible invention, and why it matters morally and practically. In particular, the chapter discusses whether to fund scientific discovery through the patent system or through other means, such as government subsidies, prizes, and so on. With respect to such decisions, the moral concerns reflected in the history of eligibility exclusions remain salient, even if their religious origins have largely been forgotten.

## I. EARLY HISTORY OF THE EXCLUSIONS FOR SCIENCE, NATURE, AND IDEAS

### A. *Western Religious Views of Science and Nature from Aristotelianism through Early Medieval Christian Theology to the Eighteenth-Century Enlightenment*

***Pre-Christian Beliefs.*** The idea of technological "invention" by humans (the subject of patent law) is deeply imbedded in religious and philosophical views of the natural world, of human understanding of that world, and of the nature of human creativity. In the Greek rational thought of the logos developed by Aristotle, "[t]he order and intelligibility of the world were due to the eternal forms that existed in its objects and made them definable. Knowledge was the perception of these forms by the mind."[7] Unlike Plato, Aristotle came to believe that the ideal world of the forms existed only in the material

[6] *See, e.g.*, Bilski v. Kappos, 561 U.S. 593, 609–11 (2010) (citing Gottschalk v. Benson, 409 U.S. 63, 67 (1972), and Diamond v. Diehr, 450 U.S. 175, 187 (1981)).

[7] Willis B. Glover, Biblical Origins of Modern Secular Culture: An Essay in the Interpretation of Western History 80 (Mercer Univ. Press 1984).

objects in which they inhered, and that the forms "had no independent, objective existence."[8] Accordingly, Greek scientific thought developed a materialist character while positing eternal, divine substances as largely beyond human sense perception and description.[9]

***Early Christian Beliefs.*** As a result of the Aristotelian vision, science in the medieval period was largely divorced from its earlier ties to religious, sacral experience of the world. "Almost single-handedly [Aristotle] had laid the foundations of Western science, logic, and philosophy. Unfortunately, he also made an indelible impression on Western Christianity. Ever since Europeans discovered Aristotle's writings in the twelfth century CE," they moved away from seeing ultimate reality as something that could be directly experienced (even if "not by reason") and established science as a distinct and secular form of knowledge of the universe.[10] Nevertheless, Christians continued to believe in a God who intervened supernaturally in the material world through miracles.[11] Over time, scientific knowledge increasingly formed the model for knowledge more generally, even if through the nineteenth century it still "did not coincide with knowledge as such."[12]

***Late-Medieval Christian Beliefs.*** In the late Middle Ages, Christian theology began to afford a very different account of nature, and thus of science, from the early medieval beliefs that were based on the earlier, Aristotelian worldview. The 1277 Council of Paris rejected the Aristotelian understanding of essences, establishing the created world as contingent and as depending on continuing acts of God's will for its existence (a doctrine which was called voluntarism).[13] As recognized by the late-medieval nominalists like William of Ockham, since God's will was not restrained by any "necessity of his own nature," the natural order that God established could not "be known by deduction from any principles whatsoever but only by observation or revelation."[14] Thus humans could observe the *potentia ordinata* "of the regularity and order observable in the world" and could thereby derive natural laws without restricting God's *potentia absoluta,* or freedom to act outside of the laws of nature (or to have created the world differently).[15] In this way, the character of nature and of morality had changed, from "law immanent in nature and particular natural objects" to "law imposed on nature by a law-giver who was beyond and outside of nature."[16] Western Christian religion was thus transformed by conscious efforts

8 Karen Armstrong, The Great Transformation: The Beginning of Our Religious Traditions 388 (Alfred Knopf 2006).

9 *See id.* At 388–91.

10 *Id.* at 392–93. *See also id.* At 421–24 (discussing various philosophers).

11 *See* Michael Horace Barnes, Stages of Thought: The Co-Evolution of Religious Thought and Science 132 (Oxford Univ. Press 2000).

12 Jürgen Huberman, Knowledge and Human Interests 3 (Jeremy J. Shapiro trans., Beacon Press 1972) (1968).

13 *See* 1 Reinhold Niebuhr, The Nature and Destiny of Man 80–83 (Prentice-Hall 1964) (1941).

14 Glover, *supra* note 7, at 83. *See also* Barnes, *supra* note 11, at 147.

15 Glover, *supra* note 7, at 85.

16 *Id.* at 91.

"to emancipate science from the shackles of mythology."[17] God was not eliminated, but rather was sought to be proved "the great 'Mechanick' who had brought the intricate machine of the universe into being."[18]

The result of these shifts away from mythological and Aristotelian thought was the development of an empirical scientific method, the purpose of which was to understand God through understanding nature. Isaac Newton thus sought to establish the existence of God through a scientific understanding that the natural world operated according to ordered and rational principles (rational design) that a creative God had established. "I see nothing extraordinary . . . for proving a Deity [from various natural phenomena as] . . . they could hardly arise from any cause purely mechanical . . . they seem to make up that Harmony in the System, which, I explained above, was the effect of Choice rather than Chance."[19] Newton viewed nature as revealing God, and thus as the means to know God and morality. "For Newton, who carried further the tendency present in [Johannes] Kepler and Descartes, Nature was the revelation of God as much as the Bible was, perhaps more so."[20]

***Enlightenment Christian Beliefs.*** By the late eighteenth century, notwithstanding changing theological conceptions and the growth of atheism, nature was understood as "one grand, interrelated design, comprehensible by rational investigation,"[21] the understanding of which would benefit all humans. During this period, invention was still largely understood in the classical sense of uncovering something in nature that had been present all along, through the mechanism of divine providence in permitting human access to such knowledge, even if Protestants more rigidly divided natural from supernatural knowledge.[22] And "[i]f the inventor was no more than God's instrument in bringing His gifts to the community, then he could at most claim user's rights over them."[23]

Thus claiming as one's own origination an invention derived proximately from nature, and seeking property rights in it, was an act of hubris and contrary to morality. The divine creations of science and nature that were revealed to humanity through the efforts of scientists, "those favored mortals . . . who share that ray of

17 Karen Armstrong, A Short History of Myth 124 (Canongate 2005). *See also id.* at 123–25 (citing Francis Bacon, The Advancement of Learning (1605)).

18 *Id.* at 125.

19 Letter from Isaac Newton to Richard Bentley (Dec. 10, 1692), in Isaac Newton's Papers & Letters on Natural Philosophy and Related Documents 289 (I. Bernard Cohen ed., Harvard Univ. Press 2d ed. 2014).

20 Richard S. Westfall, *The Rise of Science and the Decline of Orthodox Christianity: A Study of Kepler, Descartes, and Newton*, in God and Nature: Historical Essays on the Encounter between Christianity and Science 232 (David C. Lindberg & Ronald L. Numbers eds., Univ. of California Press 1986).

21 Christine MacLeod, Inventing the Industrial Revolution: The English Patent System 1660–1800, at 203 (Cambridge Univ. Press 2002).

22 *See, e.g.*, Dent, *supra* note 5, at 594.

23 MacLeod, *supra* note 21, at 198.

divinity we call genius," were intended by God to be freely available. Their divine origins as commons property imposed moral duties to freely share knowledge of nature and thus of God. Scientists were "entrusted by Providence with the delegated power of imparting to their fellow creatures that instruction which heaven meant for universal benefit; they must not be niggards to the world, or hoard up for themselves the common stock."[24] As noted by patent historian Edward Walterscheid, the medieval belief that "genius was a gift of God … largely precluded an earlier development of the concept of intellectual property. For how could one properly seek to obtain commercial value from that which was perceived to have been granted by the grace of God?"[25]

***Romantic-Era Beliefs.*** At the very end of the eighteenth century, however, beliefs regarding human invention and divine provenance of nature were beginning to shift. "[T]he Romantic reconceptualization of the creative process" during the eighteenth century had resulted "in recognition … of the principle that authors should benefit from their works," and reflected "roots in notions of individual self-proprietorship," providing the rationale for thinking of literary productions as personal property.[26] Nature and science as divine commons property similarly began to undergo an incomplete shift from divinely-granted use rights to human private property (which shift has continued for at least the last two centuries).[27] These changes were similar to the concurrent changes from use rights to absolute forms of property in land that had resulted during earlier centuries in the enclosure movement.[28] Nevertheless, it had been a "fixed pole of the debate" in Parliament in 1774 that there were no natural rights in mechanical inventions.[29]

### *B. Lockean Religious Views of the Limits of Property, Particularly for Intangibles*

Although many are aware of John Locke's theory of labor as a theoretical justification for property, few now know of the explicitly theological basis of Locke's theory. There is also an intimate relationship between the theological limits of Lockean property and intangible human "inventions."

***Locke and Christianity.*** Locke began his analysis of property from the basic moral equality of humans as God's creatures of equal station, having "an equal Right to the

24 William Cobbett, The Parliamentary History of England 999 (London, T.C. Hansard 1813).

25 Edward C. Walterscheid, The Nature of the Intellectual Property Clause: A Study in Historical Perspective 39 (2002).

26 Peter A. Jaszi, *Toward a Theory of Copyright: The Metamorphosis of "Authorship,"* 1991 Duke L.J. 455, 468 n.45, 469–71, 472 (1991).

27 *See* MacLeod, *supra* note 21, at 198.

28 *See id.*; James Boyle, *The Second Enclosure Movement and the Construction of the Public Domain*, 66 Law & Contemp. Probs. 33, 34–35 (2003).

29 MacLeod, *supra* note 21, at 198, 220–21.

use of the inferior Creatures, for the comfortable preservation of their Beings."[30] From this equal state of control over a natural God-given commons and from the typicality of the "God-given moral status" of each individual, Locke derived a moral principle of equal human regard.[31] This equal regard forces each individual to take every other individual's duty of self-preservation as having universality, anticipating the Kantian categorical imperative.[32]

Locke famously developed the concept that human labor granted the right to exclusive private appropriation of (private property in) nature, which otherwise was given by God to all humanity to share as commons property. "[T]hough the things of Nature are given in common, yet Man (by being Master of himself, and Proprietor of his own Person, and the actions or Labour of it) had still in himself the great Foundation of Property."[33] Nature was understood as commons property because it was not a human but a divine creation and was not subject initially to human control (dominion).[34] Locke based his natural law conception of property on a Biblical moral imperative for productivity, which cultivation of land achieved: "Have a lot of children! Fill the earth with people and bring it under . . . control."[35] And as with one's body, nature was God-given, which imposed inherent stewardship obligations on the uses humans could make of it.[36]

***Limits of Lockean Property for Intangibles.*** As Justin Hughes has noted, Locke's labor theory debatably may depend on an assumption of "abundance" in nature, so that everyone who labors can obtain "enough and as good" from the commons as anyone else can obtain.[37] This condition (reflecting Locke's concern for equal regard) has been called Locke's "sufficiency" or "equivalency" proviso.[38] Further, Locke found an additional limit to property rights in what has been called his "spoliation" or "waste" proviso:[39]

> The same Law of Nature, that does by this means give us Property, does also bound that Property too . . . As much as anyone can make use of to any advantage of life

30 JOHN LOCKE, TWO TREATISES OF GOVERNMENT § 87, at 243 (Peter Laslett ed., New American Library 1965) (1689).

31 JEREMY WALDRON, GOD, LOCKE, AND EQUALITY: CHRISTIAN FOUNDATIONS IN JOHN LOCKE'S POLITICAL THOUGHT 155–57 (Cambridge Univ. Press 2002). *See* RICHARD HOOKER, OF THE LAWS OF ECCLESIASTICAL POLITY § 8, at 80 (Arthur Stephen McGrade ed., Cambridge Univ. Press 1989) (1593).

32 IMMANUEL KANT, ETHICAL PHILOSOPHY: THE COMPLETE TEXTS OF GROUNDING FOR THE METAPHYSICS OF MORALS AND METAPHYSICAL PRINCIPLES OF VIRTUE, PART II OF THE METAPHYSICS OF MORALS ¶ 421, at 30 (James W. Ellington trans., Hackett Publishing 2d ed. 1994) (1785).

33 LOCKE, *supra* note 30, § 44, at 340–41.

34 *See id.* § 26, at 328.

35 Genesis 1:28.

36 *See* Sibyl Schwarzenbach, *Locke's Two Conceptions of Property*, 14 SOC. THEORY & PRAC. 141, 145–47 (1988); Roberta Rosenthal Kwall, *The Author as Steward "For Limited Times,"* 88 B.U. L. REV. 685, 691–92, 700–708 (2008).

37 Justin Hughes, *The Philosophy of Intellectual Property*, 77 GEO. L.J. 287, 297 (1988).

38 *See, e.g.*, ROBERT NOZICK, ANARCHY, STATE, AND UTOPIA 175 (1974); WALDRON, *supra* note 31, at 172.

39 *See, e.g.*, WALDRON, *supra* note 31, at 170; Hughes, *supra* note 37, at 325.

> before it spoils; so much may he by his labour fix a Property in. Whatever is beyond this, is more than his share, and belongs to others.[40]

For property in both acquired natural materials and enclosed land, to exceed the boundaries of what could be productively used was "to transform properly human liberty into license and thereby violate the highest potential of the species."[41]

Unlike physical property, however, intangible ideas are not part of a commons that can be depleted. Rather, they are part of a metaphorical area that is nonrivalrous and expands with use; intangible ideas thus arguably always meet Locke's sufficiency proviso.[42] For Locke, a person who "leaves as much as another can make use of, does as good as take nothing at all,"[43] and there can be no objection to acquisition that does not prejudice the interests of others.[44] But for the sufficiency proviso to be met with regard to property *in ideas*, the ideas subjected to appropriation must be fungible[45] and the property rights granted in those ideas must not preclude similar ideas from being used by others.

In general, the broader the idea is, the less likely it will be that any similar idea will exist for others to use, either as an idea that already lies within the intellectual commons or as a potential idea that might be added to the commons by an initial "inventor" and appropriator of it.[46] For any nonfungible ideas that are understood to preexist their discovery by humans, the commons is necessarily diminished and others are necessarily prejudiced by its exclusive acquisition and withdrawal from use by granting exclusive rights. These Lockean limits on making property out of ideas apply directly to patent law.

***Lockean Limits on Property in Science, Nature, and Abstract Ideas.*** For preexisting science and nature and for newly discovered and highly abstract ideas, "it is hard to imagine anything 'as good' that could be left for others to discover ... [Some] ideas are so fundamental ... that allowing ownership in them would violate the equality of creative liberty which the proviso also embodies."[47] As recognized by the US Supreme Court in the *Funk Brothers* case, these ideas are the "storehouse of knowledge ... free to all men and reserved exclusively to none."[48] Some ideas simply cannot be "designed around" but must be relied upon. Thus Locke's provisos implied at natural law a prohibition on the patenting of science, nature, and highly general abstract ideas. For such ideas, there are no adequate substitutes.

40 Locke, *supra* note 30, § 31, at 332.

41 Kirstie M. McClure, Judging Rights: Lockean Politics and the Limits of Consent 94 (Cornell Univ. Press 1996).

42 *See* Hughes, *supra* note 37, at 315.

43 Locke, *supra* note 30, § 33, at 333.

44 Waldron, *supra* note 31, at 172.

45 *See* Hughes, *supra* note 37, at 336.

46 *Cf. id.* at 323–25, 327.

47 Wendy J. Gordon, *A Property Right in Self-Expression: Equality and Individualism in the Natural Law of Intellectual Property*, 102 Yale L.J. 1533, 1581 (1993).

48 Funk Bros. Seed Co. v. Kalo Inoculant Co., 333 U.S. 127, 130 (1948).

Locke's spoliation proviso added another limit to any natural law of property in such ideas. Even if creating private property in some ideas would not violate the sufficiency proviso, the *legal* inability of others to use or to build off of such ideas might still violate Locke's *waste* proviso. For Locke, the waste that may result from creating property rights in functional ideas would have been a loss not so much of value to the creator or to others[49] but of the productivity that might otherwise have been made, by preventing others from using the ideas. Stated differently, Locke recognized and limited the opportunity costs to society of granting property rights.

To avoid waste in produced physical goods, Locke depended on trade in the money economy and the willingness of the property owner to sell the unused surplus goods to others without involving "tremendous reallocations of wealth toward the property holders."[50] Otherwise, possession of such property would "[deny] everyone ... the use of them by someone who has no use for them himself, or does not propose to put them to human use."[51] In contrast, private ownership of ideas may fail to account adequately for the waste that results from their non-rivalrous and non-depletable character. Market exchanges for all possible occurrences or uses of science, nature, and broad, abstract ideas would seem impossible, given their ability to be brought into existence and use virtually anywhere.

***Obligations to Freely Disseminate Scientific and Natural Knowledge.*** Sharing knowledge of God's natural laws – science, nature, and ideas – could not reduce a discoverer's own ability to employ them. The theological premise of productivity thus imposed on the discoverer a God-given duty to freely share knowledge of nature.[52] Sharing knowledge would increase cultivation by others, which would expand productivity beyond any surplus the discoverer might generate and trade.

This moral duty to share information with others also followed from Locke's effort, with Richard Hooker's help, to make a case

> for the Golden Rule: "Love thy neighbor as thyself" or "Do unto others as you would have them do unto you." ... Richard Hooker's argument cited by Locke in section 5 of the *Second Treatise* is supposed ... [t]o show that once we acknowledge that no human has a superior status, we have no choice but to treat the needs and desires of others as on a par with our own.[53]

Given equal concern and regard for others, humans would want discoverers to freely and widely share their information regarding nature, so as to better assure human self-preservation in accordance with God's purposes. The Golden Rule thus dictated creating a public domain of science, nature, and ideas. And this meant keeping such discoveries free from private property rights.

49 *See* Hughes, *supra* note 37, at 327–28.
50 *Id.* at 320.
51 Waldron, *supra* note 31, at 171.
52 *Cf.* 1 William C. Robinson, The Law of Patents for Useful Inventions § 25, at 39 (Little, Brown 1890).
53 Waldron, *supra* note 31, at 155.

Consequently, it was commonly recognized by the end of the eighteenth century that property rights in functional ideas (inventions), unlike in literary authorial ideas (published words),[54] did not arise under natural law.[55] Such invention rights could exist only by the positive act of a government – the grant of patents – and were not otherwise recognized at common law.[56] Utilitarian philosophers in the late eighteenth century therefore felt the need to articulate a call for government intervention to create private rights to inventions as incentives for their production and distribution.[57] But these positive law rights did not and – given contemporaneous moral beliefs – could not attach to the discovered science, nature, and ideas themselves. Discoveries were to be freely disseminated for mutual benefit. As aptly put almost a century later by William Robinson, "To benefit by the discoveries of his fellow-men is thus not only a natural right, it is also the natural duty which every man owes to himself and to society; and the mutual, universal progress thence resulting is the fulfillment of the earthly destiny of the human race."[58] Thus only the *applications* of discovered natural products and phenomena or scientific knowledge that the inventor had actually created could be patented.

## II. EXCLUSIONS FOR DISCOVERIES OF SCIENCE, NATURE, AND IDEAS, AND THE LINE BETWEEN INVENTIVE AND NON-INVENTIVE APPLICATIONS

Given the changing conceptions of religion and nature discussed previously, it is no surprise that the line between ineligible discoveries of nature and patent-eligible inventions has changed over time and remains contested. A brief understanding of the history helps show how the religious underpinnings have been lost over time, while the moral concerns remain highly relevant to modern disputes.

### A. *Early Eligibility Doctrine in the United Kingdom: The Watt Steam Engine Case*

In England in 1795, in the famous James Watt steam engine case, *Boulton v. Bull*, Chief Judge Eyre clearly articulated the unanimous view of all the judges that "[u]ndoubtedly there can be no patent for a *mere principle*."[59] Similarly, Judge Buller noted that "an idea or a principle alone would not support the patent."[60]

54 *See, e.g.*, Donaldson v. Beckett, (1774) 98 Eng. Rep. 257 (H.L.), *reversing* Millar v. Taylor, (1769) 98 Eng. Rep. 201 (K.B.).

55 *See, e.g.*, Walterscheid, *supra* note 25, at 203–04.

56 *See Millar*, 98 Eng. Rep. at 230–31 (Yates J.).

57 *See* Harold G. Fox, Monopolies and Patents: A Study of the History and Future of the Patent Monopoly 204 (1947).

58 Robinson, *supra* note 52, § 25, at 39.

59 *Boulton v. Bull* (1795), 126 Eng. Rep. 651, 667 (opinion of Lord Eyre, C.J.) (emphasis added).

60 *Id.* at 662 (opinion of Buller, J.).

Joseph Bramah had argued to the judges against the validity of Watt's patent, stating that the "'works of men begin'" at the point "'where the independent works of God end, who by his own secret principles and methods . . . established the elements and their properties, and stocked the universal storehouse'" of endless changes producible by different combinations and proportions.[61] In contrast, it was acknowledged that patents could issue for the invention of "manufactures" under the 1623 English Statute of Monopolies,[62] which "applied not only to things made, but to the practice of making, to principles carried into practice in a new manner, to new results of principles carried into practice."[63]

The critical issue for the *Boulton* case, on which the judges split, was whether patents could issue for practical methods (processes) of applying scientific principles that were disembodied from any particular arrangement of machinery through which those principles operated.[64] Stated differently, ineligible mere principles – which constituted scientific or natural discoveries or abstract ideas – were distinguished from principles of invention *at least* by requiring practical application and *disputedly* by also requiring additional creativity in the application of scientific or natural discoveries. As stated by Lord Eyre:

> [B]ut for a principle so far embodied and connected with corporeal substances as to be in a condition to act, and to produce effects in any art, trade, mystery, or manual occupation, I think there may be a patent . . . [The patentee] has discovered a practical manner of doing it; and for that practical manner of doing it he has taken this patent. Surely this is a very different thing from taking a patent for a principle; it is not for a principle, but for a process.[65]

In contrast, Judge Buller was concerned that most scientific principles of operation were already well known, and thus patents that were not limited to new *structures* that creatively applied scientific principles in a particular manner would withdraw useful knowledge from the public domain that otherwise was free for all to use:

> [I]f the principle alone be the foundation of the patent, it cannot possibly stand, with that knowledge and discovery which the world were in possession of before . . . [A]ll machines which are worked by steam are worked on the same principle. The principle was known before, and therefore if the principle alone be the foundation of the patent, though the addition may be a great improvement, (as it certainly is), yet the patent must be void ab initio. But then it was said, that though an idea or a principle alone would not support the patent, yet that an idea reduced into

61 MacLeod, *supra* note 21, at 220 (quoting Letter from Joseph Bramah to Sir James Eyre, Lord Chief Justice of the Common Pleas (1797)).

62 Statute of Monopolies of 1623, 21 Jac. 1, c. 3, § 6.

63 *Boulton*, 126 Eng. Rep. at 665.

64 *Compare id.* at 667 (Lord Eyre, C.J.) and *id.* at 659 (Rooke, J.), with *id.* at 482 (Heath, J.) and *id.* at 662, 664–65 (Buller, J.).

65 *Id.* at 667.

practice, or a practical application of a principle was a good foundation for a patent, and was the present case. *The mere application or mode of using a thing, was admitted in the reply not to be a sufficient ground* (a); for on the court putting the question, whether if a man by science were to devise the means of making a double use of a thing known before, he could have a patent for that, it was rightly and candidly admitted that he could not.[66]

The Court of Common Pleas in *Boulton and Watt v. Bull* split 2–2 on this critical question.[67] However, in the subsequent 1799 case of *Hornblower v. Boulton & Watt*, Watt argued that the patent was actually for a specific mechanical improvement (a manufactured steam engine), and not for a disembodied method of preheating steam that any steam engine might use. Accordingly, Watt's claimed invention of improved machines was held patentable.[68]

### B. *Early Eligibility Doctrine in the United States: Statutory Law and the Morse Telegraph Case*

***Constitutional, Legislative, and Administrative Limits on Eligibility.*** At the end of the eighteenth century, the US Constitution vested in Congress the power to grant exclusive rights to "Inventors" for their "Discoveries." Specifically, Congress was granted the power to "promote the Progress of Science and useful Arts ... by securing for limited Times to ... Inventors the exclusive Right to their ... Discoveries."[69] Discoveries of inventors had a particular meaning, distinct from discoveries of science and nature and thus from the "mere principles" discussed in *Boulton*.[70] In the terminology of the time, scientific and natural discoveries were the province of "natural philosophy,"[71] not of the "useful Arts."[72]

The exclusion from patent eligibility for mere principles (natural discoveries or abstract ideas) was also well recognized in the United States by the beginning of the nineteenth century: "[A] patent may be for a new and useful art; but it must be practical, it must be applicable and refer[a]ble to something which may prove it to be useful. A *mere abstract principle is unsusceptible of appropriation by a patent*."[73] As the Supreme Court later explained, "[t]he rule that the discovery of a law of nature cannot be patented rests ... on the more

66 *Id.* at 662–63 (emphasis added).

67 *Id.* at 658–70. *See* Jeffrey A. Lefstin, *Inventive Application: A History*, 67 FLA. L. REV. 565, 578 (2015).

68 Hornblower v. Boulton, (1799) 101 Eng. Rep. 1285, 1291 (K.B.); Lefstin, *supra* note 67, at 578.

69 U.S. CONST. art. I, § 8, cl. 8.

70 *See* WALTERSCHEID, *supra* note 25, at 125–33, 348–57.

71 *See, e.g.*, JOHNSON'S DICTIONARY OF THE ENGLISH LANGUAGE, IN MINIATURE 169 (W.P. & L. Blake, E. Lincoln, 1st Am. ed. 1804).

72 U.S. CONST. art. I, § 8, cl. 8.

73 Evans v. Eaton, 8 F. Cas. 846, 852 (C.C.D. Pa. 1816) (No. 4559) (emphasis added), *rev'd on other grounds*, 16 U.S. (3 Wheat) 454 (1818).

fundamental understanding that they are not the kind of 'discoveries' that the statute was enacted to protect."[74]

As Thomas Jefferson famously explained in 1813, ideas are non-rivalrous and thus are not proper subjects of exclusive rights *at natural law*:

> [Ideas are] incapable of confinement or exclusive appropriation. *Inventions then cannot, in nature, be a subject of property*. Society may give an exclusive right to the profits arising from them, as an encouragement to men to pursue ideas which may produce utility, but this may or may not be done, according to the will and convenience of the society, without claim or complaint from anybody.[75]

In 1790, pursuant to the patent power vested by the Constitution, Congress created a Patent Board and authorized it to grant patents to inventors who had "invented or discovered any useful art, manufacture, engine, machine, or device, or any improvement therein."[76] The Patent Board adopted various restrictions on what qualified as a *statutory* "invention" or "discovery." In particular, the Board excluded the following from patentability: the application of a machine to a new use, changes in materials of construction, changes of form, and the use of previously known implements in combination (such as using a saw and axe together).[77] Implicit in this approach to new uses was the belief that the public's knowledge of a machine entitled it to "any use of which it is susceptible."[78] By providing exclusive rights for applying existing machines to new uses or to new materials, a patent would take this right to use from the public and give it to the inventor.[79] Similarly, mere changes of form or of materials for constructing machines would simply apply the same inventive principle already known by the public, and thus would deprive the public of opportunities to make and use machines already within its grasp, even if they had not yet been constructed.[80]

In 1793, Congress codified at least one of the Board's negative rules of patentability,[81] specifying that "simply changing the form or the proportions of any machine, or composition of matter, in any degree, shall not be deemed a

[74] Parker v. Flook, 437 U.S. 584, 593 (1978).

[75] Letter from Thomas Jefferson to Isaac McPherson (Aug. 13, 1813), in THOMAS JEFFERSON, THE PORTABLE THOMAS JEFFERSON 525, 530 (Merrill D. Peterson ed., Penguin Classics 1977) (emphasis added).

[76] Patent Act of 1790, ch. 7, § 1, 1 Stat. 109, 110.

[77] JEFFERSON, *supra* note 75, at 531.

[78] *Id.* at 531.

[79] *Id.*

[80] *Id. See also* Brown v. Piper, 91 U.S. 37, 41 (1875).

[81] The precise origins of the relevant provision are in doubt. Joseph Barnes proposed the language to Congress, *see* Edward C. Walterscheid, *The Hotchkiss Unobviousness Standard: Early Judicial Activism in the Patent Law*, 13 J. INTELL. PROP. L. 103, 109 (2005), but the same language had previously been enacted in the French Patent Act of 1791. *See* John F. Duffy, *Inventing Invention: A Case Study of Legal Innovation*, 86 TEX. L. REV. 1, 36 (2007). *See generally* Sean M. O'Connor, *The Overlooked French Influence on the Intellectual Property Clause*, 82 U. CHI. L. REV. 733 (2015).

discovery."[82] Congress thus excluded from patent *eligibility* (i.e., as not "inventions" or "discoveries") things that, although novel, had been treated by the Board as not *patentable* due to constructive public possession.[83] Further, Congress thereby prompted courts to restrict patents to new principles employed by particular machines.[84] For similar machines to be patentable, they would have to operate on different principles.[85]

Nevertheless, Congress in 1793 also expanded patent-eligible subject matter to compositions of matter, adopting what is essentially the same terminology as in the modern statute: "any new and useful art, machine, manufacture or composition of matter."[86] Although the contemporaneous meaning of "art" and "useful art" is not entirely clear, at the end of the eighteenth century pure method patents (the critical subjects in *Boulton v. Watt*) were considered ineligible for protection and remained so until the late nineteenth century.[87]

***The Morse Telegraph Case.*** Under the 1836 Patent Act (which preserved the language of the 1793 Act), the Supreme Court reiterated that newly discovered scientific principles simply were not patent-eligible. Only particular configurations of machinery applying those principles constituted patentable "inventions."[88] Discussing a then-recent English patent case,[89] the Supreme Court noted, in upholding most of Samuel Morse's patent claims but rejecting one that was not limited to specific means:

> [T]he court [in *Neilson v. Harford*] at first doubted, whether it was a patent for any thing more than the discovery that hot air would promote the ignition of fuel better than cold. And if this had been the construction, the court, it appears, would have held his patent to be void; *because the discovery of a principle in natural philosophy or physical science, is not patentable.*
>
> But after much consideration, it was finally decided *that this principle must be regarded as well known*, and that the plaintiff had *invented a mechanical mode of applying it* to furnaces; and that his invention consisted in interposing a heated receptacle, between the blower and the furnace, and by this means heating the air after it left the blower, and before it was thrown into the fire . . .

82 Act of Feb. 21, 1793, ch. 11, § 2, 1 Stat. 318, 321 (emphasis added).

83 *See* John F. Duffy, *Rules and Standards on the Forefront of Patentability*, 51 WM. & MARY L. REV. 609, 624–25 (2009).

84 *See, e.g.*, Evans v. Eaton, 20 U.S. (7 Wheat.) 356, 431 (1822); *Evans*, 8 F. Cas. at 852. *See also* Walterscheid, *supra* note 81, at 108–15.

85 *See* Joshua D. Sarnoff, *The Historic and Modern Doctrines of Equivalents and Claiming the Future, Part I (1790–1870)*, 87 J. PATENT & TRADEMARK OFF. SOC'Y 371, 386–91 (2005).

86 Act of Feb. 21, 1793, ch. 11, § 1, 1 Stat. 318, 319.

87 *Compare* Corning v. Burden, 56 U.S. (15 How.) 252, 268–69 (1853) and *Evans*, 16 U.S. (3 Wheat.) at 517; with Cochrane v. Deener, 94 U.S. 780, 787–88 (1876).

88 O'Reilly v. Morse, 56 U.S. (15 How.) 62, 115–16 (1853).

89 Neilson v. Harford, (1841) 151 Eng. Rep. 1266, 1267–68.

> Undoubtedly, the principle that hot air will promote the ignition of fuel better than cold, was embodied in this machine. *But the patent was not supported because this principle was embodied in it* . . .
>
> *If the Court of Exchequer had said that Neilson's patent was for the [scientific] discovery*, that hot air would promote ignition better than cold, and that he had an exclusive right to use it for that purpose, there might, perhaps, have been some reason to rely upon it. *But the court emphatically denied this right to such a patent.*[90]

Three aspects of this discussion from the *Morse* decision are critical. First, as Professor Jeffrey Lefstin has carefully detailed, the Supreme Court in *Morse* was mistaken in its understanding of the English *Neilson* case (including a later Scottish case and an earlier English case), which had determined that a patent-eligible invention required only a new *practical* application, and not a creative and inventive application, of a scientific discovery.[91] Further, just one year earlier, in *Le Roy v. Tatham*, the Court had adopted the English *Neilson* approach and had held that applying a newly discovered or well-known natural property to a new and useful practical purpose constitutes invention (and thus patent-eligible subject matter).[92]

Second, the basis for finding eligibility in *Morse* was not the fact that the machines embodied and applied a newly discovered scientific principle. Rather, what grounded patent eligibility was the *inventive* (in other words, a creative as well as novel) *application* of that discovered principle, which was embodied by the *particular mechanical mode* invented by Morse. Stated differently, the principle of *invention* (the specified "means . . . to produce the result or effect"[93]) capable of being patented by Morse was a principle in addition to that of the discovery of the natural properties of electromagnetism; it was a principle regarding the configuration of machinery or the arranged steps of a specific process that employed the scientific discovery. The eligible "invention" was not merely the application of the scientific discovery to a new, practical use.[94]

Third, contrary to *Le Roy*, merely applying a new scientific discovery to exploit its properties in a non-inventive (but novel and practical) manner would not have justified granting a patent.[95] This is because the Court in *Morse* had implied that even a newly discovered scientific principle "must be regarded as well known."[96] Although the 1836 Act had eliminated the 1793 Act's language stating that changes to form or proportions were not discoveries,[97] the courts had continued to require a new "principle of invention" to establish patent eligibility.[98] "Invention" thus required creativity in the

[90] *Morse*, 56 U.S. (15 How.) at 115–16 (emphases added).

[91] Lefstin, *supra* note 67, at 580–92.

[92] Le Roy v. Tatham, 55 U.S. (14 How.) 156, 175–77, 188 (1852).

[93] *Morse*, 56 U.S. (15 How.) at 119.

[94] *See id.* at 132 (Grier, J., dissenting).

[95] *See* Tilghman v. Proctor, 102 U.S. 707, 726–27 (1880).

[96] *Morse*, 56 U.S. (15 How.) at 116.

[97] *See* Act of July 4, 1836, ch. 357, § 6, 5 Stat. 117, 119; Act of Feb. 21, 1793, ch. 11, § 1, 1 Stat. 318, 319.

[98] *See* Sarnoff, *supra* note 85, at 387–89; Edward C. Walterscheid, *Novelty and the* Hotchkiss *Standard*, 20 FED. CIR. B.J. 219, 255 (2010).

application of novel scientific discoveries (i.e., more than merely applying the discoveries to a new use).[99]

### C. *Later Eligibility Doctrine in the United States: Natural-Material Derivatives Cases*

Later decisions of the Supreme Court built upon *Morse*'s requirement for a creative advance *in the application* of new discoveries, imposing difficult line-drawing distinctions between creative new things and uncreative variations of newly discovered but excluded things. In 1887, in *Hartranft v. Wiegmann*, the Supreme Court articulated a standard, in the context of import duties rather than patent law, for when a new "manufacture" had been created from natural materials.[100] To qualify as a manufacture required "a new and different article, having a distinctive name, character, or use … The application of labor to an article, either by hand or by mechanism, does not make the article necessarily a manufactured article."[101] The Court in *Hartranft* rejected the idea that human or machine labor applied to improve a natural thing was sufficiently creative (transformative) to "invent" a new thing.[102]

Similarly, in 1892 in *Ansonia Brass & Copper Co. v. Electric Supply Co.*,[103] the Court denied eligibility to new uses of existing processes or products that were "analogous" to those already known (and that could include new uses suggested by scientific discovery of the properties of those existing processes or products).[104] As the Court noted, "nothing is better settled in this court than that the application of an old process to a new and analogous purpose does not involve invention, even if the new result had not before been contemplated."[105] The analogous use doctrine of *Ansonia Brass* was implicitly applied to natural discoveries by the Ninth Circuit court of appeals in *Wall v. Leck*:[106]

> [E]mployment of [a scientific discovery] in the modes or through the instrumentalities by which it is applied in nature is a mere imitation of what every man is able to perceive and reproduce as well as [t]he [patentee]. All endeavors to confine it to himself are at once futile and unjust … The laws … do not permit any man to exclusively use the conditions which are the gifts of nature, simply because he was the first one to discover its value. Not until some new instrument or method is

99 *See, e.g.*, Pamela Samuelson and Jason Schultz, *"Clues" for Determining Whether Business and Service Innovations Are Unpatentable Abstract Ideas*, 15 Lewis & Clark L. Rev. 109, 119–20 (2011) (citing Giles S. Rich, *Principles of Patentability*, 28 Geo. Wash. L. Rev. 393, 393–94 (1960)).

100 121 U.S. 609, 615 (1887).

101 *Id.*

102 *Id.*

103 144 U.S. 11, 18 (1892).

104 *See id.* at 18.

105 *Id. See also* Lovell Mfg. Co. v. Cary, 147 U.S. 623, 637 (1893).

106 66 F. 552 (9th Cir. 1895).

contrived for its direction *towards ends which it cannot naturally accomplish* does his *creative* genius manifest itself.[107]

*Leck* appears to have adopted its reasoning from William Robinson's 1890 patent treatise regarding the need for creativity in the application of natural discoveries:

> Man can discover and apply [a natural principle], but his employment of it in the modes or through the instruments *by which it is employed in nature are mere imitations* of what every man is able to perceive and reproduce as well as he. Not until some new instrument or method is contrived for its direction *towards ends that it cannot naturally accomplish*, does his creative genius manifest itself; but even these new instruments and methods effect no change within the [principle] itself, *nor do they render [the discovered principle] in any manner of his inventive act.*[108]

Nevertheless, other courts later found to be eligible useful products that were isolated and purified from natural materials. In *Parke-Davis & Co. v. H.K. Mulford & Co.*,[109] in 1911, Judge Learned Hand famously held that purified adrenaline (claimed as a product and not for its use) was an inventive new thing based on its novel commercial and therapeutic use (whether or not it reflected a change in kind rather than of degree).[110] However, the Supreme Court implicitly called this line of reasoning into question in 1935 in *American Fruit Growers v. Brogdex.*[111] There, borax-treated fruit was held not to be an eligible invention, as it did not possess "a new or distinctive form, quality, or property" (notwithstanding the added preservative function).[112] Based on *Brogdex*, the Commissioner of Patents contemporaneously indicated that the Patent Office would no longer grant patents for purified natural products like Louis Pasteur's purified yeast.[113]

In 1948, just before the modern US Patent Act was adopted in 1952, the Court held in *Funk Brothers Seed Co. v. Kalo Inoculant Co.*[114] that novel combinations of inoculants discovered to have certain non-inhibitory properties were not eligible inventions. Instead, they were "phenomena of nature,"[115] that "serve the ends nature originally provided and act quite independently of any effort of the patentee."[116] Although efforts were made to reverse *Funk Brothers* in the 1952 Act, they were inconclusive and did not adopt the specific language proposed to do so.[117]

107 *Id.* at 558 (emphases added) (citing, inter alia, 1 Robinson, *supra* note 52, § 136).
108 1 Robinson, *supra* note 52, § 136, at 195 (emphases added).
109 195 F. 95 (S.D. N.Y. 1911), *aff'd*, 196 F. 496 (2d Cir. 1912).
110 *Id.* at 103.
111 283 U.S. 1 (1931).
112 *Id.* at 11.
113 Pasquale J. Federico, *Louis Pasteur's Patents*, 86 Science 327 (October 8, 1937); U.S. Pat. No. 141,072 (1873).
114 333 U.S. 127 (1948).
115 *Id.* at 130.
116 *Id.* at 131.
117 *See* Lefstin, *supra* note 67, at 631–34.

### D. Modern American Subject-Matter Eligibility Doctrine: Lost Understandings and Consequent Confusion

Since *Hartranft*, *Ansonia Brass*, *Brogdex*, and *Funk Brothers*, the Supreme Court has not been terribly clear in specifying how, why, and on what criteria to draw the relevant lines between patent-eligible new things and processes on the one hand, and ineligible products derived from nature, or ineligible applications of laws of nature or abstract ideas, on the other. For example, the Court in 1972, in *Gottschalk v. Benson*,[118] had suggested that the line between eligible and ineligible subject matter was whether the claimed invention involved a particular (novel) machine or a (novel) transformation of matter.[119] But a few years later, the Court rejected the argument that the machine-or-transformation precedents established a limiting rule of patent eligibility,[120] restoring the focus on the question of whether the claimed invention reflects a sufficiently creative change of an ineligible discovery.[121] Similarly, in 1980, in *Diamond v. Chakrabarty*,[122] the Court permitted patents on living organisms, synthetic bacteria containing oil-degrading plasmids that were "not nature's handiwork, but [the patentee's] own," distinguishing these novel products from the *Funk Brothers* combined inoculants on the basis that they were "markedly different."[123]

More recently, the Court has applied different standards to products derived from nature and to processes applying natural discoveries. In the *Myriad* gene-patent case, the Court applied the "markedly different" standard from *Hartranft* and *Chakrabarty* to hold isolated DNA ineligible (while applying a different standard, "creat[ing] something new," to hold complementary DNA [cDNA] eligible).[124] In contrast, for processes addressing discovered natural correlations, "laws of nature," the Court in *Mayo Collaborative Services Inc. v. Prometheus Labs., Inc.*[125] adopted a two-part inquiry. The Court in *Alice Corp. v. CLS Bank Int'l*[126] summarized the *Mayo* test (when applying it to claimed applications of abstract ideas):

> [*Mayo*] set forth a frame work for distinguishing patents that claim laws of nature, natural phenomena, and abstract ideas from those that claim patent-eligible applications of those concepts. First, we determine whether the claims at issue are directed to one of those patent-ineligible concepts. If so, we then ask, "[w]hat else is there in the claims before us?" ... To answer that question, we consider the elements of each claim both individually and "as an ordered combination" to determine whether the additional elements "transform the nature of the claim" into a patent-eligible application. ... We have described step two of this analysis as

118 409 U.S. 63 (1972).
119 *See id.* at 70.
120 *See, e.g.*, Parker v. Flook, 437 U.S. 577, 588 n.9 (1978).
121 *See, e.g.*, Bilski v. Kappos, 561 U.S. 593, 610–11 (2010).
122 Diamond v. Chakrabarty, 447 U.S. 303 (1980).
123 *Id.* at 310.
124 Association for Molecular Pathology v. Myriad Genetics, Inc., 569 U.S. 576, 590, 595 (2013).
125 566 U.S. 66 (2012).
126 134 S. Ct. 2347 (2014).

a search for an "inventive concept" – *i.e.*, an element or combination of elements that is "sufficient to ensure that the patent in practice amounts to significantly more than a patent upon the [ineligible concept] itself."[127]

In summary, both *Myriad* (at least in its holding concerning isolated DNA) and *Mayo* restored the requirement for patent eligibility that there be creativity in the applications themselves. But the cases did so without clearly holding that natural, scientific, and abstract discoveries should be treated as if they were prior art and thus become part of the public domain when disclosed by the patent applicant. The Supreme Court also has created problems by failing to repudiate its prior holdings, such as *Diamond v. Diehr*, that appear inconsistent with its current approach.[128] Under current US law, natural, scientific, and abstract discoveries therefore remain protected against becoming property through claims to those discoveries or to uncreative applications of them. But the line between ineligible discoveries that cannot be patented and eligible inventions having sufficient additional creativity remains highly uncertain. Nevertheless, medical diagnostic methods based on newly discovered natural correlations – such as testing for myeloperoxidase or diagnosing birth defects based on the presence of cell-free fetal DNA – are now most likely to be held ineligible, as they tend to be based on simple recognition of the implications of the natural discovery.[129]

### E. *Comparison to Modern European Subject-Matter Eligibility Doctrine*

The prior discussion of American law stands in contrast to the current situation in Europe, where medical diagnostic inventions used outside the body are eligible subject matter, even if they are frequently found invalid on other grounds such as lack of novelty or inventiveness.[130] Under the European Patent Convention (EPC), moreover, ineligible subject matter includes "discoveries, scientific theories, and mathematical methods,"[131] but the exclusion is limited to the discovery "as such."[132] By no longer considering whether the ineligible discovery supplies the relevant creativity (the so-called contribution approach) but looking instead at whether the invention has a "technical nature" for purposes of eligibility, the EPC now largely avoids the need for similar line drawing in regard to the "as such" criterion (although it partially revisits the issue of contribution for "inventive step" analysis, and must address related

127 134 S. Ct. at 2355 (quoting *Mayo Collaborative Servs., Inc.*, 566 U.S. at 72, 80).

128 *See, e.g.*, Diamond v. Diehr, 450 U.S. 175, 188–92 (1981).

129 *See, e.g.*, Cleveland Clinic Found. v. True Health Diagnostics LLC, 859 F.3d 1352, 1360–63 (Fed. Cir. 2017); Ariosa Diagnostics, Inc. v. Sequenom, Inc., 788 F.3d 1371, 1377–78 (Fed. Cir. 2015). *See generally* Saurabh Vishnubhakat, *The Antitrusting of Patentability*, 48 Seton Hall L. Rev. 71 (2017).

130 *See, e.g.*, European Patent Convention, art. 53(c) (hereinafter EPC); Kim Treanor, *US, EU Diverge On Medical Diagnostic Patents*, Intell. Prop. Watch (May 10, 2017).

131 EPC, art. 52(2)(a) (2000).

132 EPC, art. 52(3) (2000).

line-drawing issues).[133] And as noted previously, the United Kingdom (in *Neilson*) adopted a "practical application" test of eligibility that is essentially similar to the "technical nature" test.[134]

Further, unlike in the United States (where, after *Myriad*, isolated DNA is not eligible), the European Biotechnology Directive of 1998 authorized patents on isolated human cells and genetic sequences (so long as they are not in their natural state), while banning patents on humans as a whole and cells and sequences in their natural state on deontological moral grounds.[135] And more recently, the European Patent Office Enlarged Board of Appeals held tomatoes and broccoli produced through conventional breeding techniques ("essentially biological processes") to be patent-eligible,[136] although the Administrative Council of the EPC amended the rules implementing the EPC in 2017 in an effort to overturn this result.[137]

Thus isolated genes and other products isolated from nature currently are patent-eligible in Europe, unlike in the United States.[138] These divergent results reflect not only differences of analytic approach, but also different political resolutions of the relevant moral concerns with patenting science, nature, and abstract ideas.

## III. CONTINUING RELIGIOUS AND MORAL CONSIDERATIONS

Many countries continue to prohibit patents on surgical, medical, and diagnostic methods (or, as in the United States, to prohibit enforcement of such patents against doctors and medical facilities).[139] In this context, such nations not only implicitly reject

133 *See, e.g.*, Opinion of the Enlarged Board of Appeal of 12 May 2010, in relation to a point of law referred by the President of the European Patent Office pursuant to Article 112(1)(b) of the EPC, Case No. G 003/08, at ¶¶10.3–10.13, 12.2, http://www.epo.org/law-practice/case-law-appeals/pdf/g080003ex1.pdf. *See generally* Sigrid Sterckx & Julian Cockbain, Exclusions from Patentability: How Far Has the European Patent Office Eroded Boundaries? 182–274, 310–15 (2012); Justine Pila, *Software Patents, Separation of Powers, and Failed Syllogisms: A Cornucopia from the Enlarged Board of Appeal of the European Patent Office*, 70 Cambridge L.J. 203, 204–09 (2011).

134 *See* Lefstin, *supra* note 67, at 580–92.

135 *See* Directive 98/44/EC of the European Parliament and of the Council of 6 July 1998 on the Legal Protection of Biotechnological Inventions, OJ L 213, 30.07.1998, arts. 5(1)–(3), http://eur-lex.europa.eu/legal-content/EN/TXT/?uri=celex:31998L0044.

136 *See* Opinion of the Enlarged Board of Appeal of 25 March 2015, Method for breeding tomatoes having reduced water content and product of the method, Case No. G 002/12, https://register.epo.org/application?documentId=EXBZW10W4599684&number=EP00940724&lng=en&npl=false; Opinion of the Enlarged Board of Appeal of 25 March 2015, Method for selective increase of the anticarcinogenic glucosinolates in *Brassica* species, Case No. G 002/13, https://register.epo.org/application?documentId=EXBZX31D2974684&number=EP99915886&lng=en&npl=false.

137 2017 EPO Official Journal A56, http://www.epo.org/law-practice/legal-texts/official-journal/2017/07/a56/2017-a56.pdf.

138 *See, e.g.*, Decision of the Technical Board of Appeal 3.3.04 of 27 September 2007, Case No. T 1213/04 – 3.3.04, at ¶¶ 43–57 (Bd. Appeal Eur. Pat. Off.), http://www.epo.org/law-practice/case-law-appeals/pdf/t051213eu1.pdf; Sterckx & Cockbain, *supra* note 133, at 124–29.

139 *See, e.g.*, 35 U.S.C. § 287(c); EPC, *supra* note 130, art. 52(4); Code de Propriete Intellectuelle art. L611-16 (Fr.); Patentgesetz, Dec. 16, 1980, BGBl. 1981 I at 1, last amended by Gesetzes, Sept. 1, 2017 BGBl. I at 3346, § 2a(1)2 (Ger.).

the utilitarian belief that patent incentives lead to overall increased social welfare. Patent rights are not thought to be needed to incentivize developments in medical care, and to provide such rights would be bad for society. Further, they explicitly assert it is ethically wrong to require doctors to obtain prior consent and to pay license fees in order to provide patients with known treatments. Doctors should not be prevented by patents from providing the best care they know of when treating their patients, and to do so would violate doctors' ethical duties (and would be unfair to particular patients).[140]

Although differing moral beliefs may have pointed toward the same policy of ineligibility (or unenforceability) of patents for medical-treatment method inventions, such beliefs may point in different directions for other kinds of inventions, such as uncreative applications of new discoveries of science, nature, and abstract ideas. For example, many biotechnology companies argue vigorously that patent incentives are needed to obtain investment capital, and that they will not be able to justify research and development expenditures in biotechnology absent patent protection.[141] Most people believe that patent protection does induce greater investment and leads to more invention in biotechnology than would occur without patent protection. They therefore tend to believe that such protection will benefit society (even while recognizing that this belief is subject to significant theoretical and empirical dispute regarding the need for such patents and whether they unduly inhibit sequential invention).[142] Consequentialist analyses typically lack adequate empirical data on the effects of patent rights and of potential alternatives, such as expanded government funding of research. Sometimes, the alternatives (particularly government-directed development of production) are thought to be less efficient or may appear politically or culturally infeasible.[143]

But even assuming the validity of the argument that patents for biotechnological inventions would consequentially benefit society, many people nevertheless believe that some such patents are improper for religious and deontological moral reasons. This is particularly true in regard to inventions relating to the human body or applying to public health, such as stem-cell inventions. But the moral concerns discussed herein also apply to the patenting of non-human life forms. In the 1980s, increasingly more complex modified organisms were patented, which led (in the United States and Europe, but not in Canada) to acceptance of a patent for a genetically modified "Harvard" mouse.[144]

[140] *See, e.g.*, 142 Cong. Rec. S12,023–24 (daily ed. Sept. 30, 1996) (statement of Sen. Frist); STERCKX & COCKBAIN, *supra* note 133, at 137.

[141] *See, e.g.*, Amicus Curiae Brief of Biotechnology Industry Organization at 1–2, *In re* Bilski, 545 F.3d 943 (Fed. Cir. 2008).

[142] *See, e.g.*, Sigrid Sterckx, *The Ethics of Patenting – Uneasy Justifications*, in DEATH OF PATENTS 175, 193–202 (Peter Drahos ed., Lawtext Pub. Ltd. 2005).

[143] *See generally* Joshua D. Sarnoff, *Government Choices in Innovation Funding (with Reference to Climate Change)*, 62 EMORY L.J. 1087 (2013).

[144] *See, e.g.*, *Chakrabarty*, 447 U.S. at 303 (microorganisms); *Ex parte* Hibberd, 227 U.S.P.Q. 443, 1985 WL 71986 (B.P.A.I. Sept. 18, 1985) (plants); *Ex parte* Allen, No. 86–1790, 2 U.S.P.Q.2d 1425, 1987 WL

As consequentialist justifications for expanding the scope of the patent system increasingly are challenged for a lack of empirical support, religious and deontological moral considerations likely will play a more explicit role in determining the limits of patent eligibility.[145] Religious and moral concerns also may apply to social classifications imposed by patents or to the discriminatory licensing of patents to particular groups, such as patents for the medical treatment of a particular racial group.[146] As a result, patent eligibility will remain a highly controversial subject.

It is therefore important to remember and understand the religious and moral grounding for those limits. Understanding the history may help us not only explain why such conflicts arise, but also arrive at better and more coherent theoretical solutions to the inevitable conflicts. To do so, it will be critically important to determine whether or not to treat new scientific, natural, and abstract discoveries as if they are part of the public domain. Such treatment will thereby require additional creativity in the application of those discoveries in order to generate patentable "inventions." Without such treatment, the scientific, natural, and abstract discoveries being applied may contribute the required novelty and creativity of the claimed applications.

In the end, these issues will remain political disputes among people with different world views. Religious and moral (including utilitarian) views likely will continue to limit patents or other intellectual property for natural, scientific, and abstract discoveries "as such,"[147] and possibly for uncreative applications and for many other forms of "inventions." In such cases, public-sector alternatives to fund the creative efforts to develop and disseminate such knowledge may be needed.[148]

123816 (B.P.A.I. Apr. 3, 1987) (higher organisms); Decision of the Technical Board of Appeal 3.3.8 of 6 July 2004, Case No. T 0315/03 – 3.3.8 (Bd. Appeal Eur. Pat. Off.) (discussing articles 52(1), 53(a), 53 (b), 56, 57, 83, 112(1), 116(1), 125, 164(2), and 177(1) of the EPC), http://www.epo.org/law-practice/case-law-appeals/pdf/t030315ex1.pdf; Harvard College v. Canada (Commissioner of Patents), [2002] 4 S.C. R. 45 (Can.). *See generally* Daniel J. Kevles, A history of patenting life in the United States with comparative attention to Europe and Canada (2002), https://scholarworks.iupui.edu/bitstream/handle/1805/757/Patenting%20Life%20-%20Comparative%20Study%20US%2C%20CN%2C%20EU%202002.pdf?sequence=1&isAllowed=y.

145 *See, e.g.*, Sterckx and Cockbain, *supra* note 133, at 178–93.

146 *See, e.g.*, Jonathan Kahn, *Race-ing Patents/Patenting Race: An Emerging Political Geography of Intellectual Property in Biotechnology*, 92 Iowa L. Rev. 353, 356–60 (2007). *See generally* Shubha Ghosh, *Race-Specific Patents, Commercialization, and Intellectual Property Policy*, 56 Buff. L. Rev. 409 (2008).

147 *See, e.g.*, C.J. Hamson, Patent Rights for Scientific Discoveries 20–29 (Bobbs-Merrill Co. 1930).

148 *Cf.*, *e.g.*, Vishnubhakat, *supra* note 129, at 77.

# 4

# Life-Form Patents: Proceedings in the European Patent Office and the Role of Non-commercial Parties

*Christopher Rennie-Smith*[1]

## INTRODUCTION

Those familiar with the jurisprudence of the Boards of Appeal of the European Patent Office (EPO) will know well that it contains several warnings against taking the text of a conference paper published in a post-conference volume of transactions as an accurate record of what the speaker actually said at the conference. This is beyond doubt true of the present paper as published here. At the *Patents on Life* conference I spoke on 'Judgments of the Boards of Appeal of the European Patent Office', which necessarily required me to summarise the relevant legal provisions of the European Patent Convention (EPC). The presentation I gave in Cambridge bears some resemblance to the first half of the following text. However, for the purposes of this publication I was asked to develop the paper to cover the role of non-commercial parties in EPO proceedings relating to life-form patents, an extension I was pleased and interested to add having personally seen at first hand a number of cases involving such parties.

This chapter begins by summarising the provisions of the EPC which touch on life forms and the principal decisions in this field. It then reviews the manner in which opponents of patents other than commercial parties – such as pressure groups, political parties and churches – have organised themselves as opponents. Lastly, it considers how these non commercial opponents have conducted their cases, their degree of success, whether they could do more to question the grant of life-form patents – and whether, as I consider they should, churches and other organisations with ethical credentials should actively question life-form patents.

## 1. THE EUROPEAN PATENT SYSTEM IN SUMMARY

For readers unfamiliar with patent law, there follows a thumbnail sketch of the system provided for in the EPC.

[1] I wish to record my grateful thanks to Simon Ravenscroft for much helpful editing advice. The remaining faults are entirely mine. I must also observe that I was a member of the Boards of Appeal of the EPO when many of the decisions mentioned in this article were issued and was in a few cases a member of the actual Board and thus in part responsible for those decisions.

The EPC is a treaty to which thirty-eight European states currently subscribe. It provides for the application for and grant of a patent covering as many of those states as the applicant elects to designate. Once granted, the patent becomes in effect no more than a national patent in each of the designated countries and the EPO, which administers the EPC, plays no further role with one very significant exception – the opposition to a granted patent, which can be launched up to nine months after grant by any person. Such oppositions are then heard as contested proceedings in the EPO, and if they are successful or partly successful, the patent is then revoked or amended for all its designated states. The advantage of the system is that both grant and opposition are dealt with in one central procedure; the disadvantage is that infringement is not dealt with centrally but in the national courts of the various states.[2] The attractions of this centralised system include simplicity and economy of scale, but its success (or, to be exact, the failure to cope with its popularity) has led to unacceptable delays.

The 'machinery' of the system consists principally of three bodies. First, there is the Examining Division, which examines applications for patents which after examination are either granted or refused. The examination process is a dialogue between the applicant and the EPO, although written objections may be filed by others and are usually considered.

Second, there is the Opposition Division, which considers oppositions to granted patents, which anyone may file within nine months after grant and which may maintain a patent, either as granted or in an amended form, or revoke it.

Third, decisions of both the Examining and Opposition Divisions may be appealed to the Boards of Appeal, which may uphold the original decision or substitute any decision which the lower tribunal could have made. An appeal suspends the effect of the appealed decision.

We must also mention the Enlarged Board of Appeal, which is not a second appeal stage but provides opinions on legal issues referred to it either by a Board of Appeal or by the president of the EPO. Parties to appeal proceedings giving rise to the referral are automatically parties to the Enlarged Board proceedings, and others may file written *amicus curiae* submissions after the Enlarged Board issues an invitation to do so.[3]

The Examining and Opposition Divisions are composed of EPO examiners who have graduate or higher qualifications in the relevant technical field. The Boards of

[2] Although an additional system has now been enacted by treaty which will provide for the grant of patents (Unitary Patents) covering nearly all the states of the EU and the litigation of issues, including infringement, relating to those patents in a new European court (Unified Patent Court). When, and indeed whether, this will come into operation is currently questioned by the UK's possible departure from the EU and a case pending in the German constitutional court.

[3] All Enlarged Board decisions (with a G reference number) and important Board of Appeal decisions (with a T reference number) are reported in the EPO Official Journal and referenced below by 'OJ EPO' with the year and page number. However, all such decisions, whether or not reported in OJ EPO and including those mentioned herein, can be found at http://www.epo.org/law-practice/case-law-appeals/advanced-search.html.

Appeal are composed of both former examiners and external appointees and include legally as well as technically qualified members. Proceedings before all three bodies include written submissions and the right to a hearing (called 'oral proceedings'). Oppositions and appeals must satisfy several formal criteria just to be admissible, before consideration is given to the substantive merits of the case. Thus admissibility and allowability form two distinct aspects of all oppositions and appeals.

The vast majority of patent applications and oppositions are filed by patent attorneys who then represent their clients in all subsequent stages of proceedings. Occasionally specialist patent lawyers present cases at oral proceedings. A very few patent applicants and some opponents conduct their own cases. While there are necessarily time limits in all proceedings, and while appeals are generally more formal than other proceedings, all EPO proceedings are relatively informal and certainly less formal than most national court proceedings.

The text of the EPC has been adopted and amended in three official languages – English, French and German. The language used throughout a particular case is the official language in which the patent application in question was filed, but applicants, opponents and their representatives may use any of the three languages.

The EPC as originally adopted came into force in 1973. A major revision in 2000 produced changes which took effect in late 2007. The two versions are known as EPC 1973 and EPC 2000. Both versions contain two categories of provisions, Articles and Rules. The Rules are collectively called the 'Implementing Regulations' and, as that term implies, are subsidiary to the Articles insofar as in the case of any conflict the provisions of the Articles prevail.[4]

## II. PATENTING LIFE FORMS: THE LAW

Neither EPC 1973 nor EPC 2000 contained any overall provisions about the patenting of life forms: such patents are neither expressly permitted nor expressly excluded. Accordingly, life-form inventions which meet the basic requirements by which all patent applications are assessed – that they are novel, inventive and industrially applicable[5] – are *prima facie* patentable, and in fact life-form patents were granted from the early days of the EPC. The provisions of the EPC which impinge on life-form patents are found in Article 53 and Rules 26–29.

Article 53 contains both a general ethical provision which has been invoked in a number of life-form cases and some very limited life-form exceptions. Rules 26–29 contain, under the heading 'Biotechnological inventions', some definitions which affect the interpretation of Article 53; some provisions as to biotechnological inventions which are patentable, as if there was some wider provision casting doubt on their general patentability; some provisions headed 'Exceptions to patentability'

[4] Article 164(2) EPC.
[5] Article 52(1) EPC.

which are expressed as being 'Under Article 53'; and some provisions about inventions, patentable and not, relating to the human body which are not by their language related to Article 53 or any other provision of the EPC. This inconsistent – and unfortunate – situation reflects the legislative history of these rules.

In the 1990s, animal rights campaigners who objected to genetic manipulation of animals added the patenting of such activities to their objections, fuelled in part at least by publicity surrounding Harvard University's applications in Europe and elsewhere to patent its genetically modified mouse for use in cancer research (the 'oncomouse').[6] The discussion extended to the European Parliament and led to the passage of EC Directive 98/44/EC on the 'legal protection' (in other words patenting) of biotechnological inventions (the 'Biotech Directive'). Although the Biotech Directive was a product of the European Union, of which the EPO is not a part, and although many EPC contracting states are not member states of the EU, the EPO's Administrative Council, which has the power to add or amend Rules but not Articles of the EPC, inserted a number of provisions of the Biotech Directive into the Rules of the EPC. It should be observed that the Directive had itself already 'borrowed' not just Articles 53(a) and (b) from the EPC, but also the basic requirements of novelty, inventive step, and industrial applicability.

That the new Rules are neither perfectly drafted nor wholly consistent with the rest of the EPC is not altogether surprising. Their general provisions are declaratory and of limited effect. Rule 26(1) declares that for European patents and patent applications concerning biotechnological inventions, the relevant provisions of the EPC shall be applied and interpreted in accordance with these Rules and that the Directive shall be a supplementary means of interpretation. The reference to 'relevant provisions' is curious because there are no provisions relating to just biotechnological inventions. The only common-sense meaning of 'relevant provisions' is to view it as referring to Article 53(a) and (b), which can concern biotechnological inventions. Rules 26(2)–(6) contain a number of definitions some of which are referred to below. We shall consider each of the general ethical and life-form exception provisions of Article 53 in turn, with reference to Rules 26–29 and Board of Appeal case law as appropriate, and then the remaining provisions of those Rules.

## A. *Article 53(a)*: 'Ordre Public' *and Morality*

The general ethical provision reads:

> European patents shall not be granted in respect of:
>
> (a) inventions the commercial exploitation of which would be contrary to 'ordre public'[7] or morality; such exploitation shall not be deemed to be so contrary merely because it is prohibited by law or regulation in some or all of the Contracting States[.]

[6] In the EPC, the Harvard case gave rise to two appeal decisions referred to below as *Oncomouse I* and *Oncomouse II*.

[7] The French expression is used in both the English and French texts. The German is 'die guten Sitten'.

In EPC 2000, the words 'the commercial exploitation of which' replaced the words 'the publication or exploitation of which'. This makes clear (to the writer if not to all commentators[8]) that it is neither the morality of patenting a particular invention, nor the morality of the invention itself, but the morality of the commercial exploitation of the invention which is targeted. This is consistent with the overall approach of the EPC, which is to permit inventions which fulfil the basic requirements of novelty, inventive step, and industrial applicability, and which then requires exceptions to be precisely worded. Thus what must distinguish an otherwise patentable invention in order to refuse it a patent is what will, or would, happen when it is put into commercial practice. If that will be both morally acceptable and not contrary to *ordre public*, then the exception will not apply.

The interpretations of *ordre public* and 'morality' in the decision *Plant Cells/ PLANT GENETIC SYSTEMS*[9] have generally been accepted since:

> 5. It is generally accepted that the concept of 'ordre public' covers the protection of public security and the physical integrity of individuals as part of society. This concept encompasses also the protection of the environment. Accordingly, under Article 53(a) EPC, inventions the exploitation of which is likely to breach public peace or social order (for example, through acts of terrorism) or to seriously prejudice the environment are to be excluded from patentability as being contrary to 'ordre public'.
>
> 6. The concept of morality is related to the belief that some behaviour is right and acceptable whereas other behaviour is wrong, this belief being founded on the totality of the accepted norms which are deeply rooted in a particular culture. For the purposes of the EPC, the culture in question is the culture inherent in European society and civilisation. Accordingly, under Article 53(a) EPC, inventions the exploitation of which is not in conformity with the conventionally-accepted standards of conduct pertaining to this culture are to be excluded from patentability as being contrary to morality.

The *Plant Cells* case concerned genetically engineered plant cells which were resistant to a class of herbicide and would produce plants and seeds selectively protected against weeds and fungal diseases. The Board held that seeds and plants cannot per se be contrary to Article 53(a) EPC merely because they are 'living' matter and because plant genetic resources should remain the 'common heritage of mankind'. It also held that the exploitation of the invention related to these herbicide-resistant plants was not contrary to *ordre public* or morality. In *Euthanasia compositions/MICHIGAN STATE UNIVERSITY*,[10] an Article 53(a) argument that a chemical composition for the humane killing of animals was morally unacceptable because it would be so if used for humans was rejected.

8 See Liddell and Ravenscroft, this volume.

9 T356/93 (OJ EPO 1995, 545), points 5 and 6.

10 T 866/01.

In *Breast and ovarian cancer/UNIVERSITY OF UTAH*,[11] the invention concerned a cancer susceptibility gene. One opponent argued that the absence of consent by the donors of cells used to arrive at the invention was unethical and the invention therefore immoral, but the Board decided that while such consents might be required by other legal provisions, they were not a condition of patent law. Another opponent argued that the patent would increase healthcare costs, affecting patients and their treatment by doctors, and that the dependence on the patentee by patients was contrary to human dignity. The Board held that these arguments were directed to the exploitation of the patent and not to exploitation of the invention.

In *Slaughterhouse method/STORK PMT BV*,[12] the Board held that including a human being as a feature of a claim – an observer in a slaughterhouse process – could be seen, 'however far-fetched such an interpretation may seem', as the ownership or enslavement of a person contrary to human rights and thus contrary to *ordre public*.

## B. *Rule 28*

Under the heading 'Exceptions to patentability', Rule 28(1) states:

> Under Article 53(a), European patents shall not be granted in respect of biotechnological inventions which, in particular, concern the following:
> (a) processes for cloning human beings;
> (b) processes for modifying the germ line genetic identity of human beings;
> (c) uses of human embryos for industrial or commercial purposes;
> (d) processes for modifying the genetic identity of animals which are likely to cause them suffering without any substantial medical benefit to man or animal, and also animals resulting from such processes.

Apart from the introductory words '*Under Article 53(a)*', this appears to be a straightforward list of items which may not be given a patent. However, those words would appear to make the items on this list, if commercially exploitable, contrary to *ordre public* or morality. The Rule is a dreadful piece of legislative drafting. Article 53(a) is quite clearly a provision which is open to interpretation by examiners or boards of appeal; and rightly so, since the question of what is moral or not (or contrary to *ordre public* or not) is one which needs to be considered according to the particular facts of each case and in the light of attitudes which may change over time. In short, it is a provision which gives the decision makers some discretion. Rule 28 introduces a seemingly mandatory list of items with an apparent instruction to the decision makers to consider these items as per se 'immoral' or contrary to *ordre public* or both.

[11] T 1213/05.
[12] T 149/11.

In *Use of embryos/WARF*,[13] the Enlarged Board of Appeal held that Rule 28(1)(c) forbade the patenting of a claim to human embryonic stem cell cultures which, when the application was filed, could only be prepared by a method which involved the destruction of the human embryos from which the cultures were derived, even if the method was not part of the claim. It was not relevant that after the filing date it became possible to obtain the cultures without the destruction of embryos. Since the only method of obtaining the cultures at the filing date involved destruction of embryos, the response of the Enlarged Board was not surprising as regards 'uses of human embryos'. However, whether such uses were 'for industrial or commercial purposes' might be thought a little more problematic, since obtaining the cultures clearly preceded any use of or dealing in the cultures once obtained. However, the Enlarged Board found that making the claimed product, even for further research, involved the destruction of embryos and remained commercial or industrial exploitation of the invention.

That the Enlarged Board in *WARF* considered only the listed item in Rule 28(1)(c) reflects the fact that the questions referred to it made no mention of Article 53(a), not to mention the fact that there was no previous provision such as Rule 28(1)(c). By comparison, in *Oncomouse II*[14] the Board of Appeal had to grapple with both the broader words of Article 53(a) itself and the more limited but specific words of Rule 28(1)(d) prohibiting processes causing animal suffering.

Before the EU enacted the Biotech Directive, let alone before Rules 26–29 were added to the EPC, the Board of Appeal had in the earlier case *Oncomouse I*[15] formulated this test for the application of Article 53(a) in animal cases:

> The decision as to whether or not Article 53(a) EPC is a bar to patenting the present invention would seem to depend mainly on a careful weighing up of the suffering of animals and possible risks to the environment on the one hand, and the invention's usefulness to mankind on the other.

This was the origin of the 'balancing test' which was reproduced, using another form of words, in the Directive and thus in Rule 28(1)(d) of the EPC. In *Oncomouse II* the Board decided that the effect of Rule 28 was to create two separate objections under Article 53(a) – what it called a 'Rule 28 type' Article 53(a) objection[16] and a 'real' Article 53(a) objection. In a case falling within one of the four categories of

[13] G 2/06 (OJ EPO 2009, 306). WARF stands for the patent applicant Wisconsin Alumni Research Foundation. See also Schneider, this volume.

[14] T 315/03 (OJ EPO 2006, 15).

[15] T 19/90 (OJ EPO 1990, 476).

[16] Rule 28(1)(d) is referred to here, as that is the current number of the rule in question. However, those reading the actual decision will find instead references to a 'Rule 23(d) type' Article 53(a) objection, since the then-current numbering of the Rule was 23d (and thus current Rule 28(1)(d) was then, confusingly, Rule 23d(d), but the double 'd' was not used in the Board's terminology since the expression 'Rule 23(d) type' would in principle apply to all sub-rules (a) to (d) of Rule 23d). An additional complication arises from the recent addition of sub-rule (2), discussed later, which now requires Rule 28(a)–(d) to be referred to as Rule 28(1)(a)–(d).

Rule 28(1), such as an animal manipulation case falling within Rule 28(1)(d), the Board held that Rule 28 must be applied first and, if the case failed that test, Article 53(a) need not be considered further. However, if the case passed the Rule 28 test, then in animal manipulation cases, the 'real' Article 53(a) objection required the balancing test in *Oncomouse I* to be applied. So, Rule 28(1)(d) has in effect introduced an additional test for animal cases.

The Board further held that application of the Rule 28(1)(d) test requires the assessment of three requirements: a likelihood of suffering caused to animals, a likelihood of substantial medical benefit to man or animal and, which is not expressly stated in the Rule, a necessary correspondence between the two as regards the animals in question. In other words, the test is only satisfied in the case of those animals whose suffering is shown to be balanced by a likely benefit. Thus Harvard's preferred claim to rodents was not allowed, since there was no demonstration of any benefit from the use of rodents such as beavers or porcupines, whereas its fallback claim to just mice was allowable, since the patent described the use of 'oncomice' for cancer research.

As regards rodents, the Board did not therefore need to proceed to assess the case under the *Oncomouse I* test; but it nonetheless stated that had it done so, it would have reached the same result. As regards mice, it did proceed to the *Oncomouse I* test, which allowed more factors to be considered. Under it, a decision was said to depend only 'mainly' on the balancing test, which was itself in broader terms than Rule 28(1)(d), since it weighed not just animal suffering but also possible risks to the environment, against not just medical benefit to man or animals but also the invention's usefulness to mankind.

The Board found that many of the opponents' arguments were irrelevant since they were about the morality of animal patenting and not of exploitation of the invention. Further, many of the relevant arguments were no more than assertions unsupported by evidence. The Board rejected arguments about the degree of suffering the oncomice would endure,[17] the possible use of non-animal alternatives such as cell cultures (as to which the patentee produced evidence that the oncomice were more useful), environmental risks if the oncomice should escape into the wild, the threat to evolution, and the suggestion that a patent would promote the use of transgenic mice in research and trade in animals generally. A final category of arguments that the public found the use of genetically modified animals in medical research morally unacceptable was also rejected, principally because the opponents' only evidence was unconvincing opinion polls. Thus claims to mice, as opposed to rodents, also passed the *Oncomouse I* test.

[17] Since any suffering was enough to bring Article 53(a) into play, the degree of suffering did not assist matters. Indeed, the implicit suggestion of a distinction between acceptable and unacceptable degrees of suffering probably worked to the opponents' disadvantage. I return to this later in the context of arguments by non-commercial parties.

### C. *Article 53(b): Plant or Animal Varieties, Essentially Biological Processes*

The limited life-form exceptions provision reads that 'European patents shall not be granted in respect of':

> (b) plant or animal varieties or essentially biological processes for the production of plants or animals; this provision shall not apply to microbiological processes or the products thereof[.]

#### 1. Plant Varieties

In *Transgenic plant*/NOVARTIS,[18] the Enlarged Board held that 'plant varieties' which are not individually claimed are not excluded. A claim in a patent may cover a number of plant varieties, but if no one of them is claimed as such, then the claim is allowable under Article 53(b). The Enlarged Board's reasoning was based on the historical purpose of Article 53(b), which was to allow patent protection for developments in plant technology other than as provided by plant breeders' rights, which afford protection specifically for new varieties. Further, the Enlarged Board held that this applies irrespective of how plant varieties are produced. Thus a plant variety produced by genetic engineering – as in the NOVARTIS case – was not allowable if specifically claimed, but a claim to genetically engineered plants which did not specify any individual variety was allowable.

#### 2. Animal Varieties

The expression 'animal varieties' has suffered from unnecessary confusion because different words are used in the different language texts of the EPC – three different words originally, reduced to two in EPC 2000.[19] The Board of Appeal in *Oncomouse II* observed, and described as 'absurd', that as a result the width of the exclusion would depend on the language used. It suggested that a single taxonomic category should be used as in the case of plants,[20] but it did not need to select one since the animal in question (a mouse) was in a higher taxonomic category ('genus') than 'species', which was the widest of the three alternatives (variety, species, race) then available. Four opponents had asserted that the subject of the invention – mice genetically manipulated for use in cancer research – were a new animal species because they inherited the particular characteristic of an increased probability of developing tumours, but the Board rejected this. If the opponents were correct, one could perform the claimed method of the patent upon three different species of

18 G1/98 (OJ EPO 2000, 111).

19 'Animal varieties' has always been the form of the English text, while the French has always been 'animal races' ('*les races animaux*'). The German originally read 'animal species' ('*Tierarten*') and now, like the French, reads 'animal races' ('*Tierrassen*').

20 In the Enlarged Board decision G 1/98 (OJ EPO 2000, 111), as discussed previously. Much more helpfully, Article 53(b) has always used 'plant variety' in all three languages.

mice, which would therefore all become members of one new species. However, there was no evidence that the alleged new species would be accepted as such by anyone.

3. Essentially Biological Processes for the Production of Plants and Animals

This exception is now defined by Rule 26(5) EPC:

> A process for the production of plants or animals is essentially biological if it consists entirely of natural phenomena such as crossing or selection.

In the first of two referrals in each of two parallel cases, *Broccoli*/PLANT BIOSCIENCE and *Tomatoes*/STATE OF ISRAEL,[21] the Enlarged Board had to consider the interpretation of '*consists entirely of natural phenomena*' and the extent, if any, to which a human intervention with some technical effect might render a process less than essentially biological and therefore patentable. The Enlarged Board answered the question by stating that a non-microbiological process containing or consisting of sexually crossing the whole genomes of plants and subsequent selection is 'essentially biological' and therefore excluded. It also held that a further technical step, such as the molecular marker used in the Broccoli case, which enabled or assisted the conventional steps of crossing and selection, did not take a case outside the exclusion. However, the exclusion did not apply if the process included an additional technical step which by itself introduced a trait into the genome of a plant or modified a trait in the genome, such that the plant has a trait which does not result from the mixing of the genes of the plants chosen for crossing. Therefore, the key to exclusion or not is whether the additional technical step, which must perforce be a human intervention, produces a trait which could not result in nature. The Enlarged Board added that it was not relevant whether the technical step was itself new or already known, whether it was trivial or fundamental, whether it does or could occur in nature, or whether the essence of the invention lies in it.

The later referrals *Broccoli and Tomatoes II* concerned claims to plants, plant parts and the fruits of plants. The Board of Appeal saw the difficulty that allowing a patent for products obtained by essentially biological processes would in effect allow the use of the excluded processes and thus defeat the purpose of the exclusion of those processes. The Enlarged Board dealt robustly with the Board's concerns, holding that the exclusion of essentially biological processes for the production of plants does not apply to plants or plant material produced by the process (unless the product claimed is a plant variety, in which case it is excluded as such anyway). One

[21] The Broccoli cases were G 2/07 (OJ EPO 2012, 230) and G 2/13 (OJ 2016, 61, an abridged version of the decision in *Tomatoes II*); and the Tomato cases were G 1/08 (OJ EPO 2013, 206, an abridged version of the decision in *Broccoli I*) and G 2/12 (OJ 2016, 1). The cases were the subject of two separate pairs of referrals to the Enlarged Board by Board of Appeal 3304 during the pendency of just one appeal in each case. The Enlarged Board consolidated the proceedings in each pair of referrals and thus produced two single decisions, which are usually referred to as *Broccoli and Tomatoes I* and *Broccoli and Tomatoes II*.

immediate response to this would be to suggest that a patent application should always be drafted to cover a plant or its fruits, rather than their method of production. But the Enlarged Board refuted the idea that its decision would lead to 'skilful claim drafting'; it observed that claims to plant products would still have to fulfil the requirements applied to product claims by the EPC. More importantly, perhaps, it considered that broadening the scope of the process exclusion to include the products obtained by processes for the production of plants would introduce an inconsistency in the system of the EPC, which generally allows patents for plants and plant material other than plant varieties.

Despite that decision of its own highest authority on interpretation of the law, in the light of a statement subsequently issued by the European Commission expressing the contrary view in relation to the Biotech Directive, the Administrative Council of the EPO amended Rule 28 to add a provision directly contrary to the decision of the Enlarged Board in *Broccoli and Tomatoes II*. Thus new Rule 28(2)[22] read:

> (2) Under Article 53(b), European patents shall not be granted in respect of plants or animals exclusively obtained by means of an essentially biological process.

There was no need for the EPO to react to the Commission's statement, which had no legal force and related to the Biotech Directive and not to the EPC.[23] The result was a provision which was seemingly inoperable because, since the Rules of the EPC are subordinate to its Articles,[24] Article 53(b) takes precedence over Rule 28(2). Thus on the one hand the Enlarged Board interpreted the superior Article 53 (b) so as to allow the patenting of plants and animals produced by essentially biological processes, while on the other hand the inferior Rule 28(2) said they are not to be patented.[25] The inconsistency now appears to have been resolved by the recent decision T 1063/18, which decided (unsurprisingly) that Article 53(b) as interpreted by the Enlarged Board prevails over Rule 28(2). While the full reasons for this decision are not yet available, it appears that Rule 28(2) is invalid and unenforceable.

In *Oncomouse II*, the Board of Appeal dismissed the argument that producing the transgenic mouse was an essentially biological process because the method of producing the oncomouse did not meet the definition in Rule 26(5).[26]

22 This entered into force (or purportedly so) on 1 July 2017.

23 The statement itself accepted it was not an authoritative interpretation of the Directive. After the statement was issued, and before the new Rule was introduced, the EPO (seemingly unlawfully) suspended all proceedings before it with claims relating to plants and animals produced by essentially biological processes.

24 Article 164(2) EPC.

25 There are also doubts as to the lawfulness of the provision under EU law – see Mike Snodin, 'Patentability of Plants under the EPC: Act in Haste, Repent at Leisure?', *Bio-Science Law Review* 16 (2017), 115–22.

26 'A process for the production of plants or animals is essentially biological if it consists entirely of natural phenomena such as crossing or selection.'

## D. Rule 27

Rule 27 states under the heading 'Patentable biotechnological inventions':

Biotechnological inventions shall also be patentable if they concern:

(a) biological material which is isolated from its natural environment or produced by means of a technical process even if it previously occurred in nature;
(b) plants or animals if the technical feasibility of the invention is not confined to a particular plant or animal variety;[27]
(c) a microbiological or other technical process, or a product obtained by means of such a process other than a plant or animal variety.

It is difficult to see what Rule 27 has achieved. Sub-rule (a) describes a situation covered by the previous law because there was no exception for it, so it simply spells out the previous position. Sub-rule (b) is just another way of expressing the first words of Article 53(b), and sub-rule (c) another way of expressing the latter part of that Article. In the context of the Directive itself, which was new law in that the EU had no previous legislation on biotechnological inventions, these provisions related to other provisions and at least arguably contributed to a coherent whole. However, when just 'borrowed' and inserted into the EPC, they add little if anything. Since the Articles of the EPC must take precedence over the Rules, and since Rule 27 adds nothing but alternative wording to pre-existing Articles, it seems destined to be sidelined for all practical purposes.

## E. Rule 29

Rule 29 states under the heading 'The human body and its elements':

(1) The human body, at the various stages of its formation and development, and the simple discovery of one of its elements, including the sequence or partial sequence of a gene, cannot constitute patentable inventions.
(2) An element isolated from the human body or otherwise produced by means of a technical process, including the sequence or partial sequence of a gene, may constitute a patentable invention, even if the structure of that element is identical to that of a natural element.

Sub-rule 29(1) introduces a new category of exception – the human body – but one which is neither contained in an Article as are plant and animal varieties, nor made specifically subsidiary to an Article as are the items listed in Rule 28(1). However, since Rule 26(1) requires that Article 53 be applied and interpreted in accordance with Rule 29(1), then there may arguably be little difference in legal effect between Rule 28(1)'s list of items explicitly stated as being 'under Article 53(a)', which are thus

27 As in Article 53(b), the French and German texts of Rule 27 refer to 'plant variety or animal race'.

apparently immoral per se, and Rule 29(1), which appears to be an implicit extension of Article 53 without even a reference to that Article. The inconsistency of drafting is unfortunate to say the least. At least the permissive provision of sub-rule 29(2), which allows that elements isolated from the human body may be patentable and thus gives back to the decision makers some of their discretion, does not suffer from the same degree of inconsistency.

## III. NON-COMMERCIAL PARTIES AS OPPONENTS

### A. *Background*

The vast majority of patents are applied for, granted to, and maintained by companies for whom patents are important assets or by others, such as universities, who licence patents commercially. Such patentees typically have substantial budgets for their patent programmes, and they either employ in-house patent attorneys or engage outside firms of attorneys to implement those programmes. These commercial patentees also form the vast majority of opponents to granted patents – denying a competitor the monopoly protection of a patent may be as important as obtaining and defending such protection of one's own. Thus nearly all proceedings in the EPO and other patent offices (and indeed patent infringement proceedings in courts) are conducted on behalf of commercial organisations by representatives who are trained in, and familiar with, the law and the procedures and the technical fields of the cases they handle.

Against that background, the appearance in EPO proceedings of any other, non-commercial parties, such as pressure groups, political parties, and churches, is rare. These parties are by definition opponents and not patent applicants or patentees, so they only appear in opposition proceedings and then only in cases where their interests are engaged – such as open software supporters by software patents or animal rights campaigners by genetic engineering of animals. There is thus a stronger likelihood of non-commercial parties appearing in oppositions to life-form patents than in other patent cases, although even then such appearances will be infrequent. They are also unlikely to engage patent attorneys or other professionally qualified representatives. Their knowledge and experience of patent proceedings (or any legal proceedings) may be limited, as in many cases may be the costs at their disposal.

However, the relative informality of EPO proceedings assists those unfamiliar with its proceedings. An entity of almost any type (such as a charity or a church) will be accepted as a party without any inquiry into its exact legal personality or status under the laws of its own country (in the absence of any challenge by another party (in effect, a patentee)). An officer or member of such an entity will be heard in oral proceedings as its de facto representative on the basis that a party which has no professional representative may be heard in its own right.

The EPC allows any person to oppose a patent without limitation, provided that both the opposition is filed and the opposition fee is paid within the time limit of nine months after the publication of the grant of the patent in question. 'Any person' has been generously interpreted to include not just natural and legal persons, but also nominal parties acting on behalf of others who wish to oppose but not to disclose their identities. Such a nominal opponent – often referred to as a 'man of straw' – may be a patent attorney using his name to give his client anonymity, a company which exists only to lend its name to others, or possibly a natural person who for some political or other purpose coins a fictitious identity as an opponent.

Further, a number of opponents may together oppose a patent as one opponent, paying only one opposition fee, provided they identify all the members of the group, inform the EPO of any changes and have one of their number as their 'common representative'.[28] In addition to the fee advantage for multiple opponents who act together, the costs regime in the EPO is also beneficial for non-commercial opponents. Unlike litigation in many national legal systems, an unsuccessful party is not usually ordered to pay the successful party's costs. The normal rule is that in opposition proceedings each party bears its own costs. The only exception is if the Opposition Division (or Board of Appeal in opposition appeals) orders what is called 'a different apportionment of costs' for 'reasons of equity'.[29] The case law shows that this will usually only be applied in a case of misbehaviour and not, as one might expect, in a case where one party causes wasted time and cost by advancing a hopeless case.[30]

### B. *Third-Party Observations and* Amicus Curiae *Briefs*

Although the following are not oppositions and do not give any right to participate in EPO proceedings, we should mention two further avenues available to non-commercial parties to influence patent decisions.

As mentioned previously, although the pre-grant examination of patent applications is conducted between the EPO's examiners and applicants, third parties may during examination file written observations, which invariably take the form of objections to the grant of a patent, and these are usually taken into account by the Examining Division. While it is possible that non-commercial parties may file such observations, the writer has not encountered it. It may be that non-commercial organisations are not used to monitoring the status of pending patent applications and/or do not have the resources and professionally qualified staff to do so. It is also a fact that successful third-party observations – those that cause a patent application to be refused or substantially reduced in scope – will not generate anything

[28] G 3/99 (OJ EPO 2002, 347).

[29] Article 104(1) EPC.

[30] See, for details, *Case Law of the Boards of Appeal of the European Patent Office*, 8th Edition (2016), 981–1004 (English edition), http://www.epo.org/law-practice/case-law-appeals/case-law.html.

approaching the publicity of an opposition. Against that is the fact that there is no fee or other cost, apart from preparing the observations, involved.

Similarly, but of higher profile, are the *amicus curiae* briefs which the Enlarged Board invites at the outset of proceedings before it. Those who file such briefs acquire no right to appear in the proceedings, but their views are considered (and sometimes referred to) in the decision. This provides an opportunity not only to gain publicity, but also to influence the development of the law. Typically, individual patent attorneys, both individually and as associations; companies with an interest in the questions referred to the Enlarged Board; and some non-commercial organisations file *amicus curiae* briefs.[31] The EPO's website currently[32] lists such briefs by case since 2004, and those in earlier cases could be accessed by the case number.

The writer is not aware of any briefs by non-commercial parties which are stated to have influenced the outcome of referrals to the Enlarged Board, but research might show otherwise. An example of possible influence is the *Transgenic plant/NOVARTIS* case, in which the Enlarged Board recorded that more than six hundred letters had been filed arguing against patenting of plants and animals. It specifically summarised the Greenpeace submission but later described it as interesting if irrelevant, since it raised an Article 53(a) objection when Article 53(b) was in issue. In *Use of embryos/WARF*, the Enlarged Board stated that numerous submissions had been made by third parties, of which some were made in identical form by hundreds of individuals (which suggests an unfortunate syndication of the same submission in an attempt to impress by sheer numbers). The submissions were both in favour of and against patenting of stem cell inventions.

## C. *Who Are Non-commercial Parties?*

A number of non-commercial parties have taken advantage of the relatively friendly nature of the EPO's procedures. In the area of life-form patents, environmental and animal rights pressure groups and a few political groups have been active in opposing patents. Somewhat surprisingly, especially in view of the morality criterion in Article 53(a), neither religious institutions nor non-religious ethical organisations have been anything like as active. This is unfortunate because, on moral issues, they should be able to make better and more informed cases, and moreover make them with greater moral authority than commercial parties can. We will return to this topic later. First, and by way of background to that discussion, I shall consider the conduct of cases by non-commercial parties in EPO oppositions to date.

[31] The number of such briefs can be high: 102 in *Programs for computers* (G 3/08 OJ EPO 2011, 10). Somewhat ironically, in that case the referral (by the EPO President) was dismissed by the Enlarged Board as inadmissible.

[32] Accessed 7 June 2018.

### D. *Organisation of Non-commercial Parties*

When, as mentioned, almost any entity can file an opposition in its own name and be accepted as an opponent, it seems curious that many non-commercial parties have elected to file oppositions in ways that were not only unnecessary but which have inflicted avoidable problems on themselves.

In *Admissibility of joint opposition or joint appeal/HOWARD FLOREY*,[33] the Enlarged Board held that a single opposition, even when filed in the name of two or more persons, is nonetheless a single opposition for which only one opposition fee needs to be paid; that an appeal (or opposition) filed by multiple persons must be filed by a common representative; and that if any of the common opponents or appellants withdraws from the proceedings, the EPO must be informed.

Far from discouraging multiple oppositions, the Enlarged Board's requirements seem to have encouraged non-commercial parties to indulge in filing large group oppositions when it might have been much easier to file in the name of one person or organisation. It sometimes appears that this organisational aspect of an opposition means more to a non-commercial party than the actual content of its opposition (or its likelihood of success). The publicity value of a long queue of several hundred persons filing slowly through the EPO to deposit their notices of opposition has apparently outweighed the procedural problems which might follow.

### E. *Performance of Non-commercial Parties*

The non-commercial opponents who challenge life-form patents will necessarily appear infrequently in opposition and appeal proceedings. In the case of many such parties, they may only ever take part in one opposition. In the only case (*Oncomouse II*) in which all the opponents (seventeen of them at the outset and subsequently fourteen) were non-commercial parties,[34] nearly all of them had no previous experience of patent litigation; and although half of them were professionally represented,[35] their arguments showed either a lack of awareness of the precise issues or a determination to present a particular point of view regardless of the precise issues, or possibly both. Although much of the criticism which could be made of the performance of non-commercial parties could be ascribed to their undoubted lack of experience of both the law and the procedure (and in the writer's experience the EPO goes to considerable lengths to accommodate inexperience,

[33] G 3/99 (OJ EPO 2002, 347).

[34] Five animal rights organisations, three groups against 'life patents', three political groups, one regional church, and two individuals.

[35] One representative was clearly familiar with the issues, since he was the first author of an article published while the appeal was pending – David Thomas and Georgina A. Richards, 'The Importance of the Morality Exception under the European Patent Convention: The Oncomouse Case Continues . . .', *European Intellectual Property Review* 26 (2004), 97–104. Unfortunately, he did not attend the appeal oral proceedings.

among represented as well as unrepresented parties), it is almost impossible to distinguish fully the consequences of inexperience from the consequences of the political stance and the polemical message which many non-commercial parties are resolved to take and make. Whether caused by inexperience or resolve, the results can be unfortunate, viewed objectively.[36] There are several aspects to this.

First, non-commercial parties may adopt an approach which seems at least unconventional and at worst deliberately defiant of the expected norms. Typically, commercial opponents may argue that the invention claimed in a patent is not novel or that it cannot be worked because the patent does not describe it sufficiently. The novelty objection requires a comparison with an item of prior art such as an earlier patent document. The insufficiency objection requires at least a coherent argument explaining how the patent's description is inadequate, or possibly an experiment showing that the opponent has tried and failed to work the invention. The patentee is then expected to present his case why the patent is novel and sufficient, explaining why the prior art does not anticipate his invention and how the patent's description does allow the invention to be worked. The opposing cases are presented to the tribunal, which expects them to be in the form of argument supported by evidence, and which ultimately must choose the one it finds convincing or, since total conviction is rare, the more persuasive.

That being the usual pattern, the tribunal and the patent proprietor expect an opposition on less familiar grounds such as Article 53(a) to proceed in the same way: with an argument that the commercial exploitation of the invention claimed in the patent would be contrary to morality, supported by evidence that the exploitation would not conform with the conventionally accepted standards of conduct in European society.

However, non-commercial opponents tend not to comply with the format of opposing arguments supported by evidence leading to a judgment between them. They no doubt recognise the format from the general knowledge of legal proceedings which they have as members of the public, but they do not necessarily recognise that their opposition to a patent should adopt that format. They are, to continue the morality example, convinced that the invention is immoral and appear to see the mere assertion that this is so as the main purpose of their opposition. They either fail to see or do not mind that this will not be effective – that the patentee will reply simply that he need not challenge unsupported assertions beyond formal disagreement, and that the tribunal will, adopting the classic judicial stance, ask the opponent what he relies on to justify his assertion and find against him when he cannot answer.

It may be that non-commercial parties regard patent opposition proceedings as an extension of their conventional activities, which involve convincing others of the

36 Not in absolute terms of success or failure (winning or losing an opposition), but by the lower standard of ability to engage a tribunal with a plausible and relevant case.

rightness of their cause or beliefs. Persuasion by emotive argument is at the core of their 'business'. Thus, when they wish to express disagreement with something, they are used to doing so via a debate which they hope or expect to win by force of rhetoric, rather than through the more regulated forum of a tribunal where each side presents a case, both in writing and orally, and then awaits a judgment from a neutral body.

Alternatively, it is perhaps in the nature of a pressure group, which exists to promote a cause, or of a religious institution, which exists and functions because of a set of beliefs to which all its members adhere, to consider that its views are not just correct ones but also possibly the only correct ones. It then follows that if such a party to opposition proceedings puts forward an argument that exploitation of an invention is immoral, no supporting evidence and no attempt to counter alternative arguments is required. Ethicist Audrey Chapman, writing of the United States,[37] notes the suggestion that religious critics of patenting have been disadvantaged because they make such 'prophetic' arguments – an effort to expose the roots of what is perceived to be fundamentally and systematically right or wrong – while supporters of the patent regime have the status quo on their side.[38] Chapman herself appears to squash this defence emphatically by observing that 'opting for a prophetic approach does not excuse imprecise and inadequate theological and ethical reasoning'.[39] It might be added that it also does not excuse an incorrect approach to the conduct of legal proceedings, especially when the correct approach might be more effective.

Second, the unconventional approach can produce an opposition case which is irrelevant or even counter-productive for the opponent. Many of the arguments pursued in *Oncomouse II* were irrelevant. The Board of Appeal devoted several pages of its judgment to explaining why several of the opponents (at this point appellants) had produced irrelevant arguments in several categories: (1) By a misinterpretation of the EPC, a number of arguments were not directed to the morality of exploitation of the invention. (2) The case was not about 'the patenting of animals'. (3) The Biotech Directive and the Rules 'borrowed' from it in the EPC had not introduced, as several appellants had argued, the possibility of patenting animals when such had previously not been the case.

The fourth, final category of irrelevant argument was perhaps the most surprising, as it showed an almost total misunderstanding of the judicial nature of the proceedings. One appellant argued that it was not established that the Opposition Division's

[37] And of the debate on gene patenting in general, not of the conduct of patent challenges per se.

[38] John H. Evans, 'The Uneven Playing Field of the Dialogue on Patenting', in *Perspectives on Genetic Patenting: Religion, Science and Industry in Dialogue*, ed. Audrey R. Chapman (Washington, DC: American Association for the Advancement of Science, 1999), 60–61.

[39] Audrey R. Chapman, 'Religious Contributions to the Debate on the Patenting of Human Genes', *University of St. Thomas Law Journal* 10 (2013), 650–81 (see 651; 681–82).

opinion had been 'representative'. The Board observed that it was not and that it should not have been:

> [T]he Opposition Division was under no obligation to establish any such thing. The Opposition Division was not required to form its own opinion and then somehow establish that such opinion was representative of a wider group. Quite the contrary, the task of the Opposition Division was to assess whether or not the exploitation of the invention conformed with conventionally-accepted standards of conduct in European society . . . only on the basis of the evidence placed before it by the parties in support of their arguments and with no consideration for personal opinions . . . The actual opinion or opinions of the members of the Opposition Division (or the Board) are irrelevant.

Experienced advocates, in preparing a case for presentation to a court, will consider the repercussions, both favourable and unfavourable, of a particular line of argument. *Oncomouse II* also provides a stark illustration of an argument which, although part of the opponents' very clear stand on animal rights, actually undermined itself and possibly gave the opponents a weaker position in the proceedings than if it had not been advanced. One argument deployed to demonstrate the immorality of the oncomouse was the degree of animal suffering which the opponents said was involved. Of the argument that the degree of animal suffering demonstrated immorality, the Board found the absence of any evidence of the actual degree in question 'wholly unsurprising':

> To suggest that the degree of suffering could be material is to suggest the possibility of a distinction between 'acceptable suffering' and 'unacceptable suffering'. In the Board's opinion this is not a distinction which the parties in question really wished to evoke and it is certainly not a distinction which would assist in deciding this or similar cases. Those observations are sufficient to show how it would be not only distasteful but effectively impossible for the Board (or any other decision-making instance) to make findings as to degrees of suffering.

Third, as already indicated, the unconventional approach of non-commercial parties leads to problems of evidence. Unless a fact is so well-known that a court can be asked to recognise it as established without evidence, a risky approach at the best of times, any assertion needs evidence to show it is based in fact and is not a mere hypothesis. Evidence is the everyday currency of legal proceedings, including patent oppositions. Even in the EPO, which has a relatively informal approach to evidence, there are rules and case law governing it. Non-commercial opponents are at a twofold disadvantage when it comes to evidence. Their reliance on assertions means they fail to appreciate, or at least underestimate the need for, evidence to support their arguments; and, as with arguments, they fail to assess the value of evidence, both as a means of support for and as a possible detraction from their own cause.

Again, *Oncomouse II* offers an example. Another argument of the opponents was that a patent for the oncomouse invention had been denied in Canada, by a decision of the Canadian Supreme Court.[40] It is a basic rule of evidence that if a party wishes to rely on the law of a foreign jurisdiction, the law in question must be proved as a matter of evidence. No copy of that decision was produced during the opposition proceedings. In the appeal, the Board of Appeal wrote to the parties directing them to file evidence, including national case law, in support of as yet unsupported arguments, but none of the opponents did so. The Board did not reject the argument for that reason, but it obtained a copy of the Canadian judgment for itself and found that it turned on the meaning of terms present in Canadian patent legislation but not in the EPC. It seems more likely that the opponents simply failed to understand the requirement for evidence than that they withheld the Canadian judgment knowing it would in fact undermine their argument.[41]

In contrast to the usual approach of non-commercial parties, the Article 53(a) arguments of the opponents in *Breast and ovarian cancer/UNIVERSITY OF UTAH* show that a plausible morality case can be formulated and presented, even though in this case there was again no supporting evidence. The case concerned a cancer susceptibility gene which could be used as a diagnostic tool. It will be remembered that the second of two morality arguments was that the patent would increase healthcare costs, inhibit medical treatment and research, and affect the dignity of a particularly vulnerable group of patients (those with breast cancer). The Board was of the opinion that these arguments concerned the exploitation of the patent and not of the invention. The decision records that, in an attempt to overcome that hurdle, the opponent in question argued that in this case exploitation of the invention and of the patent could be assimilated: this invention concerned breast and ovarian cancer and had a significant impact on public health, so in those special circumstances Article 53(a) should apply to the exploitation of the patent. The Board acknowledged the sensitive nature of healthcare but was not prepared to make an exception for one area to the consequences of all patents. The case does, however, show a thoughtful attempt to argue a legal issue.

### F. *Have Non-commercial Oppositions Succeeded?*

If this question is measured by the trite criterion of whether patents have been revoked on Article 53 grounds at the instance of non-commercial opponents, then the answer is no. The writer is not aware of any such 'absolute success' but does consider that achievements have been made which would not have been made otherwise.

[40] *Harvard College* v. *Canada* (Commissioner of Patents), [2002] 4 SCR 45.

[41] It is also tempting to think that they simply assumed that the refusal of a patent in another jurisdiction was in itself of persuasive value in the EPO. But that suggests that they did not even read the Canadian judgment, which did not turn on issues of morality but of statutory interpretation – whether the words 'manufacture' or 'composition of matter' were broad enough to encompass higher life forms.

One example is the role of Greenpeace in opposing plant patents. While Greenpeace cannot point to a number of revoked patents, it has challenged patents on genetically engineered plant technology so regularly that patentees in that area know they may well face at least one opponent. At the same time, the EPO has come to accept Greenpeace as a regular participant in its proceedings, and Greenpeace has learned by experience how such proceedings are conducted and, in recent years, increasingly engaged patent attorneys to present its oppositions. While some might see this as an example of a 'rebel' being drawn into the 'establishment', the result, viewed objectively, has been better (if not successful) oppositions. Parties who present their cases cogently and persuasively will almost necessarily stand a better chance than those who do not.

A second and more specific example is the *Oncomouse II* case. While at the end of the appeal hearing the opponents were understandably disappointed,[42] the fact is that their oppositions had resulted in reducing the scope of a patent originally granted for all non-human mammals to rodents and then again just to mice. The opponents would argue that the focus of their attack was always the oncomouse itself, or perhaps the iniquity as they saw it of using animals for medical research. If they could have seen the matter objectively in terms of patent law, they should rightfully have felt pleased with their achievement.

It remains the case, however, that non-commercial opponents would be more effective in opposing patents if they understood better the relevant law and procedure. In particular, they should make a more balanced appraisal of the issues in a case and present not just arguments, but evidence to support those arguments. This in turn supposes that they use EPO oppositions less as a means to deliver their point of view as in a debate, certainly less as a debate in which they appear to believe that their point of view is the only correct one, but more for what it is: a contested procedure before a tribunal which must apply the law and decide on the merits between two competing parties or sets of parties.

## G. *Could More Non-commercial Opponents Challenge Life-Form Patents?*

More particularly, and as already observed, neither churches nor other religious institutions nor any non-religious ethical organisations have been as active in patent oppositions as one might expect. On issues such as the compatibility of the exploitation of life-form patents with morality and *ordre public*, they should be able to make better, more informed cases, and moreover make them with greater authority, than commercial parties or other non-commercial parties with no or fewer ethical credentials.

[42] The writer recalls a verbal outburst by an individual from one opposing party, directed at the lady chairman of the Board of Appeal when she announced the decision, to the effect that at least a woman should have considered the patent immoral. The latent sexism in the comment seemed incongruous in the circumstances.

In the writer's view, churches and ethicists have a contribution to make to patent oppositions which has so far not been made. One church community filed an opposition in *Oncomouse II*,[43] but the writer is not aware of any larger church body – such as the Roman Catholic Church or Church of England – filing opposition.[44] Moreover, it should be expected that those whose task is to hear and judge EPO oppositions and appeals, whose own background is in either science or law, would welcome Article 53(a) cases in which constructive (and well-constructed) arguments and evidence as to what are, or are not, accepted norms in European culture are presented by, or on behalf of,[45] those who are expert in the subject.

The writer has heard it said,[46] as explanation for not opposing patents, that churches are not merely unfamiliar with the law and procedure, but that the subject-matter of patents is always highly technical in nature, with life-form patents being particularly so. While that technical complexity is undeniable, it should not prevent churches or other like bodies from opposing patents, for several reasons.

First, the objection of non-commercial parties to life-form patents will usually be centred on Article 53(a). To argue against a patent on the basis that exploitation of its invention will be contrary to *ordre public* or morality does not require detailed examination of the entire scientific and technical content of the patent and the prior art, such as a commercial opponent would undertake. The 'too technical' reason for not opposing misses the point – it is the moral, not the technical, issues which call for the attention of (non-technical) religious and ethical experts. If a churchman or ethicist can see a moral objection to the exploitation of a particular invention, then he or she needs little further scientific education to formulate an opposition.

Second, explaining the technical aspects of any patent case is usually the province of the appropriate professional, nearly always a patent attorney. While some non-commercial opponents may not have the same resources at their disposal as the commercial patentees whose patents they challenge, there is no plausible reason why a religious institution, for example, should not retain professional advisers for the limited purposes of monitoring granted patents and, if an opposition is decided upon, advising on and conducting the opposition. If the technically untutored moralist needs technical explanation, it is available.

43 The Evangelical Church in Cologne.

44 No distinction on the basis of size per se is intended (a local church congregation might present moral arguments based on religious principles as effectively as its national body). However, the larger the organisation, the more resources may be available (e.g., to engage professional advocacy services), the better the opposition case may be developed and (for what it is worth) the more gravitas the opponent may appear to have before the tribunal.

45 I am not suggesting that individual churchmen or ethicists become advocates – rather, that they or their organisations become opponents and instruct experienced representatives from their expert knowledge.

46 At the conference which prompted this book.

Third, the Article 53(a) arguments which a non-commercial opponent is likely to deploy will be arguments reflecting its own moral or ethical stance, and thus, even if a professional is used to conduct and present an opposition, the client rather than the professional will be primarily responsible for suggesting the arguments and supplying the evidence. In response, it may be asked, 'What evidence?' The answer is evidence of accepted norms in European culture which could be demonstrated by, for example, religious or polemical writings, doctrinal material or opinion polls.[47] Writings on morality should be admissible evidence as to accepted norms, just as scientific writings are admissible evidence as to the state of the art in the field of a patent. Typically, an Opposition Division or Board of Appeal is presented with a scientific review article which summarises the state of research in a particular field at its date, or with scientific textbooks as evidence of the common general knowledge in a field. By simple analogy, there is no reason why religious, ethical or other writings[48] should not be put forward as evidence[49] of prevailing moral views at a certain time.[50]

Writing of the debate on gene patenting in the United States, Chapman has observed that the religious contribution has been both sporadic and lacking in consensus;[51] she notes that there was only one church intervention in the recent *Myriad* patent case.[52] The fact that US patent law has no 'morality' provision akin to Article 53(a) possibly explains the absence of a religious contribution to life-form patent cases in that country. But it also serves to make the similar absence of such a contribution in EPO proceedings – where the morality provision invites such a contribution – all the more puzzling.

47 Opinion polls should in principle be admissible evidence of 'accepted norms', but they require great care in both preparation and presentation. In both *Plant Cells/PLANT GENETIC SYSTEMS* and *Oncomouse II*, the Boards were critical of the unrepresentative nature of the poll results relied on. It is very difficult to produce a poll of sufficient numbers to be representative, conducted in conditions which cannot be criticised, and with questions which are neither leading nor suggestive.

48 Such as sociological or philosophical.

49 Of course, careful thought should be given to the possibility of a patentee responding with writings or other evidence which suggest a different moral view.

50 In *Oncomouse II*, the Board was of the opinion that the relevant date for assessing an objection under Article 53(a) is the filing date of a patent; evidence produced later (which might include for example a review written now of the accepted moral norms of five years ago) may be taken into account provided it relates to the position at that date.

51 Chapman, 'Religious Contributions', 651.

52 An *amicus curiae* brief by the Southern Baptist Convention in *Association for Molecular Pathology v. Myriad Genetics, Inc.* (569 U.S. 576 [2013]). Chapman describes it as 'poorly written', without giving details (ibid., 682).

II

# Religious Perspectives on Life Patents

# 5

# Intellectual Property Rights and the Fundamental Right to the Commons in the Light of Catholic Social Teaching

*Monsignor Osvaldo Neves de Almeida*

## INTRODUCTION

Fourteen years ago, at the first session of the World Intellectual Property Organization's (WIPO's) Intergovernmental Committee on Intellectual Property and Genetic Resources, Traditional Knowledge and Folklore (Geneva, April 30 to May 3, 2001), the Permanent Observer Mission of the Holy See submitted a paper, issued afterward by WIPO as document WIPO/GRTKF/IC/1/7.[1] At the same time, the Holy See presented two other papers about intellectual property rights (IPR) and access to medicines to the World Trade Organization (WTO).[2] These two interventions made an important contribution to a fair interpretation and reform of Annex 1 C of the WTO Treaty (TRIPs).[3] This chapter will develop some concepts that are mainly implicit in those papers of the Holy See. Catholic Social Teaching does not address directly the subject of IPR, but it contains all the elements necessary for a moral judgement on the present system of IPR.

I will develop the chapter in three parts: First, I will present some ideas about private property and the universal destination of goods, together with some concepts about the right to knowledge. Second, I will try to propose a vision of IPR in the light of the aforesaid principles about property in Catholic Social Teaching. Third, I will examine IPR in the context of the rights of native peoples.

[1] Permanent Observer Mission of the Holy See, "Intergovernmental Committee on Intellectual Property and Genetic Resources, Traditional Knowledge and Folklore," WIPO 1st session, no. WIPO/GRTKF/IC/1/7 (Spring 2001), www.wipo.int/edocs/mdocs/tk/en/wipo_grtkf_ic_1/wipo_grtkf_ic_1_7.doc (hereinafter "Intellectual Property and Genetic Resources").

[2] "Intervention by His Excellency Mons. Diarmuid Martin to the Plenary Council of the World Trade Organization on Trade-Related Aspects of Intellectual Property Rights," June 20, 2001, http://www.vatican.va/roman_curia/secretariat_state/documents/rc_seg-st_doc_20010620_wto_en.html; "Intervention by the Holy See at the World Trade Organization: Address of H.E. MSGR. Diarmuid Martin," December 20, 2002, http://www.vatican.va/roman_curia/secretariat_state/documents/rc_seg-st_doc_20021220_martin-wto_en.html.

[3] *Id.*

## I. PRIVATE PROPERTY AND THE UNIVERSAL DESTINATION OF GOODS

### A. *Natural Right to Property*

Private property (both real estate and personal property) is generally deemed a natural right, and the same is true in Catholic Social Teaching: "Private property or some ownership of external goods confers on everyone a sphere wholly necessary for the autonomy of the person and the family, and it should be regarded as an extension of human freedom. Lastly, since it adds incentives for carrying on one's function and charge, it constitutes one of the conditions for civil liberties."[4] Small real estate property often represents the only means of subsistence for some people. Thus Catholic Social Teaching has defended the right to access to some amount of private property for everybody.[5] More recently, Pope Francis in his Encyclical Letter on the environment, *Laudato Si*, has remembered the right to property of very small farmers.[6]

### B. *Private Ownership Also of Capital Assets*

On the other hand, the same Catholic Social Teaching has constantly stressed that ownership of capital assets and means or factors of production is justified when it is beneficial for the whole society and when it is respectful of the fundamental rights of everybody, including people beyond each country's boundaries.[7] The social doctrine of the Church is not against private ownership of capital and capital accumulation, but it affirms that private property should work for the sake of all, including the poorest sections of society.[8]

4 Pope Paul VI, Gaudium Et Spes: On the Church in the Modern World (1965), ¶ 71, Vatican Website, http://www.vatican.va/archive/hist_councils/ii_vatican_council/documents/vat-ii_const_19651207_gaudium-et-spes_en.html.

5 *Id.*

6 Pope Francis, Laudato Si': On Care For Our Common Home (2015), ¶¶134–35, Vatican Website, http://w2.vatican.va/content/francesco/en/encyclicals/documents/papa-francesco_20150524_enciclica-laudato-si.html (hereinafter Laudato Si).

7 Cf. Pope John Paul II, Centesimus Annus: On the Hundredth Anniversary of Rerum Novarum (1991), ¶38, Vatican Website, http://w2.vatican.va/content/john-paul-ii/en/encyclicals/documents/hf_jp-ii_enc_01051991_centesimus-annus.html (hereinafter Centesimus Annus).

8 Pope Francis, message on "First World Day of the Poor" (2017), ¶3, Vatican Website, https://w2.vatican.va/content/francesco/en/messages/poveri/documents/papa-francesco_20170613_messaggio-i-giornatamondiale-poveri-2017.html; Laudato Si, ¶ 158; Pope Francis, Evangelii Gaudium: On the Proclamation of the Gospel in Today's World (2013), ¶ 190, Vatican Website, http://w2.vatican.va/content/francesco/en/apost_exhortations/documents/papa-francesco_esortazione-ap_20131124_evangelii-gaudium.html; Pope Francis, message "On the Occasion of the Annual Meeting at Davos-Klosters (Switzerland)" (2014), Vatican Website, http://w2.vatican.va/content/francesco/en/messages/pont-messages/2014/documents/papa-francesco_20140117_messaggio-wef-davos.html.

## C. *The Social Mortgage of Private Property*

### 1. Right of Everybody for Subsistence

All men and women of all nations are entitled to have whatever they need for their subsistence and personal advancement, taking it from all the resources available at any given time in history. The provisions protecting private property thus cannot ever lose sight of the common destiny of all goods, so much so that it has to be said that all private property is subject to a "social mortgage."[9]

### 2. Property as a Service

Private property, then, is for no one an unconditional, absolute right; rather, it is an instrument with which to achieve effective and responsible access to a portion of goods, contributing to ensure at the same time that all individuals and all families have their essential environment of freedom and just autonomy. Consequently, should there be an institutional conflict between acquired private rights and over-riding community necessities, it is for the public authorities to set about resolving it with active involvement on the part of individuals and social groups.

## D. *The Relation of IPR to Human Dignity and the Universal Destination of Knowledge*

As regards human-created goods established through labor, the intellectual component plays a fundamental role. The accumulation of knowledge is necessary to control nature and its use in service to society. It creates intellectual capital that becomes essential for the development of humankind and that constitutes a decisive element for the wealth of people and the power of nations.[10]

Thus there is another form of property that exists in our time and is no less important than land: the property of knowledge, of technology, and of know-how.[11] If at one time the decisive factor of production was land and then capital (understood as machinery and instrumental goods), today the decisive factor is ever more man himself – that is, his capacity for knowledge.[12]

The social nature of men and women is evident. Moreover, Christians, as well as those of other religions, believe that this social nature encompasses a duty of justice, solidarity, and charity. Thus intellectual property, being something that belongs to

[9] Pope John Paul II, Sollicitudo Rei Socialis: On the Social Teaching of the Church (1987), ¶ 42, Vatican Website, http://w2.vatican.va/content/john-paul-ii/en/encyclicals/documents/hf_jp-ii_enc_30121987_sollicitudo-rei-socialis.html#-27.

[10] Cf. Centesimus Annus, ¶¶ 30–44.

[11] *Id.* at ¶ 32.

[12] *Id.*

the very nature of man as a social being, relates directly to human dignity and, at the same time, has a universal destination.

For the existence of a just society, it is fundamental that attention be given to how intellectual goods are distributed. When people have access to knowledge, they can express their creativity and develop their potential; they can enter into a system of knowledge and inter-communications, putting their qualities at the service of their neighbors.

Finding a way for a practical and efficient sharing of intellectual know-how is an important act of justice.

> There is a paramount need for *a greater and more equitable sharing of resources*. This includes the transfer of science and technology. . . . It means a technology that is appropriate to the needs and best interests of the people and nations involved. But it means much more than just material sharing. There is an urgent need for a sharing of the resources of the mind and the spirit, of scientific knowledge and cultural and artistic expression.[13]

The Catholic Church "has consistently taught that there is a 'social mortgage' on all private property, a concept which today must also be applied to 'intellectual property' and to 'knowledge.'"[14] "The law of profit alone cannot be applied to that which is essential for the fight against hunger, disease and poverty."[15]

## II. A VISION OF IPR IN LIGHT OF THE AFOREMENTIONED PRINCIPLES ABOUT PROPERTY IN CATHOLIC SOCIAL TEACHING

### A. *The Classical System as a Good Balance between the Right to Private Ownership and the Universal Destination of Goods*

In its 2001 Note to the WIPO, the Holy See remembered that "[t]he raison d'être of intellectual property protection systems is the promotion of literary, scientific or artistic production and inventive activity for the sake of the common good."[16] "That protection officially attests the right of the author or inventor to recognition of the ownership of his work and to a degree of economic reward, at the same time as it serves the cultural and material progress of society as a whole."[17]

[13] Pope John Paul II, message "To the General Assembly of the United Nations" (1980), Vatican Website, https://w2.vatican.va/content/john-paul-ii/en/messages/pont_messages/1980/documents/hf_jp-ii_mes_19800822_nazioni-unite.html (emphasis in original).

[14] Pope John Paul II, message "To the Group 'Jubilee 2000 Debt Campaign'" (1999), Vatican Website, https://w2.vatican.va/content/john-paul-ii/en/messages/pont_messages/1999/documents/hf_jp-ii_mes_19990923_jubilee-2000-debt-campaign.html.

[15] *Id.*

[16] "Intellectual Property and Genetic Resources," p. 1, ¶ 2.

[17] *Id.*

Indeed, "the classical intellectual property system ... included in its original conception the notion of a social [mortgage], manifested in the substantive and time limitations on the rights granted."[18] In the case of patents, the right of allowing or forbidding the use of the invention was balanced with the duty of disclosure, "the discretion of governments with regard to the choice of the industrial sectors to be protected, the free determination of the scope of the conditions of patentability, the various options for opposing patents, and the compulsory license regime."[19]

In harmony with the social teaching of the Church, the Holy See stressed that what "intellectual property protection works for is the recognition of the dignity of man and his work, in its double dimension, namely as a medium for the expression and growth of the individual personality[,] and as a contribution to the common good."[20]

## B. *The New Situation of IPR*

### 1. The Extension of Patentable Matter

One of the characteristics of the economy of the twentieth and twenty-first century is the fluid relation between science and industry. Science has developed such knowledge on the structure of matter and on the mechanisms of life that every new scientific discovery also discloses a vast panorama of potential practical applications relevant to the many aspects of human life. Such practical applications arrive to the public through markets as new goods and services. Business has become not just the intermediary between science and people, but the main financial propeller of science, as both the seller of technological goods and the source of financial investments. Accordingly, the need to ensure a strong market position in order to increase both profits and financing has led to a race to patents. Lawyers explore and use all the possibilities offered by a given legal framework for obtaining new, broader, and stronger protections, struggling to convert all the intellectual heritage of companies into patentable matter. At the same time, the particular characteristics of scientific knowledge, for instance in the area of genetics, make it difficult to distinguish between discovery and invention, thus contributing to the enlarging of the concept of patentable matter.

Accordingly, the administrative practice of a number of industrialized countries concerning patents and the corresponding case law have evolved into accepting a broad concept of "invention" that could encompass not only novel creations of the human mind, but also discoveries. For instance, some jurisdictions accept that it is possible to grant a patent on matter that is just a replication of matter existing in nature, but has been obtained by means of a biochemical process – a sort of reverse

18 *Id.* at 4, ¶ 9.
19 *Id.* at 4–5, ¶ 9.
20 *Id.* at 1, ¶ 2.

engineering of the complex results of natural evolution. This legal development has made it possible to patent genetic components of plants, animals, and human beings that possess properties of particular usefulness.

### 2. From Invention to Assets: Erosion of the Social Purpose of Patents

During the twentieth century, especially its second half, the previously noted bond between applied science and industry has caused "intellectual property" to evolve from a reward for one's own individual ingenuity to the capital asset and production factor of businesses.[21] Actually, the capacity of companies to undertake scientific research (on their own or in association with academic bodies) and "the corresponding legal protection of the intellectual heritage that results [from it] have become one of the most important parameters governing their economic strength and their" ability to attract investment and loans.[22] Such a trend has changed the social meaning of patents and could erode the social function of IPR.

## C. *Capitalism with Limited Free Markets*

### 1. The Rights Granted by Patents, during their Life-Span, Are Stronger than those Granted by Personal Property

Patents, in fact, "grant to their owners [a] discretionary power to control or charge for acts involving the content of the patent."[23] This means that patents have "certain legal connotations that even go beyond the protection of the ownership of the material goods: the latter grants only the power to object to third-party ambitions to exercise prerogatives of ownership on goods of which one is the owner."[24] "[T]he patent, on the other hand, tends to grant throughout the lifetime of the patent the right to control [any act of any person] that entails a use of the patented knowledge, in any place within the jurisdiction or jurisdictions in which the patent is enforceable, and regardless of the subject matter affected by those acts or the social environment in which they take place."[25]

### 2. Such Rights Could Wear Away the Universal Destination of Knowledge

Inventions prior to the age of polymer chemistry and biotechnologies did not include too much scientific knowledge; rather, they involved ingenious applications of well-known chemical and physical principles. For these reasons, the

21 *Id.* at 2, ¶ 4.
22 *Id.*
23 *Id.* at 5, ¶ 9.
24 *Id.*
25 *Id.*

ban on the practical use of the know-how included in a patent meant mainly limiting imitations of the invention, and not a real ban on the alternative use of the know-how contained in it. As such, the traditional patent system did not restrain further or parallel scientific and industrial research, in contrast to what seems to be happening today in the area of biotechnology. Today's patents, instead, could bar every practical sharing of knowledge other than plain scientific teaching.

### 3. Yet Patents Harvest from All the Knowledge Heritage of Humankind

If one can generally assert that all goods have a "universal destination," such a statement should have particular significance for intellectual property, and more specifically for certain aspects of intellectual property (inventions, technological know-how, patents) *that have a universal origin*. Inventions develop in the context of a preexisting cultural environment from which the inventor benefits either directly or indirectly for his research. The innovations of the Industrial Revolution were constructed against a scientific and technical background steadily accumulated from the very beginning of humankind, which acquired its momentum during the eighteenth and the first part of the nineteenth centuries (also the time of the inception of the modern patent system). Most of those inventions, moreover, occurred within the cultural frame of European-Christian culture, and probably would not have happened without its contribution. The same could be said about almost all the successive scientific and technological developments of the nineteenth and twentieth centuries. Likewise, the new development of biotechnology was nourished in the soil of a rich and deep heritage of scientific theoretical development.

### 4. Monopoly and Profits

From a legal point of view, a patent essentially implies the right to exclude others from any commercial (public) use of a technical formula or of know-how. In strict juridical language, it cannot be said that patents per se grant an economic monopoly, because the effective exploitation of the invention could be forbidden for other reasons or the production and spread of the inventions could be governed by other, different legal contexts. But when such other factors are not present, a patent is a tool to establish a kind of monopoly during its life period.

The economic functioning of patents can allow the owners to artificially raise prices in order to maximize profits. It is believed that the availability of a time-limited exclusive right will permit the patentee to secure adequate rewards for the contribution made, through freedom to adjust prices for the products or services involved without fear of competition.

### 5. Why Capitalism with Limited Free Markets?

To summarize, today's economic system seems to concentrate on the monopolistic effect rather than the social utility of the patent, not only to ensure profit but also to draw a flow of investments toward the patent holder. Scientific, technical know-how is now considered a company's most important asset, even more important than production. Hence the monopolistic element becomes fundamental, warranting universal protection, an increase in duration, and the severe limitation of mandatory licenses.

Thus, in conclusion, IPR and particularly patents have become some of the most important tools of modern capitalism, which paradoxically works by appropriating the universal knowledge heritage of humankind without often paying the whole fair price to society, by restricting free competition and, to some extent, also by obstructing the free spread of knowledge.

## D. *The Essential Moral Framework of Patents*

### 1. The Inherent Morality of the Act of Granting a Patent

There are no social acts that do not also have a moral meaning. Patent protection could be seen as having a positive moral meaning, because patents are like a contract (*do ut des*) between the inventor and society. In fact, the inventor agrees to release and publicize all the information regarding the invention, thus contributing to the spread of useful knowledge, and society rewards him with the rights granted by the patent.

### 2. Biotechnological Patents: The Inherent Risks of Advanced Technologies, Especially Biotechnologies

However, the first moral issue concerns the invention, in itself. It is well known that granting a patent is not per se an ethical approval of the invention, inasmuch as the protected activities could also be conducted without this type of ownership coverage. The activity in itself will be moral or amoral with or without a patent. But in the absence of other relevant legal prescriptions, the granting of a patent is a moral endorsement, since patents are given for inventions that are deemed of public utility.

The latest developments in Catholic Social Teaching address the relation of humankind to science and technology.[26] Benedict XVI addressed the subject in his Encyclical Letter *Caritas in Veritate*, and Francis came back strongly to the subject in the Encyclical Letter *Laudato Si*. The reason is that a Promethean vision of humankind, by which men feel omnipotent through science and by which progress is equated with the ever increasing supply of comfort and consumer goods for the wealthiest people, is critically harming the environment, does not resolve the problem of poverty (on the contrary, it increases social exclusion), and in the long run could also harm human nature itself.

26 *Id.*; Laudato Si.

From time to time we hear news of people experimenting in the direction of an eventual symbiosis of humans and machines, or in the creation of chimeras; and this is not science fiction, for it suffices to remember how difficult it was to achieve, in 2005, a UN Declaration on human cloning.[27] Those inventions, which someday could alter the human species or which use living people (whether embryos or involving other experiments with living people), should be neither allowed nor patented, because they never could be deemed of public utility.[28] In addition to the prohibition of the activity, the non-patentability of such inventions would be a clear affirmation that their public utility could indeed never be established, while it would also discourage any use of such inventions to attract investment.

## 3. Rights and Interests Met By the Patent

The second moral issue regards the consequences of the exclusion implied by a patent. How equitable is it to exclude all the non–rights holders from activities which are of direct importance for individual or social well-being?

An evaluation of the institution at a juridical and ethical level should take into consideration its relationship with the diverse fundamental rights and interests at stake. Within the context of social life, the patent satisfies society's interest in the progress of inventive activity, as well as the right to stimulate the creation of new goods and services. At the same time, it meets the interest/right of the inventor in the acknowledgment of the paternity of his creation and the outcome of his work. Presently, with the industrial application of biotechnology, it also guarantees a service to investors by ensuring through the monopoly a prospective income on their investments, and a service to the companies by supplying a legal means (the monopoly) to attract investments.

## 4. Patents and the Universal Destination of Goods: Demand Side and Supply Side

However, the patent is only indirectly useful for the universal destination of goods. In its moral consideration of the production of goods and services, the concept of universal destination regards the creation of new market goods only as an afterthought, because its first objective is the universal enjoyment of all necessary goods and the right to a dignified existence.[29] Its primary concern regards the "demand"

[27] "General Assembly Adopts United Nations Declaration on Human Cloning by Vote of 84-34-37," *UN News Centre*, March 8, 2005, https://www.un.org/press/en/2005/ga10333.doc.htm.

[28] *Juicy Whip, Inc. v. Orange Bang, Inc.*, 185 F.3d 1364, 1368 (Fed. Cir. 1999) (quoting *Webber v. Virginia*, 103 U.S. (13 Otto) 344, 347–48 (1880)); "Facts on Patenting Life Forms Having a Relationship to Humans," *United States Patent and Trademark Office*, April 1, 1998, https://www.uspto.gov/about-us/news-updates/facts-patenting-life-forms-having-relationship-humans.

[29] Pope John Paul II, Laborem Exercens: On Human Work, ¶ 14 (1981), Vatican Website, http://w2.vatican.va/content/john-paul-ii/en/encyclicals/documents/hf_jp-ii_enc_14091981_laborem-exercens.html.

side of the basic economic equation and therefore seeks a just division of the earnings of economic activity. The patent, on the other hand, works on the side of supply and free enterprise, with a proven capacity to stimulate technological developments, but does not offer the means to a fair satisfaction of the demand side.

First, while the patent does reveal certain new techniques or know-how to society, at the same time it blocks their public use for a substantial period of time, in order to maximize the inventor's profits. Second, and of utmost importance, the monopoly's ability to stimulate technological progress is selective and partial, because in today's economic reality, investments in patented products appear attractive when directed toward market segments with great economic potential, but not when oriented toward segments with a limited market.

The question of the lack of access to medicines seems to constitute proof that the combination of private investments and patents does not promote development. As it has often been said, the increasing strength of intellectual property rights and their disengagement from any social consideration (the retreat of public investments, factual elimination of compulsory licenses, universal coverage of patents, etc.) still have not produced enough research benefiting the most elemental needs of the poorest sectors of society. Likewise, it has widened the gap between rich and poor people by linking access to new drugs to investment returns.

Consequently, it is not clear if, from a public good point of view, the economic exploitation of inventions always deserves, in every circumstance, the position of particular privilege given by patents. It is not even clear if the amount of protection conceded by the present regime necessarily corresponds to society's needs. The patents system needs to be balanced by tools conceived for the purpose of putting the invention at the service of the public when necessary. That is the sense behind the institution of compulsory licenses, which have been maintained during two centuries in both national and international laws. It should be said that the present universal patent protection cannot be deemed unfair per se if complemented by the necessary corrections.

## III. IPR AND THE RIGHTS OF NATIVE PEOPLES

### A. *The Previous Human Rights and the Constitutional Question of Rights on the Land*

#### 1. "Intellectual Property and Genetic Resources, Traditional Knowledge and Folklore," and Land

The title of the WIPO Committee and law developments relating "Intellectual Property and Genetic Resources, Traditional Knowledge and Folklore" leaves aside a question that seems prior and more basic in the discussion of all native peoples' rights: the question of the ownership of land or, at least, of the existence of some kind of sovereignty originating in their long-established presence on the land. Perhaps, therefore, the title of this subject should be "Land Property, Genetic Resources, Traditional Knowledge and Folklore." However, the vindication of the fundamental rights of indigenous people outside those rights related to the IPR legal system is beyond the scope of this discussion. Nevertheless, it is important to say that without recognizing some kind of right of native peoples to their traditional territories, it is impossible to attribute rights on the genetic resources. The basic affirmation should be that their long-dated presence in their territories, together with "ancestral concern for the soil on the part of indigenous communities[,] generates a right to its use and usufruct, and it has also to be recognized that this right extends also to the plants and animals that go with the territory."[30]

### 2. A Constitutional Law Problem

The development of the whole present legal framework of most countries started with the constitution of the nation states in the nineteenth and twentieth centuries. Such constitutional developments have tended to acknowledge only the rights of the nationalities or social groups that were already dominant in a given country, who were often colonists who arrived not long before, or their descendants. The pattern of real property rights did not recognize the rights of native peoples who were in the land many centuries before the arrival of the colonists or of the dominant group or culture. The IPR legal system generally developed after the constitutional stabilization of European countries, simply ignoring the rights of native peoples because their rights or even their very existence were ignored by constitutional law. The vindication of indigenous peoples' rights is already concretized in many political international declarations and in some international legal instruments, but still it has not been incorporated into the constitutional law of most of the countries themselves.

It is worth mentioning here that in the twenty-first century at times of the "conquest" or "colonization" of the Americas, Catholic missionaries in the "new" continent, like Montesinos e Las Casas, and Catholic theologians in Europe, mainly in Spain – Vitoria, Suarez, and others – were strenuous defenders of the rights of native peoples, including their "natural right" to some kind of sovereignty or ownership of the land.[31] Recently, Pope Francis has explicitly referred to such rights.[32]

## B. *Biological Resources, Native Populations, and Traditional Knowledge*

### 1. Biological Resources and Indigenous Communities and Territories

Biological research has highlighted the great social and economic usefulness of a large range of natural resources, both vegetal and animal. It has found ways to

30 "Intellectual Property and Genetic Resources," pp. 2–3, ¶ 5.
31 See William J. Bennett, *America: The Last Best Hope, Vol. 1: From the Age of Discovery to a World at War* (Tennessee: Thomas Nelson, 2006), 10–11.
32 Laudato Si, ¶ 94.

develop their potential and has obtained new patents to protect the ownership of the findings. However, most of these biological resources are located in "territor[ies] inhabited since time immemorial by native communities."[33] Moreover, it often "happens that those native communities already have some knowledge and make use of some of the biological properties" of the natural resources.[34] In addition, their biological environment is closely associated with their culture and "constitutes an integral factor of their identity and social cohesion."[35] In any case, those indigenous communities lack the ability to oppose patents obtained by the biotech industry, because there is no room in the patent law for their demands, because the patent jurisdiction in which patents are granted is in a country different from their own, or simply because they have no legal or economic means to litigate.

### 2. Biological Resources That Are Part of a Human's Genetic Heritage

The Holy See's Permanent Observer Mission has made clear at WIPO that

> [o]ther biological material susceptible to industrialization forms part of the genetic heritage of men and women, and especially the members of those native communities which, owing to their peculiar lifestyles, have in the course of generations developed specific genetic features. Any attempt at economic exploitation of such resources has to be strictly regulated, in order always to ensure full respect for personal dignity and freedom, which includes the right to be fully informed on a given project, the right to fair participation in the benefits, and also the right to object to the use of resources derived from their own bodies.[36]

### 3. Traditional Knowledge

In addition, the Holy See has emphasized there are "innovative or creative activities developed and refined throughout history, which are [akin] to the social manifestation of the work of several generations and [reflect] a genius peculiar to communities, peoples[,] or families."[37]

> The intellectual manifestations of tradition or folklore nevertheless deserve recognition, because they correspond in all respects to the substantive concepts that afford entitlement to "classical" protection of intellectual property, as on the one hand they constitute a means of constructing and projecting the identity of the members of the community concerned, and on the other are a common asset of that same community, which has grown with small, anonymous contributions over a great many generations.[38]

33 "Intellectual Property and Genetic Resources," pp. 2, ¶ 5.
34 *Id.*
35 *Id.* at p. 3, ¶ 5.
36 *Id.*
37 *Id.* at p. 1, ¶ 3.
38 *Id.* at pp. 1–2, ¶ 3.

### C. *Different Sets of Rights at Stake*

"In the case of the copyright and patent law [in place] before the 1980s, the two types of legal claim that the intellectual property system had to balance against each other were the right of the author or inventor to recognition of [the] authorship of [his] work and to remuneration on the one hand, and on the other hand the interest of society in stimulating intellectual innovations that were of general usefulness."[39] However, "[t]he new legal panorama created . . . by the extension" of the concept of patentable matter and by the linking of intellectual property to business "includes a wider range of rights and interests."[40]

Now, different overlapping sets of rights are at stake:

(1) "[t]he rights of the native populations that have developed the traditional knowledge and the expressions of folklore[,] or who occupy the territories from which the genetic material comes";
(2) "the right of the countries to the resources associated with biological diversity, the right of the inventor or discoverer to remuneration for any intellectual value that he may have added, the possible rights and interests of companies";
(3) "society's right to or interest in the stimulation of inventive activity and the development of science and the arts"; and finally
(4) "a more general right of all mankind to be assured that the products of scientific progress will serve everyone equally and not only the sectors with the greatest acquisitive potential."[41]

As the Holy See has said, "The ethical challenge to be met is therefore that of reconciling the various rights and interests at stake in such a way that the legitimate economic interest does not compromise higher values such as the social function of inventions and knowledge and the rights of the peoples with which the knowledge and resources originate."[42]

### D. *Acknowledging the Intellectual Property Rights of Native People as a Way for Redressing Old Injustices*

The challenge of acknowledging innovative forms of IPR could provide a very positive way of redressing the human rights' deficit affecting native peoples. Actually, it could be unfair and sometimes impossible to scrub out the rights of other populations that have settled the same lands, once the native people have been conquered or submitted. Likewise, it is almost impossible to delete national governments and boundaries. Instead, the "achievement of equitable economic

39 "Intellectual Property and Genetic Resources," pp. 3, ¶ 7.
40 *Id.*
41 *Id.* at p. 4, ¶ 7.
42 *Id.*

participation of native populations in the benefits deriving from the commercial exploitation of biological resources"[43] seems feasible and easier. Biological resources, among other things, are also economic assets belonging to native populations.

### E. *The Creation of New Legal Concepts*

New legal concepts are needed to "ensure certain minimum protection requirements for those sectors whose rights and interests are not fully acknowledged in systems now in force."[44] Paramount among such new concepts seems to be the "acceptance of the institution of the [previous] informed, free consensus of persons, peoples and States[,] as a prerequisite of patenting and/or defense of industrial secrecy or trade secrets relating to biological resources."[45]

In addition to the informed and free prior consensus, "in the case of human genetic resources obtained from populations or individuals," there is a need for "integration in the legal institutions of intellectual property, in the appropriate manner, of international [rules and guidelines] already existing that relate to biomedicine and human rights."[46] Paramount "in the case of the protection of [T]raditional [K]nowledge [is] the promotion of effective means of ensuring respect for the collective ownership of traditional knowledge and full recognition and enforcement of the rights resulting from existing [community legal] systems, also where appropriate going beyond the actual territorial scope or national borders."[47]

## IV. CONCLUSION

### A. *The Utopia of a True Free Market*

In order to ensure that patents serve the universal destination of goods and the common good, a new legal theory should be developed, especially around the elements concerning the public availability of the invention, such as the duration terms of the monopoly, the right price of the licenses (compulsory or not), the possible exceptions to the patent, the means of technological transfer, and the creation of alternative means of industrial knowledge protection. A fair calculation of the monopoly range, in addition to ensuring a bigger share of inventions, would limit the excessive concentration of profit made by the inventor, and it would moderate arbitrary pricing practices by allowing early competition. This could be

43 *Id.* at 6, ¶ 10.
44 *Id.* at 5, ¶ 10.
45 *Id.* at 6, ¶ 10.
46 *Id.*
47 *Id.*

seen as utopian, but if we want to have a free economy that works for the common good, it is necessary.

What is at stake is not the patent in itself, but its present economic effects in relation to the general right to development and distribution. The tendency, at present, is toward a preponderance of the twin concepts of "investments/monopoly." There is need therefore for a discussion *de iure condendo* (i.e., about law in a transitional stage[48]) – a discussion that questions, from an economic, social, and legal point of view, the fairness of the present system and the advisability and possibility of modifying the present juridical formulation of the patent.

### B. *A Fair and Holistic View and Application of Present Patent Law while the New Legal Theory Is Developed*

In a broader perspective, international and national laws must be seen as a unity hierarchically constructed under the basic conventions on human rights, which express the inalienable features of human dignity. For such unity of law and also for the "*pacta sunt servanda*"[49] principle, other international agreements such as the Convention on Biological Diversity and its successive protocols should be deemed binding – together with the WIPO conventions, TRIPs, and other international instruments of IPR – and the different precepts should be coordinated and harmonized. In some cases, a mandatory connection of such international agreements with TRIPs should be favored, so that any possible patent exclusions based on these conventions are not considered undue obstacles to international commerce.

Last but not least, there is an urgent need to renovate and revitalize the diverse multilateral systems concerned with the transfer of know-how and technology, also with appropriate and generous funding by rich countries, accompanied by the harmonization of patents with an adequate legal system for sharing knowledge and technology.

Also, and for all the reasons here presented, it would be desirable that "case law and administrative practice . . . abide by prudent and restrictive rules in relation to the extension of [a patent's] scope, allowing such extension only in cases of proven social usefulness."[50] In summation:

> there is an urgent need to preserve the possibility of invoking where appropriate the "social encumbrance" mentioned earlier through the application, subject to respect for the rule of law, of the moderating elements devised by legal science and practice, such as compulsory licensing and the exclusion of protection for reasons of public policy and morality in the case of patents, or reasonable exceptions to copyright.[51]

[48] "*De iure condendo*," Oxford Reference, http://www.oxfordreference.com/view/10.1093/acref/9780195369380.001.0001/acref-9780195369380-e-511.

[49] "Treaties shall be complied with." International Judicial Monitor, *General Principles of International Law*, http://www.judicialmonitor.org/archive_0908/generalprinciples.html.

[50] "Intellectual Property and Genetic Resources," p. 5, ¶ 9.

[51] *Id.*

# 6

# Human Rights and Life Patents: Lessons from the Church's Social Teaching and Engagement in the United States

*Stephen M. Colecchi*

The use of patents on living matter, including human genes, DNA, and stem cells, as well as microorganisms, plants, and animals, has grown exponentially. At the same time, the Catholic Church's teaching and societal engagement with these "signs of the times" have developed in response. Life patents are a form of intellectual property protection that is increasingly being enshrined and strengthened in bilateral and multilateral trade agreements. In this chapter, I will address this phenomenon from my perspective serving the Catholic Church in the United States. My contribution to the discussion of life patents will explore two questions: What lessons can be learned from the Church's engagement with life patents as they touch the rights of persons who are poor, indigenous, or marginalized? And how can the Church appropriately defend the rights of persons at the margins of the global economy, ensuring their just and fair treatment, and their access to the life-saving benefits of the objects of life patents?

My experience is one of a practitioner who applies Catholic Social Teaching to pressing issues. For more than a decade, I have coordinated staff support to the Committee on International Justice and Peace of the United States Conference of Catholic Bishops (USCCB). This committee of bishops and expert consultants applies Catholic Social Teaching to international policies of concern to the bishops of the United States in an effort to inform and influence the actions of our government. The mandate of the committee is to assist "the [US] bishops, both collectively and individually, in advancing the social mission of the Church on international justice and peace through policy development, advocacy, education, outreach, and acts of ecclesial solidarity. The committee advises the bishops on international public policy issues, especially integral human development, human rights, religious freedom, and peace."[1]

This chapter will describe and explore key assets the Church in the United States brings to these questions, and it will close with an analysis of the effectiveness of these efforts, offering recommendations and directions for future work. My reflections will not touch on the substantial biomedical ethical questions raised by life

[1] *Who We Are: Committee on International Justice and Peace*, http://www.usccb.org/about/international-justice-and-peace/who-we-are.cfm.

patents, especially the patenting of human genetic material and the use of human stem cells and embryos. Earlier in my career, I coordinated Catholic advocacy efforts for the Diocese of Richmond at the Virginia General Assembly. In that capacity, I successfully helped lead an effort to oppose a law purportedly designed as a "prohibition on human cloning." This bio-industry-promoted bill would have permitted unfettered experimentation on human embryos, requiring their destruction. Such a utilitarian objectification of human life carries terrible moral risks for society.

My sole focus here will be on the social justice implications of life patents for communities and nations at the margins of our world, specifically drawing on USCCB's experience of engaging US policy makers. Pope Francis writes in his Apostolic Exhortation *Evangelii Gaudium*, "[A]ll of us are asked to obey [the Lord's] call to go forth from our own comfort zone in order to reach all the 'peripheries' in need of the light of the Gospel."[2] This preferential option for the poor requires answers to an essential moral question: what is the impact of life patents on persons, communities, and nations who live on the world's peripheries?

The impacts on those at the margins are likely to be profound and not entirely positive. In his encyclical *Laudato Sí*, Pope Francis warns that biotechnology is among the abilities that have "given us tremendous power." "More precisely," together with other capacities, it has "given those with the knowledge, and especially the economic resources to use [it], an impressive dominance over the whole of humanity and the entire world." In this regard, many share the conviction that "'contemporary man has not been trained to use power well,' because our immense technological development has not been accompanied by a development in human responsibility, values and conscience."[3]

There are two areas in which life patents on plants, animals, and human genetic material are particularly critical: agriculture and medicine. USCCB has engaged in policy development and public advocacy on intellectual property provisions related to life patents in trade agreements that would impact both agriculture and medicine. This engagement continues today.

Despite the current US administration's expressed skepticism regarding multinational trade agreements, such agreements remain an important vehicle for intellectual property provisions related to life patents. Although President Trump withdrew from the Trans-Pacific Partnership (TPP), his administration remains committed to renegotiating or withdrawing from the North American Free Trade Agreement (NAFTA) with Canada and Mexico, and seems open to continuing talks with the European Union on the proposed Transatlantic Trade and Investment Partnership (TTIP). The administration has indicated a receptivity to bilateral trade

[2] Pope Francis, *Evangelii Gaudium*, no. 20 (November 24, 2013), http://w2.vatican.va/content/francesco/en/apost_exhortations/documents/papa-francesco_esortazione-ap_20131124_evangelii-gaudium.html.

[3] Pope Francis, *Laudato Sí*, nos. 104, 106 (May 24, 2015), http://w2.vatican.va/content/francesco/en/encyclicals/documents/papa-francesco_20150524_enciclica-laudato-si.html.

agreements and a particular interest in protecting intellectual property rights. USCCB brings three assets to its work on life patents: the teaching of the Church; relationships with the Church in other nations; and on-the-ground experience in developing nations, especially through the work of the Conference's relief and development agency, Catholic Relief Services (CRS). I review each of these below.

## I. THE TEACHING OF THE CHURCH

The teaching of the US bishops reflects and builds upon the teaching of the Universal Church regarding life patents and intellectual property rights. While acknowledging the right to private property, the bishops go on to articulate its limits in service to the common good. Citing both Pope Paul VI and Pope John Paul II, the bishops wrote in their 1986 landmark Pastoral Letter, *Economic Justice for All*:

> The common good may sometimes demand that the right to own be limited by public involvement in the planning or ownership of certain sectors of the economy. ... "Private property does not constitute for anyone an absolute or unconditioned right. No one is justified in keeping for his exclusive use what he does not need, when others lack necessities." Pope John Paul II has referred to limits placed on ownership by the duty to serve the common good as a "social mortgage" on private property ... Therefore, as Pope John Paul II has argued, "One cannot exclude the socialization, in suitable conditions, of certain means of production." The determination of when such conditions exist must be made on a case-by-case basis in light of the demands of the common good.[4]

This teaching lays the foundation for the limits the bishops later specifically acknowledge on the right to intellectual property.

In their 2001 statement, *Called to Solidarity with Africa*, the US bishops addressed the question of life patents explicitly within the context of trade agreements. They wrote:

> The negotiation and implementation of trade agreements, including those involving intellectual property rights and patenting of life forms (plants, seeds), should be undertaken in consultation with African governments and civil society to protect the rights and dignity of all parties. In this way, U.S. trade relationships with African countries can help to promote more equitable terms of exchange, greater development, and broader political and economic participation with African societies, thus strengthening Africa's self-reliance.[5]

[4] United States Catholic Bishops, Economic Justice for All: Pastoral Letter on Catholic Social Teaching and the U.S. Economy no. 115 (1986), http://www.usccb.org/upload/economic_justice_for_all.pdf.

[5] United States Catholic Bishops, A *Call to Solidarity with Africa* (November 14, 2001), http://www.usccb.org/beliefs-and-teachings/what-we-believe/catholic-social-teaching/upload/statement-a-call-to-solidarity-with-africa-2001-11-14.pdf.

Embedded within this statement are explicit concerns for equity, participation, and human development, and an implicit concern for the power imbalance between the United States and the developing countries of Africa as trade agreements are negotiated. These concerns came into clearer focus as the US bishops addressed particular trade agreements.

A 1993 letter of the Conference of Bishops to the US Trade Representative regarding NAFTA emphasized the "preferential option for the poor" of *Economic Justice for All* and lamented that "although [the proposed trade agreement] goes into considerable detail about protections for investors and intellectual property, it says very little about workers and their rights."[6] In these early years of engagement, the US bishops acknowledged the presence of intellectual property provisions in trade agreements but did not focus on the moral problems that those protections would pose directly. Instead they focused almost entirely on what were then perceived as more direct threats to labor rights and the environment. For example, in 1997, in connection with so-called fast-track legislation that grants the president the authority to negotiate trade agreements and constrains Congress to a simple up or down vote of the resulting agreement, then archbishop Theodore McCarrick wrote on behalf of the International Committee:

> Workers and the environment are directly impacted when trade is opened up and should receive no less attention than, for example, intellectual property rights. The bill before the House establishes as a "principal negotiating objective" the "strong enforcement of intellectual property rights, including through accessible, expeditious and effective civil, administrative, and criminal enforcement mechanisms." We have no precise formula to offer and no position on, e.g., whether trade sanctions are necessary to assure enforceability, but we do believe that workers and the environment are priority, not secondary concerns, and should receive no less protection in trade negotiations than intellectual property rights.[7]

The same letter, however, did emphasize the importance of "equitable and sustainable development," "the needs of poor countries," and "broad participation in economic decision-making by the people of each country." These considerations would come to be applied directly to intellectual property rights in later years.

As trade ministers gathered in Cancun, Mexico, for the fifth World Trade Organization Ministerial Conference in 2003, the US bishops continued to amplify their focus on poor countries, specifically noting a concern for access to life-saving

[6] Letter of Msgr. Robert N. Lynch, General Secretary, United States Catholic Conference (precursor of USCCB), to Ambassador Mickey Kantor, US Trade Representative (June 29, 1993), http://www.usccb.org/issues-and-action/human-life-and-dignity/global-issues/trade/letter-to-us-trade-representative-kantor-from-msgr-lynch-on-nafta-1993-06-29.cfm.

[7] Letter of then-Archbishop Theodore McCarrick, Chairman, Committee on International Policy (precursor of the Committee on International Justice and Peace), US Catholic Conference, to Congress (November 5, 1997), http://www.usccb.org/issues-and-action/human-life-and-dignity/global-issues/trade/letter-to-congress-from-cardinal-mccarrick-on-fast-track-authority-1997-11-05.cfm.

pharmaceuticals. They reaffirmed their concern for the equal participation of governments and citizens of poor nations:

> We also hope that any agreements reached on sharing life-saving drugs will truly benefit the people of poorer countries who are ravaged by HIV/AIDS, malaria, TB and other treatable diseases. In addition, given the widening economic inequality between wealthier and poorer nations, it is especially important that governments as well as citizens of developing nations have a full place at the negotiating table with equal stature, rights and opportunities in Cancun.[8]

## II. RELATIONSHIPS WITH THE CHURCH IN OTHER NATIONS

A second asset that the bishops in the United States bring to policy debates on intellectual property's effect on the marginalized is their relationships with the Church in other nations. For decades, the bishops of the United States have fostered relationships of solidarity with the bishops and local Churches in other nations. These relationships have helped inform the US bishops' positions on the international policies of the United States, especially in regard to developing nations.

A particularly relevant dialogue ensued between US and Central American bishops shortly after the United States-Central American Free Trade Agreement (commonly referred to as CAFTA) had been signed but not yet ratified by Congress. That dialogue resulted in a 2004 Joint Statement on CAFTA by the Bishops' Secretariat of Central America (SEDAC) and the chairmen of the Domestic and International Policy Committees of USCCB. This statement, informed by the experience of the Church in the developing countries of Central America, directly addressed the issue of intellectual property. The bishops in their joint statement argued:

> The treaty will have effects on intellectual property rights. The proposed legal framework could jeopardize the right of Central American countries to exercise proper stewardship of their natural resources. If the use of generic drugs is limited, this would lead to further increases in the cost of health care.[9]

It is interesting to note that the statement expressed concerns both for the developing nations' right to "exercise proper stewardship of their national resources" and for their access to affordable pharmaceuticals. This two-pronged concern will eventually become

[8] Statement on the Occasion of the 5th World Trade Organization Ministerial Conference by Bishop John H. Ricard, SSJ, Chairman, Committee on International Policy, USCCB (September 10, 2003), http://www.usccb.org/issues-and-action/human-life-and-dignity/global-issues/trade/statement-by-bishop-ricard-on-the-5th-world-trade-organization-ministerial-conference-2003-09-10.cfm.

[9] Joint Statement on United States Central American Free Trade Agreement (US-CAFTA) by the Bishops' Secretariat of Central America (SEDAC) and the Chairmen of the Domestic and International Policy Committees of the United States Conference of Catholic Bishops (USCCB), July 21, 2004, http://www.usccb.org/issues-and-action/human-life-and-dignity/global-issues/trade/joint-statement-on-us-central-american-free-trade-agreement-2004-07-21.cfm.

three-pronged, when concerns for agriculture are added. The moral bottom line for the bishops signing the joint statement is clearly declared: "The moral measure of any trade agreement should be how it affects the lives and dignity of poor families and vulnerable workers whose voice should receive special attention in this discussion."

In 2005, USCCB hosted a meeting of Andean and US bishops in Washington, DC. Representatives of the social justice ministries of the bishops' conferences of Bolivia, Colombia, Ecuador, Peru, and the United States expressed both hope and concern in connection with the proposed United States-Andean Free Trade Agreement in these words:

> In an increasingly interdependent world, of which this trade agreement is an example, all our people should have reason to hope that greater cooperation between us will offer real solutions to the perennial and stubborn problems of poverty, hunger and lack of opportunity. . . . We are concerned that developments in international trade will not unleash the true potential for economic growth, poverty alleviation and integral human development that all people of good will hope for.[10]

They went on to express similar concerns to those articulated in the 2004 Joint Statement on CAFTA, listing those as follows:

> As pastors, the bishops underlined important issues such as safeguarding the dignity of rural life, intellectual property rights, worker rights, the environment, the importance of citizen participation, especially for the poorest, as well as the importance of articulating a broader development agenda.

Later that year, in September 2005, an even larger gathering of Latin American, Canadian, and US bishops issued a communiqué on trade policies. At that gathering, concerns regarding intellectual property rights were taken up within a broader critique of the market. Although the bishops hoped that "increased global trade will unleash creativity, initiative and economic growth," they also expressed some fears, including fears related to "intellectual property rights":

> The market has its own logic, but it does not have its own ethic. It requires what the Church calls a strong juridical framework of social institutions that can humanize it and channel the considerable energies of the market for the common good. This moral and juridical framework needs to ensure that intellectual property rights, access to technology and information, transparency and civil society participation, agricultural policies, labor standards, and environment regulations that promote sustainable development are each addressed in ways that promote the common good of all, especially the poor.[11]

[10] Communiquè of Andean and U.S. Bishops (February 10, 2005), http://www.usccb.org/issues-and-action/human-life-and-dignity/global-issues/latin-america-caribbean/colombia/communique-of-andean-and-us-bishops.cfm.

[11] Joint Communiqué of the Latin American and North American Bishops (September 8, 2005), http://www.usccb.org/issues-and-action/human-life-and-dignity/global-issues/trade/joint-communique-of-the-latin-american-and-north-american-bishops-on-trade-agreements-2005-09-08.cfm.

All of these engagements and relationships with the Church in the Western Hemisphere led to a fuller development of the US bishops' engagement in intellectual property rights. In a letter to the US Trade Representative regarding the US-Andean Free Trade agreement, then bishop Thomas Wenski, USCCB International Chairman at the time, gave a fuller articulation of the US bishops' concerns:

> We are also concerned about intellectual property rights provisions with regard to pharmaceuticals. The Church locates intellectual property rights within the broader framework of the common good and believes these rights should be balanced with the needs of the poor. We are concerned about the U.S. proposal to extend new patent rights to already patented chemical products if new uses are discovered for them. We are also concerned about the request for periods of "data exclusivity" in which generic drug manufacturers would be prevented from marketing their products. Both proposals may well increase the price of, and therefore access to, essential medicines.
>
> In addition, the bishops of the Andean region minister extensively among indigenous groups. Out of respect for their cultural heritage and in view of their economic development, we hope that the U.S. will give careful attention and deference to the proposal made by the Andean governments that requires commercial agreements to honor the patrimony of these indigenous communities, and to share the benefits of any commerce with groups in which traditional knowledge originated.[12]

In the context of the renegotiation of NAFTA, the chairmen of the relevant committees of the US and Mexican bishops' conferences issued a joint statement in November 2017, in which they declared:

> Intellectual property rights provisions regarding pharmaceuticals and agriculture are particular concerns. The Church locates intellectual property rights within the broader framework of the common good and believes these rights should be balanced with the needs of the poor for access to medicines and to food, as well as the intellectual, economic, and political recognition of traditional knowledge. An expansion of NAFTA's market exclusivity for pharmaceuticals would jeopardize the health of millions as they struggle to support their families.[13]

[12] Letter to Ambassador Robert Portman, US Trade Representative, from Bishop Thomas Wenski (November 28, 2005), http://www.usccb.org/issues-and-action/human-life-and-dignity/global-issues/trade/upload/letter-to-us-trade-representative-portman-from-archbishop-wenski-on-andean-nations-free-trade-agreement-2005-11-28.pdf.

[13] Renegotiating NAFTA: Rebuilding our Economic Relationship in Solidarity, Mutual Trust, and Justice – A Joint Statement of the Chairmen of the Committee on Domestic Justice and Human Development and the Committee on International Justice and Peace of the United States Conference of Catholic Bishops and the Chairmen of the Pastoral Social Committee of the Conference of the Mexican Episcopate on the North American Free Trade Agreement's Renegotiation Process (November 15, 2017), http://www.usccb.org/issues-and-action/human-life-and-dignity/global-issues/trade/joint-statement-on-nafta-11-15-2017.cfm.

Concerns about the effects of intellectual property provisions in bilateral and multilateral trade agreements became regular features of USCCB trade policy, largely as a result of USCCB's relationships with other bishops' conferences in the Western Hemisphere. Although USCCB has not taken positions for or against entire trade agreements, it regularly raises concerns with respect to intellectual property rights and the common good, especially as regards the poorest and most marginalized peoples, in its meetings and official correspondence with US policy makers.

## III. ON-THE-GROUND EXPERIENCE

The Catholic Church has broad and deep experience combating poverty. Catholic Relief Services (CRS) is the official overseas relief and development agency of USCCB. CRS has experience as an implementer of US foreign assistance in more than one hundred countries around the world. Its programs address emergency response, HIV and AIDS, health, education, civil society capacity, food security, agriculture, emergency relief, and peacebuilding. At one time CRS directly addressed the impacts of trade policy on developing nations. Today, CRS regularly contributes its experience to help USCCB to take the lead on trade policy.

In October 2003, CRS developed recommendations on CAFTA that were sent to the US Trade Representative. In addition to focusing on access to medicines and medical interventions, CRS drew upon its agricultural experience in developing countries. Regarding patents on genetically modified plants, the CRS report noted:

> Partners in developing countries have expressed concern with the potential loss in biodiversity, increase in prices, and commercial sanctions associated with granting of such patents, as well as the possibility of losing access to resources such as seeds, which have been developed over years. CRS believes that trade agreements should not insist on intellectual property rights that would seriously inhibit farmers' ability to use genetic plant resources such as seeds for food and agriculture.[14]

Citing the Holy See's intervention at the World Trade Organization (WTO) in Cancun in 2003, which declared that "the intellectual property rights system must exist not only to protect creative and innovative impetus but also and primarily to serve the common good of the human family," CRS went on to formulate five guidelines:

> Ensure that the implementation of the agreement regarding production and international trade in essential medicines genuinely allows poor people access to such medicines, especially in the fight against HIV/AIDS.

[14] Catholic Relief Services, Recommendations to the U.S. Trade Representative on the Central American Free Trade Agreement (October 2003), http://www.usccb.org/issues-and-action/human-life-and-dignity/global-issues/trade/recommendations-from-catholic-relief-services-to-us-trade-representative-2003-10.cfm.

> Grant patents for the minimum time and under the minimum conditions necessary to provide incentives for innovation.
>
> Prohibit private interests from patenting agricultural products and processes developed over time by rural producers, without their consent and fair compensation.
>
> Require that any marketed product constituted of genetically modified varieties be appropriately labeled and that countries importing such products be fully informed about their genetically modified content.
>
> Adopt strategies and programs to help transfer these technologies affordably to Central American countries and other developing countries, to the extent such technologies are judged safe and requested by these countries.[15]

The experience of CRS in both the agricultural and health sectors in developing countries has been an invaluable asset to the development of USCCB policy on intellectual property rights. If anything, USCCB could more regularly and seamlessly integrate a concern for the agricultural impacts of intellectual property rights on developing countries based on the experience of CRS. USCCB may have at times given the mistaken impression that the moral focus is on pharmaceuticals alone.

## IV. FOUR KEY PRINCIPLES

Based upon USCCB's articulation of Catholic Social Teaching, its relationships with the Church in developing nations, and the on-the-ground experience of the Church and CRS, four interrelated principles emerged in the bishops' moral framework on trade: respect for the rights of indigenous people, concern for a proper balance in recognizing intellectual property rights, questions related to dispute resolution mechanisms that may give "unfair advantage" to commercial interests, and the need for participation. These principles were expressed in a 2015 joint letter of the Chairmen of USCCB's Committee on International Justice and Peace and the Committee on Domestic Justice and Human Development:

> Indigenous People: Catholic bishops throughout the world minister extensively among indigenous groups. Out of respect for their cultural heritages and in view of their economic development, we hope that the United States will give careful attention to requirements that commercial agreements honor the patrimony of these indigenous communities, and share equitably the benefits of any commerce with groups in which traditional knowledge and natural resources originated.
>
> Intellectual Property Rights: We are also concerned about intellectual property rights provisions with regard to pharmaceuticals. The Church locates intellectual property rights within the broader framework of the common good and believes these rights should be balanced with the needs of the poor.

[15] *Id.*

> Dispute Resolution Mechanisms: We question the merits of requiring sovereign parties to international treaties to agree to binding international arbitration as the forum for dispute resolution. Such a path may lead to unfair advantages for commercial interests willing to exploit the rules of the arbitral system and may result in the weakening of important environmental, labor, and human rights standards.
>
> Participation: It is critical the people have a voice in decisions that touch their lives. Human dignity demands transparency and the right of people to participate in decisions that impact them.[16]

Likewise, the November 2017 joint statement of the US and Mexican bishops' conferences on the renegotiation of NAFTA reinforced both a concern for the potential impacts of intellectual property rights on vulnerable persons and the principle of maximizing the participation of people in the decisions that impact their lives and livelihoods.[17]

## V. EVALUATING USCCB ENGAGEMENT

How might we summarize USCCB engagement on intellectual property rights and poor persons, communities, and nations? Church teaching is clear: the right to life is foundational in Catholic teaching. In his encyclical *Pacem in Terris*, Saint John XXIII taught: "Man has the right to live. He has the right to bodily integrity and to the means necessary for the proper development of life, particularly food, clothing, shelter, medical care, rest, and, finally, the necessary social services."[18] Access to nutritious food and adequate medical care is essential for preserving human life and promoting integral human development. Life patents are a form of intellectual property protection with profound social justice implications for poor and marginalized communities, especially where food and medical care are concerned. There are the obvious issues of access to these essential goods, but there also exist significant concerns for protecting the rights of indigenous communities to fair and just compensation when their rich plant, animal, and human genetic patrimonies are "discovered" or "exploited" by corporations. In many cases, indigenous communities have bred plants and animals, using traditional methods, over hundreds of years. In other cases, they have used herbs and traditional lore for medicinal purposes. The genetic patrimony and traditional knowledge of indigenous communities deserve recognition, respect, and fair compensation if utilized by others.

16 Letter to Congressional Leaders on Trade Principles, from Archbishop Thomas Wenski and Bishop Oscar Cantú (January 26, 2015), http://www.usccb.org/issues-and-action/human-life-and-dignity/global-issues/trade/letter-to-congressional-leaders-from-bishop-cantu-archbishop-wenski-on-trade-2015-01-26.cfm.

17 Renegotiating NAFTA, *supra* note 13.

18 Saint John XXIII, *Pacem in Terris*, no. 11 (April 11, 1963), http://w2.vatican.va/content/john-xxiii/en/encyclicals/documents/hf_j-xxiii_enc_11041963_pacem.html.

At the same time, the protection of intellectual property rights is important for stimulating and rewarding research and discovery, but in ways that serve the common good of all, especially of persons at the margins of our global economy. With the persistent and ongoing impasse in the Doha Development Round negotiations of the WTO, the United States has pursued bilateral and multilateral free trade agreements (FTAs) and continues to do so. The engagement of USCCB on intellectual property rights and the common good has primarily been in connection with these FTAs. A great deal is at stake for persons who are poor or marginalized. It is important to evaluate the effectiveness of these efforts.

First, USCCB has been able to articulate a consistent, principled and generally holistic set of concerns, including (a) acknowledging the importance of intellectual property rights serving the common good, especially of those who are poor or marginalized; (b) protecting the rights of indigenous communities related to their traditional knowledge and resources, and insisting on both fair compensation and access to benefits flowing from these sources; (c) raising questions about dispute resolution mechanisms that may give corporate interests unfair advantages when governments attempt to protect the common good of their people against the private good of corporate interests; and (d) defending the rights of persons who will be impacted by trade agreements and their intellectual property rights to participate in the decisions that impact their lives.

The reliance of USCCB on its relationships with the Church in other nations and the experience of the Church and CRS in other countries give enhanced credibility to the voice of USCCB in public policy debates. Public officials in both the administrative and legislative branches of the US government give a respectful hearing to the concerns expressed by USCCB.

Yet we still need to ask: how effective are these efforts? USCCB's initial decision not to take positions on overall trade agreements, or even "fast track" legislation, has been a double-edged sword. On the one hand, it has allowed USCCB to remain principled and above sometimes highly polarized and politicized debates. On the other hand, it has limited USCCB's ability to affect the final provisions of trade agreements, including those related to intellectual property. The passage of fast-track legislation, in particular, allows major pharmaceutical and agricultural corporations to influence negotiations over intellectual property rights with little input from impacted populations.

Given the enormous inequality in the economic and political power of the various nations that are parties to these negotiations, the United States has disproportionate influence that often serves the interests of US corporate entities, sometimes to the detriment of poor or vulnerable communities and nations. Since fast-track legislation limits Congress to an up or down vote on an entire trade package, it is impossible to use the legislative review process to change intellectual property rights provisions that fail to take into account the common good and the welfare of poor and marginalized persons.

USCCB's partnership with the Church in other nations, particularly developing countries, has been a very positive feature of its engagement on intellectual property rights provisions within FTAs. One wonders, however, if the Church has been able to leverage fully these relationships to protect the rights of poor and vulnerable people. Multinational corporations continue to exert more influence on intellectual property rights within FTAs than the Church and civil-society organizations exert. Corporate access to the US government and the governments of other nations seems to be more direct and effective. Persons and communities who are poor continue to be largely marginalized during the negotiations of bilateral and multilateral trade agreements. USCCB, in partnership with local Churches and civil societies of nations that are parties to FTAs, will need to adopt more effective strategies to enhance transparency, accountability, and participation.

Currently, two multilateral FTA negotiations will test Church leadership in the United States and other nations: the renegotiation of NAFTA and the negotiation of the Transatlantic Trade and Investment Partnership (TTIP). NAFTA involves the United States, Mexico, and Canada; TTIP involves the United States and the nations of the European Union. Taken together, these two agreements will encompass a significant portion of the world's economy. In addition, it is quite possible that the United States will seek to advance bilateral trade agreements with inevitable implications for intellectual property rights provisions.

It will be critical for the Church in the United States to continue engaging policy makers on the intellectual property rights provisions of both NAFTA and TTIP, so that these instruments serve the common good by promoting innovation and protecting the needs and rights of persons and communities at the margins. This will need to be done in a manner that compensates for the immense power imbalances between richer and poorer nations, and between multinational corporations and civil society.

In a major shift in policy, the chairmen of the international and domestic committees of USCCB made clear that they would have opposed the Trans-Pacific Partnership (TPP), even if the Untied States had not withdrawn from it, due in part to profound apprehensions related to intellectual property rights. In a May 2017 letter to the US Trade Representative that spelled on key moral principles related to trade agreements, the chairmen wrote:

> Although the United States Conference of Bishops (USCCB) has not historically taken positions for or against particular trade agreements, we were prepared to oppose the Trans-Pacific Partnership (TPP) due to grave concerns regarding access to medicines, the vulnerability of small agricultural producers, and the troubling requirement that sovereign parties to international treaties agree to binding international arbitration as the forum for dispute resolution. The President's decision to withdraw the United States from further consideration of the TPP is to be commended, therefore, and we look forward with hope to negotiations of future trade

deals that promote integral human development and offer fundamental protections of human life and dignity.[19]

Such an approach may indeed strengthen the ability of USCCB to influence negotiations on future trade agreements.

The question is not one of being for or against the protection of intellectual property rights. As the Holy See said in February 2015 in a statement on pharmaceuticals and intellectual property by Archbishop Silvano M. Tomasi, then Permanent Observer of the Holy See to the United Nations in Geneva:

> While fully respecting the right to private intellectual property, the Holy See urges a creative and innovative approach, with full use of the flexibilities allowed under the Trade Related Intellectual Property instruments, so that the right to health for all people without any form of discrimination can be fully guaranteed and implemented. We are convinced, therefore, that concern for the protection of intellectual property rights, while legitimate in itself, must be seen within the wider perspective of promoting the common good, building global solidarity and prioritizing the life and dignity of the world's most vulnerable people.[20]

The same could be said for life patents in general and for their specific impacts on persons and communities who are poor or marginalized, especially in the areas of agriculture and medicine – two areas that touch on fundamental human rights. USCCB and the broader Church must find ever more effective ways to impact FTAs and WTO negotiations so that intellectual property is protected in ways that serve the common good and promote the rights of poor and vulnerable people. As Pope Francis said in quoting the bishops of the United States on the power dynamics related to climate change, "[G]reater attention must be given to 'the needs of the poor, the weak and the vulnerable, in a debate often dominated by more powerful interests.'"[21] A similar power dynamic must be overcome if life patents are to serve both the legitimate need to support innovation and the needs of all God's children in the human family.

19 Joint Letter to U.S. Trade Representative Lighthizer Regarding Moral Principles and the Evaluation of Proposed Trade Deals (May 12, 2017), http://www.usccb.org/issues-and-action/human-life-and-dignity/global-issues/trade/joint-letter-to-trade-representative-lighthizer-on-trade-2017-05-12.cfm.

20 Archbishop Silviano M. Tomasi, Statement of the Holy See to the Social Forum of the UN Human Rights Council (February 18, 2015), reprinted in *Archbishop Silviano Tomasi Calls for Innovative Approach to Medicine Access*, Zenit (February 19, 2015), https://zenit.org/articles/archbishop-silvano-tomasi-calls-for-innovative-approach-to-medicine-access/.

21 *Laudato Sí, supra* note 3, at no. 52.

# 7

# Intellectual Property and Genetic Sequences: A Jewish Law Perspective

*Michael J. Broyde and Steven S. Weiner*

## I. OVERVIEW AND PROBLEMS

### A. *Intellectual Property: A Balancing Act*

Since Johannes Gutenberg's printing of the Bible in the mid-fifteenth century, the question of authors' exclusive rights to their creative works has confronted publishers and writers, judges and moralists. Similarly, and also beginning in the Early Modern time period, with the emergence of increasingly sophisticated commercial manufacturing industries and ultimately the Industrial Revolution, societies have faced a demand to provide some degree of exclusive legal protection to inventors of new and useful machines and technologies.[1]

The rationale for awarding a measure of exclusive protection to creative works and useful inventions is fairly easy to articulate. Creating a truly innovative and successful invention or work of authorship typically requires a disproportionately large investment of costly resources (time, capital, etc.) at very high risk, as contrasted with the fairly modest, low-risk investment that enables *copying* a new work or technology once it has been demonstrated to be effective and brought to market. Thus, absent some form of exclusive protection for the original author or inventor – what we nowadays call a "barrier to entry" prohibiting would-be competitors from copying the same design – the high-risk investment necessary to develop breakthrough technologies and works would be economically irrational. Society as a whole would then be deprived of the great benefits delivered by artistic and technological progress. This rationale is very nicely captured in Article One (section 8, clause 8) of the United States Constitution (hereinafter the "IP Clause"), which succinctly explains the congressional power to grant exclusive patent and copyright protection as follows (emphasis added): "*To promote the progress of science and useful arts*, by securing for limited times to authors and inventors the exclusive right to their respective writings and discoveries."

[1] The Venetian Patent Statute of 1474 is often regarded the earliest codified patent system in the world.

Note that the IP Clause expressly requires Congress to limit these exclusive rights ("for limited times"). Implicitly, the Constitution thus recognizes that patent and copyright protection should strike a balance between two values. The first is the societal benefit of providing sufficient economic incentive to motivate authors and inventors (and their stakeholders) to invest and to risk the resources necessary to create valuable new works and inventions. The second is the societal cost of monopolizing those new works and inventions – resulting in monopoly (higher) pricing and also restricting third-party competitors from adapting such works and inventions in their own, (possibly) improved offerings. Federal patent law thus strikes "a careful balance between the need to promote innovation and the recognition that imitation and refinement through imitation are both necessary to invention itself and the very lifeblood of a competitive economy."[2]

This careful balance means that, on one hand, an original author or inventor enjoys exclusive rights to his work or invention for a limited period of time – in essence, a lawful monopoly period intended to provide sufficient economic returns justifying the investment risked – while on the other hand, after that time period, the work or invention ultimately enters the public domain and becomes freely available to third-party competitors, so that society can more fully reap the benefits of that intellectual property at competitive prices and through further-improved products produced by third parties.[3]

In a sense, the fundamental nature of the balance between intellectual property exclusivity as an innovation incentive or impedance applies whether we are talking about the creation of a new written work of authorship – or imagery, music, or video, for that matter – or whether we are talking about technology for generating new materials or building new machines, or for creating new life or new forms of life. In each case, the underlying policy issue is whether legally conferred monopoly rights are a socially beneficial, appropriate financial inventive to motivate the investment necessary in order to yield new and valuable creative works, or whether such monopolization of creative work for a relatively few "lucky" inventors would on balance impede overall innovation and creative productivity.

Naturally, it can be argued by policy makers and stakeholders that the balance between incentivizing more creative research (by granting patents more liberally) versus encouraging the benefits of competitive freedom should be perceived as tipping differently in particular technological or artistic areas. In order to strike the best balance for society between innovation incentives and competitive freedom, US (and other modern) patent and copyright regimes have limited the extent of exclusive rights granted not only by duration of the grant, but also by defining certain areas

[2] Bonito Boats Inc. v. Thunder Craft Boats, Inc., 489 U.S. 141, 146 (1989).

[3] See, e.g., FED. TRADE COMM'N, TO PROMOTE INNOVATION: THE PROPER BALANCE OF COMPETITION AND PATENT LAW AND POLICY (2003), https://www.ftc.gov/reports/promote-innovation-proper-balance-competition-patent-law-policy; Richard A. Posner, *Intellectual Property: The Law and Economics Approach*, 19 J. ECON. PERSP. No. 2, 57–73 (2005).

in which such forms of intellectual property protection are not available. For example, the US patent statute broadly declares as patent eligible "any new and useful process, machine, manufacture, or composition of matter, or any new and useful improvement thereof."[4] Nevertheless, under judicial doctrine established long ago by the US Supreme Court, "laws of nature, natural phenomena, and abstract ideas" cannot be patented.[5] As the Supreme Court has explained, these exceptions are essentially motivated by a public policy determination that granting even time-limited exclusive rights to such "basic tools of scientific and technological work" would create a cost to society – in terms of lost opportunities to freely exploit those basic natural laws or abstract concepts in third-party products – that decisively outweighs the societal benefits of any creativity that would be incentivized by granting such patents.[6] As explained at the end of Part III of this chapter, even as the Jewish tradition is comfortable averring that, theologically, all belongs to God, the Jewish legal tradition would not consider such ideals and ideas in a patent dispute of this type.

Interestingly, for virtually three decades following the establishment of the US Court of Appeals for the Federal Circuit in 1982 as the court with exclusive appellate jurisdiction for patent matters, the judicial exceptions to patentability were construed and applied in relatively narrow fashion by the Federal Circuit. For example, broad patent claims for software concepts and even for so-called business methods were widely upheld and enforced.[7] However, over roughly the last five years, the US Supreme Court has weighed in on patent matters far more actively and frequently, taking a relatively expansive view of the judicial exceptions to patent eligibility and reversing the Federal Circuit's position in important respects. For example, in its

4 Subsequent provisions impose various substantive requirements such as novelty and non-obviousness. However, section 101 broadly defines the limits of what *categories* of invention are potentially patent eligible.

5 Diamond v. Diehr, 450 U.S. 175, 185 (1981), and see earlier cases cited there.

6 Mayo Collaborative Services v. Prometheus Laboratories Inc., 566 U.S. 66, 71 (2012) ("Monopolization of those tools through the grant of a patent might tend to impede innovation more than it would tend to promote it."). Interestingly, in denying the patentability of natural laws and phenomena, the US Supreme Court characterizes this doctrine as a policy-based, "judicially created" eligibility requirement linked to 35 U.S.C. § 101, instead of simply asserting that merely "discovering" a previously unrecognized natural phenomenon is not a novel invention as required under 35 U.S.C. § 102. Perhaps this is because § 101 itself opens by offering patent protection to "Whoever invents *or discovers* . . . any new and useful process." The Court may not have wished to categorically assert that discovery of existing *but unknown* phenomena per se lacks novelty and cannot be patented. Indeed, one can argue strongly that Einstein's "discovery" of Relativity and Newton's "discovery" of gravity were far more deeply innovative and novel than an improved mousetrap. Thus, instead of invoking lack of novelty, the Supreme Court has treated "laws of nature" as a category of discoveries deemed patent ineligible by courts for policy reasons.

7 See, e.g., notably State Street Bank and Trust Company v. Signature Financial Group, Inc., 149 F.3d 1368 (Fed. Cir. 1998) (rejecting any "business method" exception to patentability, and finding patent eligibility for a partnership structure of mutual funds [implemented on a "data processing system"] because the claimed invention "produces a useful, concrete and tangible result" – in that case, a calculated share price).

well-known *Alice* decision, the Supreme Court applied the exclusion for "abstract ideas" to invalidate a patented business method on a computer-based escrow system – essentially a (purportedly) new application of routine computer technology.[8] Although the Court did not categorically declare "business method" inventions ineligible for patent protection, it nevertheless reformulated and applied the "abstractness" exception. Indeed, it did so in a manner that not only doomed the particular business method patent before it, but also led very quickly to a dramatic, widely observed decrease in the issuance and successful enforcement of such patents, as well as for patents on certain types of broadly claimed software inventions.[9]

An important, contemporary controversy – very much at the heart of our present discussion – is how the policy balance of incentives should be assessed in the context of intellectual property protection for genetic sequences. We examine the current state of that controversy under US patent law in the next section.

### B. *IP Protection for Genetic Sequences: A Contemporary Controversy*

As noted previously, for the first thirty years of its existence, the Federal Circuit applied the judicial exceptions to patentability in a relatively narrow manner. One example that is especially pertinent for our present purposes was the Federal Circuit's holding that newly discovered – but naturally occurring – gene sequences could be patented specifically in their isolated form, based on the reasoning that such sequences do not occur naturally in isolated form.

However, in *Myriad Genetics*, the Supreme Court reversed the Federal Circuit's decision and established that *naturally occurring* gene sequences cannot be patented.[10] Merely isolating genes that are found in nature – even if doing so required great R&D efforts – does not make those genes patentable. Strikingly, the Supreme Court declared that "[g]roundbreaking, innovative, or even brilliant discovery does not by itself satisfy the Section 101 inquiry."[11]

Around the same time, the Supreme Court in *Mayo Collaborative Services v. Prometheus Laboratories*[12] used similar reasoning to invalidate a patent for the scientific discovery that the proper dosage of a certain drug for GI disorders could be determined for an individual patient by testing the patient's level of concentration of a particular metabolite (after an initial dose of the drug). The patent in question basically covered the method of initially administering the drug, testing the patient's levels of the metabolite, and then adjusting subsequent doses for that patient based

8 Alice Corp. v. CLS Bank Int'l, 573 U.S. 208, 134 S. Ct. 2347 (2014).

9 See, e.g., Jasper L. Tran, *Two Years After* Alice v. CLS Bank, 98 J. Pat. & Trademark Off. Soc'y (2016), http://papers.ssrn.com/sol3/papers.cfm?abstract_id=2798992.

10 Association for Molecular Pathology v. Myriad Genetics, Inc., 569 U.S. 576 (2013).

11 *Id.* at 591.

12 566 U.S. 66.

on whether the metabolite level was above or below a concentration level that the inventors had discovered. Because technology to test for concentrations of that metabolite was well known and not novel in any way, the patented invention essentially amounted to the specified correlation between levels of the metabolite and an effective dose of the drug – in other words, a law of nature. The Court held that the patented process "add[ed] nothing specific to the laws of nature other than what is well understood, routine, conventional activity" and therefore was not patent eligible.[13]

While *Mayo* did not involve an invention based on genetic discovery per se, the underlying logic in *Mayo* subsequently led the Federal Circuit to deny patent eligibility in a widely watched biotech case, *Sequenom*, involving the scientific discovery that cell-free fetal DNA (cffDNA) can be found naturally in the blood of pregnant mothers.[14] Well-known techniques (e.g., PCR amplification) can be employed to amplify the cffDNA present in a sample of maternal blood and to detect the presence of particular DNA sequences of interest. The discovery of cffDNA in maternal blood was thus a breakthrough, allowing doctors to perform prenatal genetic diagnostics by means of a relatively safe and simple blood draw, instead of a more dangerous, invasive procedure like amniocentesis. Despite the acknowledged scientific merit and practical medical value of the invention, and although the natural phenomenon of cffDNA in maternal blood had been previously unknown and undiscovered, the Federal Circuit concluded that the rules and reasoning of *Mayo* compelled finding the patent invalid, because the patent's claims were essentially directed to exploiting a natural law with purely conventional techniques. The Supreme Court denied certiorari, despite numerous passionate amicus briefs from both industry and academia on behalf of the petitioner.[15]

It is noteworthy that Supreme Court patent eligibility jurisprudence focuses so categorically on whether a claimed invention involves "natural" (even if previously undiscovered) phenomena.[16] Thus the Court in *Myriad* held that naturally occurring gene sequences are not patent eligible; at the same time, it also held that complementary DNA (cDNA) derived from a corresponding natural gene sequence (simply by removing the "introns" or non-coding regions) *is* patent eligible. The Court rejected the argument that simply because "the nucleotide sequence of cDNA is dictated by nature, not by the lab technician" claims to cDNA should

[13] *Id.* at 82.

[14] Ariosa Diagnostics, Inc. v. Sequenom, Inc., 788 F.3d 1371 (Fed. Cir. 2015), *cert. denied*, 136 S. Ct. 2511 (2016).

[15] See *Amici Support Certiorari in* Sequenom v. Ariosa, Patent Docs, May 2, 2016, http://www.patentdocs.org/2016/05/amici-support-certiorari-in-sequenom-v-ariosa.html.

[16] It has been suggested that original, non-naturally occurring DNA sequences might also be protectable under *copyright* law. Christopher M. Holman, *Copyright for Engineered DNA: An Idea Whose Time Has Come*, 113 W. Va. L. Rev. 699 (2011). Copyright necessarily could apply only (if at all) to non-natural sequences, because it only protects "*original* works of authorship." 17 U.S.C. § 102 (emphasis added).

not be patentable. It reasoned that "the lab technician unquestionably creates something new when cDNA is made. cDNA retains the naturally occurring exons of DNA, but it is distinct from the DNA from which it was derived. As a result, cDNA is not a product of nature and is patent eligible under section 101."[17] This distinction may seem superficial, at first blush. However, as further explained in *Mayo*, the Court's view is that while the "law of nature" exception to patentability is motivated by public policy considerations – in particular, by an assessment that exclusive rights should not be granted to "basic tools" and "building blocks" of science and research – the exception functions as a bright-line, absolute rule. In the Court's words:

> Courts and judges are not institutionally well-suited to making the kinds of judgments needed to distinguish among different laws of nature. And so the cases have endorsed a bright line prohibition against patenting laws of nature . . . and the like, which serves as a somewhat more easily administered proxy for the underlying "building block" concern.[18]

Consider this hypothetical. Suppose there are several alternative treatments for a particular cancer. Each treatment has severe side effects, and each fails to succeed on average for a significant percentage of patients. Now, imagine that a team of medical researchers determines through careful experimentation and insightful data analysis that there is a strong correlation between certain genetic sequences and the probability of success for each alternative treatment. In other words, patients who have sequence 1 will do much better with treatment A, whereas patients who have sequence 2 will do much better with treatment B. Unfortunately for our researchers, because this correlation is a "natural phenomenon," and because treatments A and B are conventional (as is genetic testing technology), under *Mayo* and *Sequenom*, this valuable scientific discovery apparently cannot be patented. Trying to patent a "personalized medicine" method in which treatment A or B is administered, depending on the results of a patient's genetic test fails for exactly the same reasons as the patent claims in *Mayo* and *Sequenom*.

Our hypothetical is quickly becoming reality. Recently, intellectual property researchers presented an analysis showing a "spike in patent rejections" for personalized medicine inventions in the United States in the wake of *Mayo*.[19] Is this good public policy? Does it strike an optimal balance between incentives to innovate and freedom to compete? Many voices in the biotech industry consider *Mayo* and *Sequenom* terribly ill-advised from a public policy perspective. While that point is certainly debatable, for our present purposes, we wish to highlight the fact that the Supreme Court itself largely avoids the policy question by characterizing the "law of

17 569 U.S. at 595.
18 *Mayo*, 566 U.S. at 89.
19 Heidi Ledford, *US Personalized-Medicine Industry Takes Hit from Supreme Court*, NATURE, 17 August 2016, http://www.nature.com/news/us-personalized-medicine-industry-takes-hit-from-supreme-court-1.20436.

nature" exception as a "bright line prohibition" – a convenient "proxy" for the original underlying policy concerns, and justifiable on grounds that courts "are not institutionally well-suited" to make more fine-grained assessments of the social tradeoff in granting exclusive rights for some natural discoveries versus others.

We have emphasized herein the reluctance of courts to make fine-grained, case-specific policy assessments under US patent law, because we shall claim in the following sections that a *halakhic* theory of intellectual property can lead us in a very different direction with respect to these challenging issues.

## II. DEVELOPING A *HALAKHIC* (JEWISH LAW) THEORY OF INTELLECTUAL PROPERTY

In attempting to establish a *halakhic* theory of intellectual property, one must first confront the fact that there is little or no discussion of intellectual property per se in the early legal sources, such as the Talmud, that traditionally form the basis of a typical *halakhic* analysis. This fact is not difficult to explain. As we noted at the outset, what we now think of as copyright and patent systems only began to emerge in the late fifteenth century – roughly a millennium after compilation of the Talmud, and centuries after composition of the definitive, medieval Jewish legal codes and commentaries, including Alfasi, Maimonides, Tosafot, and Tur. As such, there is quite naturally no real discussion of intellectual property as such within those early, classic texts of Jewish law.

Is it possible to somehow identify a distinctively *halakhic* theory of patent law? We believe so. Despite understandable lacunae regarding intellectual property in the early classic sources, Jewish lawmakers in the early modern period nonetheless confronted and addressed many new developments in commercial law, including intellectual property and related matters. While we are not aware of any rulings under Jewish law (at least until the establishment of the modern State of Israel) specifically on patent matters,[20] many *halakhic* responsa were written regarding copyright-style protection for published works of authorship following the advent of the printing press.[21] The same is true for disputes regarding local monopoly rights, such as exclusive licenses to distribute whiskey or to slaughter beef in a particular

[20] A likely explanation is that the rabbinate could adjudicate disputes among Jewish businessmen, but generally had no jurisdiction or authority over members of the larger non-Jewish community. Because there were significant markets *within* the Jewish community for books (e.g., for printed works of Jewish scholarship), it was not uncommon for copyright disputes among competing Jewish publishers to come before rabbinical courts. In contrast, an exclusive right to manufacture or use a new type of industrial machine would, by its nature, impact businessmen and markets far beyond the Jewish community, and so rarely if ever did disputes over patent infringement come before parochial rabbinic authorities for resolution. Interestingly, a well-known nineteenth-century Jewish responsum on a question of *copyright* law involving Jewish publishers of rabbinic works does mention (secular) patent law – but only by way of analogy, to support a strong *copyright* regime under Jewish law. *Shoel uMeishiv*, first ed., vol. 1, responsum 44.

[21] We discuss several examples of these as follows.

community.[22] Our claim is that analyzing the underlying legal reasoning of these responsa allows us to develop a legal theory with meaningful application to protection of genetic inventions.

The very notion of protecting technological innovation as personal property is radical from the perspective of classical Jewish property law. The major codes uniformly rule that intangibles (*davar she'ein bo mamash*) such as "the aroma of an apple" cannot be purchased like tangible articles of property. Instead, one who wishes to sell, for example, the right to occupy a particular building *must* characterize the transaction as selling a limited interest in the (tangible) building itself, and not as a stand-alone property right separate and apart from ownership of the building.[23]

An intellectual property monopoly, like a patent, is even more profoundly intangible.[24] A patent on the design for a new type of machine bestows on the patent owner the power to exclude anyone else from building, using, or selling a machine of similar design. Such a sweeping power to restrict what others can do with their *own* tangible resources cannot be conceptualized as an "interest in" an item of tangible property. Rather, the point of a patent or copyright is a monopoly that restricts the freedom of what others can do with their own private resources, even absent any involvement with tangible property of the patentee.[25]

Nevertheless, as mentioned previously, *halakhic* decisors[26] from the early modern period onward enforced certain legal monopolies, such as copyright-style protection for publishers and exclusive licenses to distribute whiskey. On what grounds did they do so, given the utterly intangible nature of such rights?

Two very different rationales were offered. Some authorities argued that prevailing *secular* rules of commerce are *halakhically* controlling in such circumstances, under the pragmatic principle that "the law of the sovereign is the law" and/or that mercantile practices widely and generally adopted by businessmen become authoritative.[27] While the same principles can likewise be invoked to recognize patents *halakhically*, fundamentally this approach would not move us any closer to

22 We also analyze examples of these as follows.

23 Rambam, *Laws of Sale* 22:13–14; Tur and Shulchan Aruch, *Choshen Mishpat* 212:1–2.

24 This point was recognized and well-articulated in an important responsum of Maharshal involving a monopoly right to sell whiskey. We discuss this responsum in more detail as follows.

25 In this way, the concept of patent rights also differs from – and is much more abstract than – classic nuisance doctrine, which restricts a property owner's activities on his own land to the extent those activities would tend to physically harm neighboring landowners and their property.

26 "Decisors" is the common term used in English for *poskim*, who are those individual Jewish law authorities – frequently not even authorized formally by any ecclesiastical body – who issue decisions of Jewish law that are widely accepted by the community.

27 *Pitchei Tshuva* CM 212:1; *Aruch HaShulchan* CM 212:3. It is worth briefly reflecting on why Jewish commercial law includes an extensive, built-in, unapologetic deference to the "law of the land." One possibility is that Jewish law may view the essential function of commercial law as setting common "rules of the game" that participants can rely on. The rules thus need not be "religiously correct" – rather, they become "correct" precisely because they are conventionally followed. Alternatively, Jewish tradition may have wished to relieve its followers of a choice-of-law problem in which they

a *distinctively halakhic* theory of patent law – because it simply borrows and imports a reigning secular patent regime wholesale.

However, many other prominent authorities argued that legal monopolies could be *halakhically* rationalized as instances or extensions of Talmudic protection against unfair competition. The Talmud records,[28] for example, that the owner of a local business is entitled to block outsiders (but not local residents) from opening a competing business in his vicinity; similarly, a fisherman may not cast his net in the vicinity of another fisherman's net already present. *Maharshal*, an Early Modern contemporary of *Shulchan Aruch*, argues in a lengthy responsum[29] that a municipally-granted monopoly right to sell whiskey could be thought of in similar *halakhic* terms. Crucially, the Talmudic cases cited are well outside the bounds of normal Talmudic property law. A businessman does not "own" his prospective customers, nor does a fisherman own the free-swimming fish that he might prospectively catch if not for the intrusion of a competing net. Rather, these Talmudic rulings can best be understood as equitable doctrines enacted to protect society against unfair competition.[30]

faced one set of rules (*halakha*) for transactions solely among Jews while facing an inconsistent set of rules (secular) for transactions involving Gentiles: the solution was to adopt the more universal (i.e., secular) rules. Either way, for those areas of commercial law in which moral and religious values of the Jewish tradition were at stake (such as labor law, usury law, and a few other areas), *halakha* could nevertheless declare that secular law should not supersede traditional doctrine in such realms. For more on this, see Michael J. Broyde, Public and Private International Law from the Perspective of Jewish Law, in THE OXFORD HANDBOOK OF JUDAISM AND ECONOMICS 363–87 (Aaron Levine ed., Oxford University Press 2010).

28 Bab. Talmud Bava Batra 21b.

29 *Responsa of Maharshal* nos. 35 & 36; *Responsa of Chatam Sofer* (vol. 5) *Choshen Mishpat* 79.

30 A responsum from R. Ezekiel b. Judah Landau (*Noda' Bi-Yehuda, Hoshen Mishpat* 2:24) is sometimes mentioned in this context as furnishing another basis for *halakhic* copyright-like monopoly. However, upon closer inspection, that responsum does not deal so much with intellectual property rights, but more with ownership and usage rights with respect to *tangible blocks of movable type*. The question involved a dispute between an author and a printer about whether the author was entitled to recover some portion of the printing costs he paid to the printer from the revenues subsequently earned by the printer who re-used (some) of the same tangible type blocks to print and sell *other* books, but only used blocks of type corresponding to text that was *not* original to this author. This was fundamentally a question of tangible property rights, and so R. Landau fit it logically within classic *halakhic* doctrine regarding one who benefits from *tangible* property at another's cost. It is unclear to us if this responsum has true import for intellectual property theory.

Another source sometimes suggested for *halakhic* copyright protection is a contemporary ruling that forbids copying of purchased recordings, on the basis that when a producer sells her work, she withholds the right to copy it. When a purchaser of, for example, a tape copies the recording, he thus commits theft: a purchaser may not simply do whatever he wishes with his article of purchase, because the right to make further copies is (implicitly or explicitly) contractually excluded from the sale. See R. Zalman Nechemia Goldberg, *Copying a Cassette without the Owner's Permission*, Tehumin 6, pp. 185–207. But even assuming the *halakhic* efficacy of a post-sale restriction of this nature, this approach is simply inapplicable to most practical cases involving infringement of biotechnology, because a would-be competitor/infringer can typically learn about the invention by reading freely available publications (e.g., scientific papers, published patent applications, or regulatory filings) and without relying on purchasing tangible articles subject to post-sale restrictions.

The significance of a theory of patent law that conceptualizes patent rights as a variety of equitable protection against unfair competition, as opposed to a form of personal property, is that the particulars of the doctrine are more tightly coupled to considerations of public policy, fairness, and social welfare. For example, under the Talmudic doctrine of unfair competition, whether a competing business may lawfully be opened depends on the extent of harm to competitors and social value of the products or services being sold. Competition is thus more liberally permitted for teaching Torah, for important scholars, for visiting merchants selling perfume and the like to women, for wholesalers versus retailers, and for out-of-towners versus locals.[31] Such distinctions would seem curious in the context of basic personal property rights, but make perfect sense in the context of an equitable, social doctrine of unfair competition.

A thousand years after the Talmud's completion, the prominent *halakhic* authority R. Moses Sofer (Hungary, 1762–1839, widely known as *Chatam Sofer* after the title of his works) applied the policy principles underlying Talmudic unfair competition doctrine to the new, modern context of copyright infringement.[32] Invoking Maharshal's earlier analysis, and staying true to the underlying Talmudic principles of unfair competition, R. Sofer boldly drew new lines and endorsed strong copyright-style protection for many types of written religious works. Revisiting the Talmudic rule that seemingly accorded little competitive protection to teachers of Torah – a rule which some generalized to other religious services – R. Sofer argued that the Talmud plainly spoke of contexts in which increased competition would in fact "increase wisdom" by fostering even greater scholarly devotion. However, where unfettered competition would realistically destroy the incentive to provide (for example, religiously) necessary products or services, R. Sofer argued that Talmudic law called for limits on competition.

Therefore, instead of broadly encouraging free competition among suppliers of merchandise and services for religious purposes, Chatam Sofer endorsed strict (time-limited) copyright-style protection for authors and publishers of religious works that would likely never see the light of day without the economic incentive of at least limited-duration monopoly protection. He was even willing and able to draw a relatively fine distinction for works directed at broader, general audiences (such as standard prayer books), which he observed did not need monopoly protection in order to be economically viable (because much greater sales volumes allowed recovery of fixed costs even with modest, competitive profit margins). R. Sofer likewise updated the classical rule limiting anticompetitive protection to local neighborhoods, reasoning that for contemporary purposes, the market for published books was essentially global (i.e. embracing Jewish communities around the globe), and that, indeed, without legal protection extending beyond a publisher's own local

[31] Bab. Talmud Bava Batra 21b *et seq.*

[32] *Shut Chatam Sofer* (vol. 5) *Choshen Mishpat* 79.

neighborhood, there was no economically viable model for publishing works of scholarship.[33]

Contrast this flexible, policy-guided, contextually sensitive *halakhic* approach to intellectual property – and from no less an authority than the famously conservative R. Sofer[34] – with the US Supreme Court declining to draw policy-based lines based on the social impact of protecting particular scientific discoveries and instead endorsing "a bright line prohibition against patenting laws of nature" because "[courts] and judges are not institutionally well-suited to making the kinds of judgments needed" to draw such lines.[35]

Notably, patents did not emerge organically from classical property law under secular, non-Jewish legal systems either. The very term "patent" – short for "letters patent" – traditionally denotes a type of published legal instrument issued by a sovereign and granting an exclusive right or title.[36] Nevertheless, while modern patent law remains a creature of statute, not common law, modern secular jurists comfortably conceptualize patent rights as a species of property ("intellectual property"). The *halakhic* formalism that regards intangibles as explicitly outside the bounds of conventional property law would position patent protection more squarely within an equitable framework. Application of patent law to specific inventions would then arguably be more consistently and thoroughly subject to considerations of public policy.[37]

Interestingly, the US Supreme Court faced an opportunity just last year to squarely address whether patents should be treated as private property or should instead be characterized as public rights. In *Oil States v. Greene's Energy Group*, this question arose in the high-stakes context of a constitutional challenge to the IPR (inter-partes reexamination) procedure employed by the United States Patent and Trademark Office (USPTO) to review and invalidate issued US patents challenged by third parties. Particularly in the last few years, since passage and implementation of the America Invents Act (AIA), IPR proceedings have been used to invalidate numerous US patents – and with attractively high speed and low cost from the perspective of patent challengers.[38] Because the USPTO is an administrative agency and not an Article III judicial body, the petitioner in *Oil States* contended that

33 *Id.*

34 R. Sofer famously punned, in a very different, polemical context, that "everything new is Biblically forbidden." https://en.wikipedia.org/wiki/Moses_Sofer#Influence_against_changes_in_Judaism.

35 See *supra* text at note 18.

36 *Letters patent*, https://en.wikipedia.org/wiki/Letters_patent; *Patent*, https://en.wikipedia.org/wiki/Patent.

37 A more flexible, context-sensitive approach by courts to intellectual property adjudication would also better accommodate industry-specific tuning of intellectual property rights, as has been advocated by a number of leading legal scholars. See, e.g., Dan L. Burk and Mark A. Lemley, *Policy Levers in Patent Law*, 89 VA. L. REV. 1575 (2003). To date, such proposals have gotten very little traction, as US courts consider industry-specific adaptations to be the exclusive province of Congress – where political considerations often present obstacles to legal change.

38 See, e.g., *IPR Statistics Revisited: Yep, It's A Patent Killing Field*, www.patentattorney.com02/08/2017.

USPTO IPR proceedings violate the constitutional rights of US patent owners by extinguishing their lawful private property without adequate due process. Respondents countered (among other arguments) that patents are more properly viewed as public rights created by federal statute for the instrumental purpose of promoting the progress of useful arts.[39] In a 7–2 decision,[40] the Supreme Court upheld the constitutionality of IPR proceedings, declaring patents to be a "public franchise" and hence a type of public right.

Nevertheless, it remains to be seen whether *Oil States* will empower US courts to take a more flexible, public policy–oriented approach toward patent eligibility. According to a very recent, bipartisan letter signed by (among others) Senator Chris Coons (D-DE), chair and ranking member of the Senate Judiciary Subcommittee on Intellectual Property, and Representative Doug Collins (R-GA-9), ranking member of the House Judiciary Committee:[41] "Today, U.S. patent law discourages innovation in some of the most critical areas of technology, including artificial intelligence, medical diagnostics, and personalized medicine." The letter announces a bipartisan, bicameral framework for significant reform of US law for patent eligibility, in response to decisions like *Mayo* and *Sequenom*. As one signatory to the letter stated, "[L]eaders in the fields of biologics research and diagnostics will deliver the cures of tomorrow … [only] if we can protect those innovations with the patent protection that rewards the risks and investment necessary to discover the next great idea." However, this legislative framework aims to define "a closed list" of exclusive categories of statutory subject matter which alone should not be eligible for patent protection. While such a list might capture a good, current snapshot of policy priorities, it is not the same as dynamically bringing public policy interests to bear on a case-by-case basis as our society continues to evolve.

## III. TOWARD A JEWISH/*HALAKHIC* PERSPECTIVE ON IP PROTECTION FOR GENETIC SEQUENCES

As we saw in section I, secular US law for genetic patenting focuses on the unpatentability of "laws of nature." Discovering natural genetic sequences or phenomena is not a patentable invention – even if such sequences or phenomena were previously unknown, and even if the discovery required great ingenuity and enables much smarter diagnosis and/or targeted use of existing technologies. This judicial exception to patentability has been rationalized by the US Supreme Court as a way of preserving free availability of the basic "building blocks" of scientific research.

39 See, e.g., Professor Dmitry Karshtedt's concise and informative summary of the oral argument in *Oil States*, available at https://patentlyo.com/patent/2017/11/engaging-history-property.html.

40 Oil States Energy Services, LLC v. Greene's Energy Group, LLC, 584 U.S. ___, 138 S. Ct. 1365 (Apr. 24, 2018).

41 "Thom Tillis, U.S. Senator for North Carolina," https://www.tillis.senate.gov/public/index.cfm/press-releases.

The Court has resolved not to draw distinctions among different types of natural laws by assessing how critical the free availability of particular discoveries is, or how socially beneficial it might be to incentivize such discoveries through patent protection. Courts are "institutionally" bad at making such assessments, and a "bright line" test is best.

In contrast, as we saw in section II, a *halakha*-inspired approach to intellectual property as an equitable, policy-oriented social doctrine facilitates context-sensitive line drawing by courts. If courts were to adopt such an approach to intellectual property, in place of a personal-property theory, they might feel freer, or even mandated, to replace the bright line between natural and non-natural discoveries with a more context-sensitive assessment of the social benefits of intellectual property incentives for particular scientific discoveries (or types of discoveries) of genetic properties – whether purely "natural" properties or not.

Beyond offering a theory of intellectual property that is more open to case-sensitive policy analysis, can a *halakhic* perspective offer specific value judgments that might inform policy in the secular, contemporary context of genetic intellectual property? We will next review several *halakhic* considerations that may provide helpful insights.

### A. *Balancing Competitive Freedom against Incentives to Innovate*

On one hand, the Talmud's liberal permission of competition in the realm of Torah teaching (based on the principle that "competition among scholars increases wisdom") may logically weigh against strong patent protection for scientific advances in a modern secular society that values and honors scientific progress in a manner not unlike the way in which Talmudists valued and honored Torah scholarship. Indeed, the Supreme Court's pronouncement that the judicial doctrine against patenting laws of nature is concerned with maintaining free competitive access to the "basic building blocks" of scientific research is very much in the spirit of fostering "increased wisdom" through free "competition among scholars." Nevertheless, we have also seen R. Sofer's sensitivity to the policy counterargument that protection against unchecked competition provides otherwise inadequate incentives for production and distribution of content that is important for scholars and society. Ultimately, R. Sofer saw fit to update the *halakhic* rules for unfair competition and to draw more nuanced legal lines for certain types of content meriting greater protection in light of the economic realities and social needs of his time.

### B. *Balancing Social Welfare against Private Property*

Saving a person's life/health is permitted at the expense of someone else's property. By Rabbinic enactment, one who acts to save a third party (not himself) in mortal danger need not even reimburse the property owner.[42] This might suggest that from a *halakhic* perspective, intellectual property should not be strongly protected for potentially life-saving technology (genetically based or otherwise). Here again, however, this policy consideration only goes so far, and faces the counterargument that eviscerating intellectual property protection would leave inadequate incentives for development of new life-saving technologies in the first place.[43]

### C. *Ethical Danger in Treating Life as Instrumental?*

*Halakha*, like most religious systems in today's world, values human life as sacred. By itself, however, this value need not necessarily weigh against strong intellectual property protection for many common types of new genetic technology. Personalized medicines based on discoveries about the natural sequence and function of human DNA; bioengineered therapeutic compounds like monoclonal antibodies that bind to particular disease-causing targets; new organisms such as genetically modified bacteria or algae that efficiently output useful compounds; and even gene enhancement therapy, which introduces new genetic material into a human individual to compensate for anomalies in that individual's personal genome – these are all examples of genetic technologies that can be (and are being) created and used to greatly enhance the quality of human life without mistreating human individuals as if they were merely tools.

Likewise, Jewish tradition would not look askance on the use of genetic engineering to produce individuals in order to be of specific assistance to others in need of help. Consider the case of an individual dying of leukemia, in need of a bone marrow transplant, who agrees to participate in a cloning experiment with the hopes of producing another like him or her who, in suitable time, can be used to donate bone marrow and save the life of a person (and even more so, the donor). The simple fact is that Jewish law and tradition view the donation of bone marrow as a morally commendable activity, and perhaps even morally obligatory such that one could compel it even from one's child.[44] Jewish law and ethics see nothing wrong with having children for a multiplicity of motives other than one's desire to "be fruitful and multiply." Jewish tradition recognizes that people may have children to help take care of them in their old age and accepts that as a valid

42 Bab. Talmud *Bava Kama* 117b; Rambam *Laws of Battery and Damage* 8:14. No such enactment was necessary to motivate saving oneself, but the Rabbis did not want fear of potential financial liability to deter or delay anyone in a position to save a third party's life.

43 This counterargument of course did not typically apply in the classic context of a rescuer who damages tangible property in the course of trying to save a particular victim in jeopardy.

44 See J. David Bleich, *Compelling Tissue Donations*, 27 Tradition 4, 59–89 (1993). The rationale for this is that such donations (which are not really donations according to Jewish law, as they can be compelled) are neither statistically harmful nor particularly painful, and thus one who engages in this activity fulfills the biblical obligation not to stand by while their neighbor's blood is shed. This activity is compulsory activity in the same way one must jump into the water to save one who is drowning, if one knows how to swim and such activity poses no danger.

motive.[45] The same is true for a couple that conceives a child with the hope that the new child will be a bone marrow match for their daughter who is dying of leukemia and is in need of bone marrow from a relative. There is no basis to assert that one who bears a child in order to save the life of another is doing anything other than two good deeds – having a child and saving the life of another.[46] While the popular press may condemn this conduct as improper and instrumental use of human life, Jewish tradition would be quite resolute in labeling this activity as completely morally appropriate. Having a child is a wonderful, blessed activity; having a child to save the life of another child is an even more blessed activity. Such conduct should be encouraged rather than discouraged. The fact that genetic engineering can help us derive important functional value from living organisms – even from people – should not in itself be seen as ethically improper, if there is no likelihood of harm to the participant. This is consistent with deep respect for the sacred nature of life and the moral obligation to help others.

### D. *Ethical Risks to Social Equality and Human Rights*

Some have argued that enabling individuals to enhance their personal genome through genetic engineering may give rise to serious moral problems of social inequality. Such enhanced individuals will achieve success more easily than those who remain un-enhanced. For example, studies show that people who are tall and physically attractive are more likely to be hired and promoted than people who are short or unattractive. Although Western democratic societies can accommodate a certain degree of inequality, the difference in prospects between the enhanced and the unenhanced could become so pronounced and decisive that serious social instability would ensue.[47] Taken to the extreme, enhancements could be installed by manipulating gene lines, resulting in social advantages that are inherited by succeeding generations. This could eventually create a political system dominated by a genetic aristocracy or "genobility" who possess an unfair – and arguably immoral – lock on wealth, privilege, and power. The threat of inequality is amplified if technology to confer such genetic advantages is eligible for patent protection, such that it is only accessible by those who can afford to purchase access under license of the patent owner.

As intriguing as this position is for many ethicists, a strong counterargument is that many medical discoveries – not only in genetics – initially accrue to the benefit of the wealthy, and for a time allow certain advantages only to those with better access

45 See *Yevamot* 64a; *Shulhan Arukh, Even ha-'Ezer* 154:6–7; and Yehiel Michel Epstein, *Arukh ha-Shulhan, Even ha-'Ezer* 154:52–53.

46 The birth of the child itself is a fulfillment of the *mitzvah* to be fruitful and multiply, and the donation by the child of bone marrow or blood or other replenishable body serums that can save the life of another – particularly of a parent – is a second good deed.

47 Maxwell J. Mailman, *The Law of Above Averages: Leveling the New Genetic Enhancement Playing Field*, 85 Iowa L. Rev. 517 (2000).

to health care. Insisting that developing manufactured insulin to treat diabetics was unethical because the initial beneficiaries were wealthy patients who could pay for insulin[48] cannot be correct, either as a matter of *halakha* or sound public policy. Instead, we hope and expect that life-saving treatments – like manufactured insulin – that may be initially expensive will ultimately become widely available through typical industry progress. That is a much wiser policy than halting medical progress and depriving society altogether of new cures. "Solving" the need for more equitable allocation of access to medical care by blocking the development of genetic technologies (as some advocate[49]) seems to contravene the *halakhic* mandate to cure and prevent illness wherever possible.[50] Maximizing long-term benefits to society at large is always a factor in the intellectual property policy balance and motivates – for example, a (twenty-year) limit on the enforceable lifetime of all patents. Social equality seemingly furnishes little reason to treat genetic patents any differently than other patents in the medical realm.

However, some ethicists go further, and express the fear that society will mislabel genetically engineered individuals (e.g., clones) as non-human and engage in activities tantamount to murder or enslavement, by treating these individuals as organ sources, or as individuals to be experimented upon, or as forced labor. This is really a fear of what might occur through abuse of genetic technology, and not a necessary consequence of its appropriate use. It is quite plausible from a *halakhic* perspective to support temporary restrictions against human cloning, for example, as an ethical precaution until would-be practitioners of this technology – and society at large – can be adequately educated, sensitized, and regulated so as to unfailingly accord full and complete human dignity to the products of human cloning.[51] Similar considerations might militate against granting patents for such technology. However, this type of prophylactic rule, which argues that inherently permitted activity should for the time being be prohibited or discouraged in light of current ethical challenges, is not the same as asserting as a categorical rule of Jewish law that

48 See Seale Harris, Bunting's Miracle: The Story of the Discovery of Insulin (Lippincott 1946).

49 See George J. Annas, *The Man on the Moon, Immortality and Other Millennial Myths: The Prospects and Perils of Human Genetic Engineering*, 49 Emory L.J. 753 (2000).

50 For a fine volume on this topic, see The Orthodox Forum Proceedings VI: Jewish Responsibilities to Society (D. Shatz and C. Waxman eds. 1997).

51 See, e.g., *Cloning*, Pittsburgh Post Gazette, March 1, 1997, at A1:

> Rabbi Moshe Tendler, professor of medical ethics, Talmudic law and biology at Yeshiva University in New York, sees other potential good use for human cloning. In theory, the Orthodox scholar might permit cloned children when a husband cannot produce sperm. But he believes that the danger of abusing the science is too great to allow its use [at present]. As a Jew, he lives in the historical shadow of the Nazi eugenics program, in which people with "undesirable" traits were weeded out of society, forbidden to have children and ultimately killed .... "The Talmud says that man has to learn to sometimes say to the bee, 'Neither your honey nor your sting.' Are we good enough to handle this good technology? Of course we are, if we can set limits on it. And when we can train a generation of children not to murder or steal, we can prepare them not to use this technology to the detriment of mankind."

such conduct is prohibited. Rather, it is at most only a temporary safeguard, prohibiting that which is intrinsically permissible only until conditions change.

### E. *Playing God?*

Is the possibility of humans "playing God" through genetic technology fundamentally troubling for Jewish law? As the late Lord Immanuel Jakobovits stated, speaking for the Jewish tradition:

> We can dismiss the common argument of "playing God" or "interfering with divine providence." Every medical intervention represents such interference. In the Jewish tradition this is expressly sanctioned in the biblical words: "And he [an attacker] shall surely cause him [his victim] to be healed."[52] The Talmud states: "From here we see that the physician is given permission to heal."[53]

This articulation of the Jewish view is deeply rooted in Jewish law and ethics. The world was not created a perfect place – people are responsible for their own conduct and condition and need not be accepting of the conditions of nature around them. Indeed, people are charged with improving on the handiwork of the Creator. The classical code of Jewish law states simply:

> Jewish law gives the doctor the license to heal, and it is a good deed, and within the category of life saving activity. One who withholds medical treatment is a spiller of blood [a murderer].[54]

In the Jewish tradition, people were put on this earth to "improve the world in the image of the divine,"[55] and not to accept the perilous condition of the world, whatever it might be. Tampering with nature is part of the human mission in the Jewish tradition – curing illness is one facet of that mission. Mastering nature and using it to make better people is no less a fulfillment of this religious mandate than the healing of the sick with leeches or antibiotics.

### F. *Religious Value in Preserving the "Natural Order"?*

Interestingly, unlike current secular patent law, *halakhic* concerns over "tinkering" with life forms are not exclusively focused on naturally occurring genetics – quite to the contrary. According to Nachmanides (Catalonia, Spain 1194–1270, also known as *Ramban* – an acronym for R. Moses ben Nachman), one of the most prominent Jewish legal authorities and theologians of the medieval period, the Biblical prohibition of *kilayim* (interbreeding different species) focuses precisely on "engineering"

[52] Exodus 21:19.
[53] *Will Cloning Beget Disaster?*, The Wall Street J., Friday, May 2, 1997, at A14.
[54] *See* Shulchan Aruch, Yoreh Deah 336:1.
[55] This exact phrase *letaken olam bemalchut shadai* is taken from the daily alenu prayer, which is recited thrice daily in the traditional prayer liturgy.

new forms of life. Nachmanides suggests[56] that the Bible forbids interbreeding different species of animals or plants for two reasons:

a. Unnaturally crossbreeding species implies that God "failed to complete the creation of all that was needed in the world" and needs the crossbreeder's assistance, as it were, to properly finish the job.
b. Crossbreeding is rejected because it generally fails to produce fertile offspring – or in some cases cannot generate any offspring at all.

Nevertheless, while Nachmanides saw a degree of universal immorality in efforts to create new species of life, R. Judah Loew ben Bezalel (Prague, died 1609; known better as "Maharal") evidently disagreed and took a far more limited view to the ethical implications of crossbreeding:

> Regarding those who are aghast of the interbreeding of two species – certainly, this is contrary to Torah which God gave the Jews, which prohibits inter-species mixing. Nonetheless, Adam (the First Person) did this, creating [mules] that were fit to exist in order for the world to be complete; and even though the Torah [later] forbad such acts [for Jews], the Torah's laws are a separate matter. Indeed, the world was created with many species that are prohibited by the Torah to be eaten, and the world is completed by their existence. Inter-species breeding was not prohibited because of sexual immorality . . . As we already noted, the laws of the Torah are one matter, and the ways of completing the world are a separate matter [and permissible for Gentiles].[57]

Indeed, Rabbi Loewe nearly states that such conduct by general society is good – after all, we all use donkeys and eat nectarines.[58] Thus, according to R. Loew, while Jewish law prohibits Jews from interbreeding species of animals and plants, this is not because such innovations are inherently or broadly unethical. Rather, these acts are technically prohibited to Jews under the regulations of the Torah for unknown reasons, very much like unkosher food. Unlike Ramban, R. Loew saw in breeding of mules a desirable completion of the world (albeit technically not permissible for Jews to perform) – not an insult to the perfection of God's creation.[59]

56 Ramban's Commentary to the Torah, Leviticus 19:19

57 Maharal *Be'er HaGolah* chapter 10.

58 And Jewish law permits this enjoyment. Such conduct was prohibited by Jewish law because it was not part of the Divine mission for the Jewish people. Jewish law is not a general ethical category governing the conduct of all, but its scope and application is limited to Jews, not merely jurisdictionally, but even theologically. This point of view would seem apparent from the general attitude that the Jewish tradition takes to a number of proselytizing issues. For more on this, see Michael Broyde, *Proselytizing and Jewish Law*, in SHARING THE BOOK: RELIGIOUS PERSPECTIVES ON THE RIGHTS AND WRONGS OF PROSELYTISM 45–60 (John Witte, Jr., and Richard C. Martin eds. 1999).

59 Bear in mind that even according to Ramban, it does not necessarily follow that any typical bioengineering techniques would implicate the *kilayim* prohibition, legally speaking, even for a strictly observant *halakhist*. The Biblical prohibition is concerned with sexual reproduction among animals, and also with certain forms of agricultural crossbreeding. We do not suggest for

For our present purposes, it is critical to ask: Is the Ramban's view about *kilayim* merely a parochial religious value? Or can Ramban's notion of a moral obligation to give deference to God's natural order offer meaningful insight to a secular, technologically advanced society? We are inclined to claim that, yes, it does. From a modern, secular perspective, Ramban's view can be taken as a reminder to be ever-vigilant against excessive hubris in technological development – and particularly in the realm of engineering new variations or forms of life, where lack of appropriate caution and respect for the attendant safety risks carries the potential to literally unleash environmental catastrophe.[60]

Clearly, even a strict *halakhic* society – and much more so a secular one – would be foolish and mistaken to overgeneralize from the Ramban's explanation of *kilayim* and categorically forfeit bio-technological opportunities to fulfill the divine mandate of healing the sick. Technology that helps modify the human genome so as to eliminate or compensate for sickle cell anemia or Tay-Sachs, for example, would without question have extraordinarily positive *halakhic*, ethical, societal, and economic value. Genetic cures can potentially be more permanent, and thus more effective, than conventional pharmaceutical alternatives. What we nevertheless find highly enlightening and relevant for our purposes from the Ramban's explanation of *kilayim* is the ethical value of healthy self-awareness and caution as we step across the frontier of modifying life's natural blueprints. The importance of healthy caution and respect for the power potentially unleashed (perhaps unpredictably) when tinkering with life holds true in determining patent eligibility for inventions, just as it does for other legal and regulatory aspects in the biotechnology field.

It is worth noting that the dominant voices in modern Jewish law – the consensus view of the last five hundred years – treat the basic approach of Rabbi Loewe as the one that is dominant and ought to be followed. Nachmanides's (Ramban's) view serves to remind us all of the perils of ethical hubris, but the Jewish tradition has for centuries treated the violation of the natural norm as proper when it extends life, improves the human condition, or otherwise benefits life on this earth.[61]

## IV. CONCLUSION

While the classic texts of Jewish law for the most part preceded the rise of intellectual property law by centuries, we have argued here that a distinctly *halakhic* theory of

a moment that traditional *halakhic* rules of *kilayim* implicate modern biotech methods for gene sequencing and splicing. We cite Ramban's views on *kilayim* here only because the ethical principle he finds in *kilayim* has powerful resonance, by way of analogy, as far as exercising caution and avoiding hubris as the biotechnology field advances.

60 Seth D. Baum and Grant S. Wilson, *The Ethics of Global Catastrophic Risk from Dual-Use Bioengineering*, in 4 Ethics in Biology, Engineering and Medicine 59–72 (2013), online at http://sethbaum.com/ac/2013_BioengineeringGCR.html.

61 For more on this, see Michael J. Broyde, *Genetically Engineering People: a Jewish Law Analysis of Personhood*, 13 St. Thomas L. Rev. 877 (2001).

intellectual property can nonetheless be derived from the application of Talmudic unfair competition principles to intellectual property questions by *halakhic* authorities beginning in the Early Modern period. We further contend that because this Talmudic doctrine is equitable in nature and operates to protect broad social interests, it provides *halakhic* authorities with a more flexible, context-sensitive model for intellectual property law as contrasted with the personal property basis undergirding much of secular, contemporary patent and copyright law. Finally, we suggest that if contemporary courts were to adopt such a model for intellectual property, *halakha* could offer various insights regarding policy considerations and value judgments pertinent to determining patent eligibility criteria for genetic inventions.

# 8

# Intellectual Property, Islamic Values, and the Patenting of Genes

*Mohammed El Said*[1]

## INTRODUCTION

The past few decades witnessed unprecedented levels of technological progress in many areas, leading some to refer to this as the 'age of technology'. One particular area where such noticeable developments have occurred is medical technologies and innovations. New medical technologies, gene therapy, and personalized medical treatments that were once viewed as futuristic in nature are a reality today. Concurrently, such developments also raise dilemmas for many due to their extra-territorial nature and cross-border impact and their interaction with established universal human rights principles such as the rights to life, health, and access to medicines.

Such developments have been praised by many because of their impact in preserving good health, curing diseases, and prolonging life expectancy levels. At the centre of these developments emerges the debate about interplay between the process of learning, scientific progress, and knowledge creation. At the heart of the debate arises the question: what are the values and principles that should drive such developments? Should the developments be triggered by purely capitalist, social, and economic factors? Or do theological values also have a role in influencing and shaping the developments? More specifically, who owns the knowledge leading to the creation of these innovations?

This chapter will examine the above questions in more detail. In addition, the chapter will analyse to what extent such developments and their utilization are compatible with social, theological, and local beliefs. More specifically, the chapter will present Islam's stance on a number of contemporary issues including the protection of intellectual property (IP) and the issue of gene patenting in light of established Islamic bioethics and principles. It will conclude by highlighting the urgent need for Muslim states to take a proactive role to address these issues through legislation and public instruction.

1 Mel-said@uclan.ac.uk. The author would like to thank Professor Thomas Berg for his valuable feedback on this chapter. The usual disclaimer applies.

## 1. THE SCIENCE AND RELIGION DEBATE: ISLAMIC PERSPECTIVES ON LEARNING

The debate about science and theology in general preceded Islam by centuries. As noted by Thomson, 'the subject was argued by the ancient Greeks and scholars of the Middle Ages and the Enlightenment'.[2] However, the relation between science and religion was not always harmonious. To the contrary, at various times and in different places, science was seen to be incompatible with religious values. For example, in 1633, Galileo was tried for heresy by the Roman Inquisition, which forced him to recant his views and live out his days under house arrest.[3] Today, a number of Islamic and Jewish scholars largely reject the theory of natural evolution of new species, particularly for humans.[4] Differences continue to prevail between scientific claims and religious principles. However, such a debate is no longer confined to philosophers and clergymen, but extends to also include the circle of technological gurus and medical geniuses.

Islam has placed a focus on knowledge creation and scientific discovery from its early days. From its outset, Islam encouraged learning and examination of the surroundings as a way of affirming humans' existence and acknowledging the creator's ability.[5] In doing so, Islam encapsulated several mechanisms aimed towards promoting learning and knowledge-creation with the objective of achieving personal satisfaction, social welfare, and public benefit. In this regard, the Quran explicitly grants higher recognition and appreciation to those who pursue knowledge: 'Are those who and those who do not know to be reckoned alike?'[6] Moreover, the first verse of the Quran states, 'Read! In the name of your Lord who has created (all that exists). He has created man from a clot (a piece of thick coagulated blood). Read! And your Lord is the Most Generous. Who has taught (the writing) by the pen. He has taught man that which he knew not.'[7] However, such activities were and

[2] K. Thomson, Introduction, in *The Religion and Science Debate: Why Does it Continue?*, edited by H. Attridge (New Haven, CT: Yale University Press, 2009), at 1.

[3] It was only in 2000 that former Pope John Paul II issued a formal apology for the church's treatment of Galileo.

[4] See generally Nidhal Guessoum, 'Islamic Theological Views on Darwinian Evolution', *Oxford Research Encyclopedia of Religion* (May 2016), http://religion.oxfordre.com/view/10.1093/acrefore/9780199340378.001.0001/acrefore-9780199340378-e-36; Salman Hameed, 'Bracing for Islamic Creationism', *Science* 322 (2008): 1637–8; Saouma BouJaoude, Anila Asghar, Jason R. Wiles, Lama Jaber, Diana Sarieddine, Brian Alters, 'Biology Professors' and Teachers' Positions Regarding Biological Evolution and Evolution Education in a Middle Eastern Society', *International Journal of Science Education* 33.7 (2011): 979–1000; Nidhal Guessoum, *Islam's Quantum Question: Reconciling Muslim Tradition and Modern Science* (London: I. B. Tauris, 2011).

[5] For example, 'See ye not how Allah hath made subservient unto you whatsoever is in the heavens and whatsoever is in the earth, and showered you with His blessings both manifest and hidden' *Surat* Luqman: verse 20. This *Surat* highlights the need for man to exploit the resources available for welfare of mankind.

[6] *Surat* Az-Zumar: verse 9.

[7] *Surat* Al Alak: verse 96:1–5. The *Hadith* of the Prophet (PBUH) also emphasizes the value of knowledge. As the following *Hadith* narrated by Abu Hurrairah notes: 'When a man dies, his acts come to an

remain subject to a number of well-established guidelines and principles, as this chapter will discuss.

One fundamental pillar which the Islamic regime of principles relied upon in the process of knowledge acquisition and promotion is the treatment and recognition of knowledge (in its wider meaning) as a public good. Public goods are characterized by their non-rivalrousness and non-excludability. In other words, the consumption of a public good by one individual does not limit the ability of other individuals to simultaneously consume that same good (non-rivalrous), and no individuals can be excluded from consuming the good whether or not they paid the costs of its production (non-excludability). Within this meaning, essential sectors within any state which are related to the provision of vital services to society are considered as public goods, such as education and public health.[8]

In order to institutionalize the process of learning within the Islamic society, a number of learning institutions were founded and supported. The mosque was the first such institution that served both religious and educational purposes. Although originally created for worshiping purposes, it was gradually expanded to also include a school (*Madrassa*) and a library. As Wani and Maqbool explain, 'The mosques, which during the early decades of Islam formed the nerve centres of all political, social, religious, and educational activities, housed valuable libraries comprising books on religion, philosophy, and science.'[9] Moreover, universities and public libraries increasingly became more widely available to the masses in major cities and centres of learning such as Baghdad, Damascus, Cairo, and Cordoba.[10] Such state-run activities were largely supported by various private funding mechanisms such as *Waqfs* and personal patronage.[11]

Incentivizing learning and knowledge creation was therefore always part of the Islamic ethos. Finding solutions to dilemmas encountered in medicine, astrology, chemistry, mathematics, and other subjects needed a proactive and continuous effort. Although this effort was largely influenced by the need to fulfil one's duties towards the creator and society, it was also viewed as the key to ensuring the

end, but three, recurring charity, or knowledge (by which people benefit), or a pious son, who prays for him (the deceased)' (Sahih Muslim: Book 12, Hadith 4005).

8 See generally Ramazan Kilinç and Carolyn M. Warner, 'Micro-Foundations of Religion and Public Goods Provision: Belief, Belonging, and Giving in Catholicism and Islam', *Politics and Religion*, 8(4) (2015), 718–44.

9 Zahid Ashraf Wani and Tabasum Maqbool, 'The Islamic Era and Its Importance to Knowledge and the Development of Libraries', *Library Philosophy and Practice*, paper 718 (2012), at 1, https://digitalcommons.unl.edu/libphilprac/718/.

10 For more, see R. M. Elayyan, 'The History of the Arabic-Islamic Libraries: 7th to 14th Centuries', 22 *International Library Review* (1990), 119–35, at 126.

11 Nizamoglu explains, '*Waqf* means "charitable foundation". The term comes from an Arabic root, with the sense of "to stop", "to hold back", "to stand still and to tie". It means also one person religiously and socially devoting all or some parts of his assets to the public in order to meet some of their needs for the sake of The God.' See Cem Nizamoglu, 'The Institution of Waqf as a Solution to the Economic Crisis', *Muslim Heritage*, available at http://www.muslimheritage.com/article/institution-waqf-solution-economic-crisis.

development and progress of the expanding Islamic nation.[12] Notably, Islam's encouragement of learning in general was not confined to the religious sciences (*aloom al fiqh*).[13] In fact, as a religion, Islam declared that it is the duty of the believer to contribute to the welfare of society – as a whole, regardless of its components, Muslim or otherwise – through utilizing and optimizing God's creations on earth. This resulted in the period often referred to as 'the golden age of Islam', whose general theme was the utilisation of scientific innovations to serve religious duties and social needs.[14] In other words, by treating knowledge as a public good and institutionalising the learning process, Islam adopted and encouraged an 'open educational' model aimed towards achieving human development and social welfare.

## II. INTELLECTUAL PROPERTY PROTECTION AND ISLAMIC VALUES: CONVERGENCE OR CONFLICT?

Today's global knowledge governance is influenced by a number of factors. One important factor is the current global regime of intellectual property (IP) protection and governance.[15] IP refers to the legislative framework designed to award legal monopoly (protection) for the creations of the mind (intellect) for a specific period of time. This monopoly protection is granted in exchange of the disclosure of the invention to the society. Examples of such rights include patents, trademarks, copyrights, geographical indications, trade secrets, industrial models and designs, and others.

With the establishment of the World Trade Organization (WTO) in 1995, IP protection became one of the important foundations of the global economic regime. The Agreement on Trade-Related Aspects of Intellectual Property Rights (the TRIPS Agreement), which is one of the three main pillars of the WTO, represents the most comprehensive agreement ever concluded in the field of IP protection. WTO member states must incorporate the Agreement's minimum standards of protection and obligations within their national legal regimes and must accept being subjected to the WTO's Dispute Settlement Procedure.[16] This applies to all

12 Remarkably, Islam's adoption of the 'scientist/inductive approach' was clearly manifested in a number of verses from the Quran, embedded by the frequent encouragement of reflection, observation, and free thinking. Various Quranic verses (*surats*) directly allude to such affirmation, such as 'why do they not ponder' and 'in them there are signs for those who reflect' (*Surat* al-Ra'd: verse 3).

13 For more on meaning and definition of knowledge under Islam, see Franz Rosenthal, *Knowledge Triumphant: The Concept of Knowledge in Medieval Islam* (Lieden: Brill, 2007).

14 See, e.g., Jim Al Khalili, *Pathfinders: The Golden Age of Arabic Science* (London: Allen Lane, 2010); A. R. Momin, *Islam and the Promotion of Knowledge* (New Delhi: Institute of Objective Studies, 2009), p. 47.

15 Broadly speaking, this refers to various multilateral and regional trade agreements and bilateral free trade agreements (FTAs) and investment agreements (BITs), in addition to various international declarations, conventions, and treaties.

16 The TRIPS Agreement is Annex 1C of the Marrakesh Agreement Establishing the World Trade Organization, signed in Marrakesh, Morocco, on 15 April 1994.

countries, regardless of their religious, economic, and social backgrounds and regimes.[17]

A major criticism of the TRIPS Agreement concerns the 'one size fits all' approach it adopts.[18] In other words, it imposes a unified regime on all member states regardless of their backgrounds and beliefs. This uniformity also resonates with today's global technological landscape, where local particularities are often ignored. As Raquib explains, the 'modern technological paradigm seriously neglects and downplays values associated with a religious worldview'.[19]

Islam in general recognizes the concept of private property ownership in its various forms. Moreover, it also acknowledges the need to reward an individual for his/her effort regardless if such effort was physical or mental in nature.[20] In addition, one of the main objectives of *Sharia* is the preservation of property and/or the individual's wealth. However, Islam's affirmation of private property and ownership rights includes some limitations. In essence it views humans as custodians and vicegerents (trustees) on earth and says that all property – in its tangible and/or intangible form – belongs to the divine Creator.[21] Such custodianship also extends to knowledge, learning, and science, in which humans are expected to be holding knowledge on basis of trust (*Amanah*), not on the basis of monopolisation and use for self-interest.

As such, Islam views personal ownership of property as a 'way of achieving largely social ends by using an individual's entrepreneurial capacity and ingenuity'.[22] Based on this, the right of ownership is not absolute but is rather subject to a number of limitations and conditions.[23] There are a number of restrictions which apply against this right, and

17 Under TRIPS Article 66.1, least developed countries have a transition period, which has been extended, to implement protections.

18 For more on this and on the history of the TRIPS Agreement, see Mohammed K. El Said, *Public Health Related TRIPS-Plus Provisions in Bilateral Trade Agreements: A Policy Guide for Negotiators and Implementers in the Eastern Mediterranean Region* (World Health Organization, Regional Office for the Eastern Mediterranean, 2010), http://applications.emro.who.int/dsaf/dsa1081.pdf.

19 Amana Raquib, *Islamic Ethics of Technology: An Objectives (Maqasid) Approach* (Kuala Lumpur: The Other Press, 2015), xxi–xxii.

20 It was reported that the prophet PBUH stated that 'all that belongs to Muslim is forbidden to his fellow Muslim, his life, his property and his honour'. Even the Prophet Mohammad himself acknowledged the respect for the private property in his last Farewell Sermon, 'Verily your blood, your property are as sacred and inviolable as the sacredness of this day of yours, in this month of yours, in this town of yours.'

21 Nevertheless, Islam appreciates private property ownership. The Quran states in *Surat* al-Baqra: verse 1885:

> And eat up not one another's property unjustly (in any illegal way e.g. stealing, robbing, deceiving, etc.), nor give bribery to the rulers (judges before presenting your cases) that you may knowingly eat up a part of the property of others sinfully.

22 Saliem Fakir, 'The Notion of Public Goods, Sustainable Development and Islam', Paper for AWQAF Foundation of South Africa, AWQAF Seminar on Islam and Sustainable Development (World Conservation Union)-SA (1 September 2002), at 9.

23 For instance, Ibn Taymiya states that 'nobody could raise his building to obstruct light or air of his neighbors. Similarly, it was not allowed to install a flour mill or a leather tanning factory in the

in practice the line between what is permissible and what is prohibited will often depend on the context, nature, and exercise of the right under question in each case.

There is a growing literature dedicated to discussing the issue of IP protection under Islamic principles.[24] Most commentators agree that there is no explicit reference in the *Sharia* on how to treat IP-related matters. Since current regimes of IP protection are a modern concept, discussing their compatibility with Islamic principles requires the utilisation of a number of legal and interpretive tools.

The main primary sources of guidance under Islam are the Quran and the *Sunnah*. In the absence of clear and explicit reference to a certain matter under either the Quran or *Sunnah*, a number of principles are utilised to derive a legal ruling (*Hukum*) concerning the matter under discussion. Indeed, as observed by Dahlen, '*Fiqh* is (or, at best, can be) precisely a dynamic reading of revelation which affects the transformation of event into process, giving it ongoing significance'.[25] In line with this and as a general rule, the principles applicable to property ownership in general may also apply to IP ownership, subject to a number of limitations as will be seen.

In light of the aforementioned, the concept of IP as known under today's modern legal regimes needs to be evaluated against a number of principles, since there is no explicit reference to it under the Quran or *Sunnah*.[26] Based on this, current IP protection regimes raise a number of interesting – if not problematic – issues from an Islamic perspective.

One main criticism made by Muslim scholars against IP protection is the monopolistic nature of the rights it bestows upon its holders. IP, as a type of private property, creates a right that enables its owner or holder to preclude others from making, selling, distributing, importing, or dealing with the property subject to protection for a certain period of time. Such a monopoly creates a conflict with some established Islamic principles such as *Ihtikar*.

*Ihtikar*, under Islamic principles, refers to monopoly,[27] which means controlling and preventing the use of something by others. Commonly known under Islamic

residential areas as they could pollute the environment. People could not slaughter animals in the streets or houses to avoid unhygienic conditions.' Ibn Taymiya, *Public Duties in Islam: The Institution of the Hisba*, translated from Arabic by Muhtar Holland (The Islamic Foundation, Islamic Economic Series (3), 1982), at 145.

24 See, e.g., Mohammed El-Said, 'The Compatibility of Modern Intellectual Property Protection Norms with Islamic Principles: Lessons from History', 4(1) *World Intellectual Property Journal* (2012), 121–8; Mohamed Ahdash, *Copyright in Islamic Law* (The Islamic Texts Society, 2016); Silvia Baltrametti, 'The Legality of Intellectual Property Rights under Islamic Law', in *The Prague Yearbook of Comparative Law* 2009, edited by T. Mach et al. (Prague, 2010), 55–94.

25 Ashk Dahlen, *Islamic Law, Epistemology and Modernity: Legal Philosophy in Contemporary Iran* (New York: Routledge, 2003), at 43.

26 *Sunnah*, also spelled *Sunna*, refers to the body of traditional social and legal customs and practices of the Islamic community. Together with the Quran and *Hadith* (recorded sayings of the Prophet), it is a major source of the Islamic law.

27 In Arabic, the term *ihtikar* comes from the word *hakr*, which means collecting and controlling goods. *Ihtikar* is used by scholars of Islamic jurisprudence to refer to the privilege to collect and control of goods in an effort to anticipate the need for price increases. In other words, it means the process of

jurisprudence (*fiqh*),[28] the term refers to the practice of hoarding, or the 'process of monopolizing products in order to raise prices'. As an established rule, all monopolistic practices are explicitly prohibited in Islam. Although the Quran does not explicitly refer to *Ihtikar* (save for reference to hoarding of silver and gold), the *Hadith* of Prophet Muhammad PBUH makes critical reference to such practices, stating that the 'monopolist is a sinner'.[29] Applying this general principle, it is problematic to confine knowledge (as a public good) to the contours of IP protection. As Momin explains in this regard, 'Islam disapproves of concealing or withholding knowledge and emphasises its dissemination wide and far'.[30]

IP protection could be viewed from an Islamic perspective in another aspect: the Objectives (*Maqasid*) of the *Sharia*.[31] In summary, the Objectives under *Sharia* aim to preserve five main elements: religion, life, intellect, wealth, and progeny. Evaluation of a certain matter will be taken in accordance with these goals in the absence of a clear instruction or legal rule (*Hukum*) from the Quran or *Sunnah*. However, in case of a conflict between these objectives, the higher objective will be given precedence and priority over others. In accordance with this, preservation of religion ranks first, then preservation of life, then progeny, which is higher than preservation of health and intellect subsequently.

In line with this, IP would be evaluated for its compliance with the provided objectives (*Maqasid*). However, one should be cautious when undertaking such an analysis and should avoid generalization. As mentioned, IP rules cover a wide range of different rights (trademarks, patents, copyrights, and others), which serve different purposes, and therefore any analysis against the objectives referred to previously would have to be made case by case.

So let's take two IP-related examples, apply the aforementioned test, and see how the outcome may differ. One important area of growing global concern nowadays involves pharmaceutical patent protection and provision of medicines to the poor.[32] Prior to the TRIPS Agreement, countries had full discretion and autonomy to decide if they wanted to incorporate protection for pharmaceutical patents within their national legal regimes.[33] This was no longer available as an option under the TRIPS

monopolizing *ihtikar* products to result in price increases. The Quran does not mention *ihtikar*, but it makes reference to hoarding gold and silver.

28 The Quran, the *Sunnah*, *Qiyas* (analogical reason), and *Ijmā'* (scholarly consensus) constitutes the four sources of Islamic jurisprudence (*uṣūl al-fiqh*). When a situation calls for a new rule and none is to be found in the primary sources, nor in *Ijmā'* or *Qiyas*, it becomes permissible to resort to *Maslaha* (the public interest, see *infra*).

29 Shaykh Al-Islam Ibn Hajar *Al-Asqalani, Fath Al Bari Fi Sharh Sahih Al Bukhari* (1986), at 408.

30 See Momin, *supra* n.14, at 44.

31 As Raquib explains, *Maqasid* may be translated to mean 'aims', 'objectives', or 'end goals' under Islamic jurisprudence. Raquib, *supra* n.19, at xvii.

32 For a detailed analysis, see Mohammed El-Said, 'TRIPS-Plus, Public Health and Performance-Based Rewards: Schemes Options and Supplements for Policy Formation in Developing and Least Developed Countries', 31(3) *American University International Law Review* (2016), 373–444.

33 See El Said, *supra* n.18.

Agreement, which makes it compulsory upon all member states to award legal protection to pharmaceutical patents. Based on this, if a certain medicine subject to patent protection was widely unavailable as a result of its expensive price, and patients' lives were in danger because of their inability to purchase the needed medicine, this practice would contravene one of the highest objectives of *Sharia*, protecting and preserving life. Allowing IP protection in such a case without applying any restrictions or exceptions would be highly improper from an Islamic perspective.

The answer to the same question may be different in another scenario and with a different type of IP. For instance, take the example of a certain trademark. Assume that we are confronted with a question about purchasing a certain luxury product – a watch, for example – that has a famous trademark attached to it, and many members of society could not afford it as a result of its high price. In this case the *Sharia* objective of preserving the private ownership and property of the trademark owner – who invested heavily in the development and marketing of the trademark – will prevail, since consumers will have access to different ranges and types of alternative watches if they wish. In this case, the objectives of *Sharia* would be permissive towards trademark protection, since protection would lead to the preservation of wealth without hurting the wider interest of the society.

In line with this, it is evident that the test applied here will be based on the merits of each case. Further guidelines complement the above rules. Where certain benefit and harm clash, the prevention of harm would be given priority over the desired achievement of the benefit. Raquib elaborates by explaining that 'the greater of two benefits should be prioritized; collective interests should be preferred over individual benefits; lesser harm should be opted for over greater harm; individual harm should be endured to prevent collective harm and necessities should be assessed accurately.'[34]

One primary principle governing the various social and economic practices within any Islamic society is the public's *Maslaha* (or the public interest). In line with this, the purpose of legal principles in Islam shall be compatible with the public's *Maslaha* as well as with the Objectives (*Maqasid*) of the *Sharia*.[35] This is the case since an Islamic economic system can only be understood in terms of the social objectives it sets out for every Muslim individual living in that society.[36] Supplementing this position is the principle of *la darar wa la dirar*: in other words, the necessity to refrain from causing harm to oneself or others. In explaining the reasoning behind *Masalah*, Sardar states:

34 Raquib, *supra* n.19, at 86.

35 The *Maslaha* prohibits individuals from making a profit which results in the abuse or detriment of public interests. For more, see El-Said, *supra* n.24.

36 See Fakir, *supra* n.22.

> Traditional Muslim scholars developed a whole array of *Maslaha* categories, some of which required direct evidence from the Quran and the Sunnah while others could lead to binding legal sanctions on the basis that they clearly promote a noted ethical criterion such as preservation of life, property, promotion of Islamic mores and sound reasoning of the Sharia.[37]

In other words, where the grant of IP protection conflicts with the public interest, the public interest shall prevail and the grant shall be treated as null. The provided example of the accessibility of life-saving medicines will help clarify this point. As highlighted, the right to life is sacred and is well preserved under Islamic rules and objectives, thus forming a fundamental and recognized aspect of the public interest.[38] So where the high price of certain treatment or medicine precludes access to it by those who need it, arguably under Islamic jurisprudence the private property right of the manufacturer would be overruled because it conflicts with the *Maslaha* and the prevalence of the right to life, and the patent on such a medicine would be subject to limitations. Other scholars would even go further and argue that if 'certain people are in urgent need of lodging in the house of a certain person, there being no other accommodation available to them[,] . . . it is his *duty* to lodge them. Likewise, if they needed him to lend them cloths to keep them warm or utensils for cooking or tools for building or drawing water, he should provide the same free of charge.'[39] Ibn Taymiya establishes that 'it is obligatory when need arises to provide one's personal services free of charge: e.g to teach knowledge, to give legal advice, to give evidence, to arbitrate', and so forth.[40] This line of reasoning assumes that the welfare and common benefit of the society shall override any other considerations.

Notably, several conditions should be taken into consideration when applying the public interest principle, *Maslaha*:[41]

- To start with, the *Maslaha* (public interest) being sought should not contradict a recognized established rule (*Hukm*): when a ruling has been legislated by the *Sharia*, there can be no *Maslaha* to be considered for any ruling that is against it, as the (new) *Hukm* would then be a cancelled *Maslaha*.

37 Ziauddin Sardar, *Islamic Futures: The Shape of Ideas to Come (Islamic Futures and Policy Studies)* (London: Mansell, 1985), 113, cited in Raquib, *supra* n.19, at 49–50.

38 *Surat* al-Māidah, verse 32 states:

> If anyone slays a human being, unless it be [in punishment] for murder or spreading corruption on earth, it shall be as though he had slain all mankind; whereas, if anyone saves a life it shall be as though he had saved the lives of all life, it shall be as though he had saved the lives of all mankind.

39 Ibn Taymiya, *supra* n.23, at 52–53.

40 Ibid., p 54.

41 Yusuf Sani Abubakar, 'Corporate Social Responsibility of Islamic Financial Institutions: A Look from the *Maqasid Al-Shariah* (Purpose of *Shariah*) Approach', 7 *Business and Economics Journal* (2016): 255, at 3.

- There can be no *Maslaha* to create a new Islamic duty *(Ibadah)*,[42] or to add any element (*Rukn*) in an established *Ibadah*, or to eliminate some parts or obligations of any *Ibadah*.
- The *Maslaha* should be in line with definitive references and not based on speculative principles.
- When a rule (*Hukm*) is ordained, the *Maslaha* of that rule should be greater than the negative impact of it, neither equal nor less.
- The *Maslaha* should be a general and comprehensive *Maslaha*, not restricted and specific to some individual or group.
- The *Maslaha* should be compatible with the *Sharia* standards of *Maslaha*, and realistically understood.

As seen from the provided discussion, the context and objectives of the *Sharia* (*Maqasid*) are very important when determining the compatibility of IP protection with Islamic values and principles. Awarding an IP monopoly in certain cases may conflict with some established Islamic principles. Analysis of the merits and facts of each case should be undertaken accordingly.

## III. GENE PATENTING AND ISLAMIC ETHICS

Considerable global attention in recent years has turned to the impressive developments made in the gene industry. Such attention was largely triggered by the unrivalled progress made in this area during the past few decades, which culminated in the successful completion in 2003 of the Human Genome Project (HGP).[43] HGP opened the door for a whole new world of opportunities, in terms of treatment of diseases and increased interest by private companies wanting to reap the benefits of the commercialization of gene-related activities and technologies on a large scale. Governments, non-governmental organizations, and scholars have highlighted the implications of these developments with both optimism and scepticism.

The progress made in the field of gene therapy carries with it huge potential to improve humanity's welfare and cure endemic diseases. This is so because many diseases are caused internally by the interaction of genes. Decoding and understanding how genes function within the body are therefore essential to finding the suitable treatment for curing diseases. For instance, gene therapy and genetic engineering developments are triggering impressive results in terms of finding long-term treatment for Parkinson's and Alzheimer's diseases.

Other national initiatives are currently underway in a number of countries aiming to identify the population's genome map through genome sequencing and to develop precision-medicine solutions, tailored to the individual

[42] *Ibadeh* refers to the various established duties and acts of devotion or services in Islam such as canonical prayer (*Salat*), charity (*Zakah*), fasting (*Sawm*), and pilgrimage to Mecca *(Hajj)*.

[43] See 'All about the Human Genome Project', https://www.genome.gov/10001772/.

patient.[44] As Ghaly et al. explain: 'Genomics, therefore, is an appealing approach to any healthcare system that aims to unlock the role of DNA in medical care, and if implemented on a larger scale, to any nation interested in understanding heritable risk factors for common and rare diseases in its population, and in the development of long-term public health policies.'[45] Undoubtedly, these developments – which are highly controversial by their nature – have triggered the need to develop new understanding about the role of bioethics and its compatibility with social and religious norms. The latter is particularly important for Muslim states because Islamic ethics largely 'holds a religious worldview and draws its resources mainly from, religious texts'.[46] This emanates from the public nature of Islamic principles: the effect and impact of Islamic teachings and norms is not confined to spiritual teachings but rather extends to regulate the daily activities of society as well. In conclusion, serving the society and the advancement of its welfare is imperative in order for the individual to attain his/her spiritual fulfilment.

This debate stems from several considerations. The first is legal, focusing on the legality of patenting genes in general. A second is ethical and theological, emanating from the fact that these activities raise sensitive issues concerning the preservation of life and the protection of the public interest.

## A. *Legal Considerations*

The legal controversy surrounding gene patents is explained by the answer to the following fundamental question:[47] are genes considered to be 'laws, phenomena, or products of nature' and thus unworthy of patent protection, or otherwise?[48]

Before answering this question, it is important to understand the impact resulting from granting a patent for a certain gene. Correa sums the situation up by explaining

[44] Such as establishing sequencing in clinical settings, as well as the implementation of preventive programs such as premarital genetic screening, preimplantation genetic diagnosis and neonatal screening for genetic metabolic disorders.

[45] See Mohammed Ghaly et al. (eds.), *Genomics in the Gulf Region and Islamic Ethics: A Special Report in Collaboration with the Research Center for Islamic Legislation and Ethics* (2016), at https://www.imperial.ac.uk/media/imperial-college/institute-of-global-health-innovation/Islamic-Ethics-Report-English-(1).pdf.

[46] See Hassan Chamsi-Pasha and Mohammed Albar, 'Western and Islamic Bioethics: How Close Is the Gap?', 3(1) *Avicenna Journal of Medicine* (January–March 2013): 8–14. See generally Abdulaziz Sachedina, *Islamic Biomedical Ethics: Principles and Application* (New York: Oxford University Press, 2011).

[47] A gene patent refers broadly to the exclusive legal right granted to a specific sequence of DNA (a gene). Gene patent refers to patent granted for a specific gene sequence and its chemical composition. A gene patent is also a patent issued for a specific DNA sequence of a specific organism. See 'Patenting Genes: Pros and Cons', *Genetics Generation*, http://knowgenetics.org/patenting-genes-pros-and-cons/.

[48] A large number of jurisdictions exclude products of nature from legal protection, and therefore no patent could be claimed for such.

that if patent protection is awarded for a certain gene, 'nobody could use the patented gene even for functions not discovered or disclosed by the patent owner. An absolute protection of this kind is likely to discourage further research and prevent other possible uses of a patented gene until the patent expires.'[49] Accordingly, the award of a patent is likely to result in a monopolistic stance, and discouragement of research and development in that area.

The issue of gene patenting has been subject to legal debate for some time, particularly in developed countries. The experience of the United States is important, since it was among the first developed countries to engage in such activities at the national level systematically. In the United States, gene patents were granted as early as 1980, albeit with some levels of concern.[50] However, greater criticism started to surface years later, with growing fears about the negative impact such a grant may have on the development of research and clinical trials on genetic diseases. By 2006, a number of unsuccessful initiatives to introduce legislation banning gene patents were made. No legislation was introduced, however; the judicial system was left to deal with this controversial issue.

The US Supreme Court in 2013 finally addressed the issue with its landmark decision in *Association for Molecular Pathology v. Myriad Genetics*.[51] The Court unanimously invalidated a patent for an isolated gene sequence, declaring that 'separating a gene from its surrounding genetic material is not an act of invention'.[52] This forbade the patenting of isolated gene sequences in the United States – although the decision gave the biotech industry a partial victory by upholding patents on cDNA, the artificial chemical from which elements that do not code for genetic information have been removed.[53]

On the other hand, European Directives have addressed the issue explicitly but have taken a mixed (even conflicting) approach on the topic. For example, Article 5, Section 1 of the European Directive 98/44/EC explicitly stipulates that '[t]he human body, at the various stages of its formation and development, and the simple discovery of one of its elements, including the sequence or partial sequence of a gene, cannot constitute patentable inventions'. However, Section 2 of the same Article establishes that '[a]n element isolated from the human body or otherwise produced by means of a technical process, including the sequence or partial sequence of a gene, may constitute a patentable invention, even if the structure of that element is identical to that of a natural element'. Based on this, naturally

49 See Carlos Correa, 'Intellectual Property and Access to Science', The South Centre, Research Paper 69 (July 2016), at 7.

50 Jacob Sherkow and Henry Greely, 'The History of Patenting Genetic Material', 49 *Annual Review of Genetics* (November 2015), 161–82, at 171.

51 569 U.S. 576 (2013). The invalidated patent covered isolated DNA sequences that indicated a fourfold increase, or greater, in the likelihood of developing breast or ovarian cancer.

52 Ibid.

53 More recently, the High Court of Australia also invalidated patents on isolated gene sequences – and on cDNA as well – in its landmark decision in *D'Arcy v. Myriad Genetics*, [2015] HCA 35 (7 Oct. 2015).

occurring genetic sequences, whether of human or other origin, remain patent-eligible under the European Union (although human embryos are ineligible).[54]

Looking more globally, a number of countries have taken a clear legislative position against patenting gene sequences. For example, the 1996 Brazilian Industrial Property Code (No. 9.279, 14 May 1996) excludes from patentability living beings or 'biological materials found in nature', even if isolated, including the 'genome or germplasm' of any living being (article 10.IX).[55]

The TRIPS Agreement does not deal with the issue of gene patents explicitly. However, many observers believe that the language of Article 27.3(b) of the Agreement, if interpreted flexibly, allows countries to exclude gene patents.[56] However, a number of countries no longer enjoy such flexibility, because they have acceded to bilateral or regional 'TRIPS-plus' agreements.[57] The TRIPS Agreement's failure to address this issue explicitly and the rise of TRIPS-plus obligations in this field led a number of scholars to criticize the current global IP regime for its vagueness.[58]

### B. *Ethical and Theological Considerations*

Beyond the legal realm, the idea of granting patents on genes has also been criticized as 'absurd', 'unethical', and 'immoral'.[59] This stems from the fact that such activities have been equated with monopolistic practices, stifling innovation and increasing the costs of research substantially, and consequently harming the public interest.[60] Critics have also expressed fears about the potential side effects and abuse on human beings' genome arising from the various affiliated experimentation activities on

54 Paul Cole, 'Patentability of Genes: A European Union Perspective', *Cold Spring Harbor Perspectives on Medicine* (2014), http://perspectivesinmedicine.cshlp.org/content/5/5/a020891.short.

55 See Correa, *supra* n.49, at 4.

56 Under Article 27.3, members may exclude from patentability:

> (b) plants and animals other than micro-organisms, and essentially biological processes for the production of plants or animals other than non-biological and microbiological processes. However, Members shall provide for the protection of plant varieties either by patents or by an effective sui generis system or by any combination thereof. The provisions of this subparagraph shall be reviewed four years after the date of entry into force of the WTO Agreement.

57 For example, Article 16.7 (1) of the US-Singapore FTA states that 'each party may exclude inventions from patentability only as defined in Arts. 27.2 and 27.3(b) of the TRIPS Agreement'. For more on TRIPS-plus agreements, see Mohammed El-Said, 'From TRIPS-minus to TRIPS to TRIPS-plus: Implications of IPRs for the Arab World', 8 *Journal of World Intellectual Property* (2005): 53.

58 See, e.g., Cydney A. Fowler, 'Ending Genetic Monopolies: How the TRIPS Agreement's Failure to Exclude Gene Patents Thwarts Innovation and Hurts Consumers Worldwide', 25(5) *American University International Law Review* (2010): 1073–105.

59 See Nuno Pires de Carvalho, 'The Problem of Gene Patents', 3 *Washington University Global Studies Law Review* (2004): 701.

60 Resnik explains that 'since human DNA patenting uses market rhetoric to describe human body parts, it does treat human beings as incomplete commodities and, therefore, may *threaten* human dignity by taking us further down the path of human commodification'. David Resnik, 'DNA Patents and Human Dignity', 29 *Journal of Law, Medicine and Ethics* (2001): 152–65, at 152.

genes. In the mid-1990s, the American Medical Association (AMA) amended its Code of Ethics to recommend that doctors should not patent medical procedures because these patents compromise patient care. Since that time, the AMA (with other medical and scientific groups) has issued numerous statements that gene patents should not interfere with appropriate medical care and the development of better medical treatments and technologies, and declaring unethical any limitations on the dissemination of medical knowledge.[61] It is no secret that the arguments against patents 'on life' remain widely alive today among academies, practitioners, and religious figures.[62]

An important question in this context is what is Islam's position on research activities related to genes? And who would be entitled to own the knowledge created in the process?

In order to answer this question, we need to combine two novel matters under Islamic jurisprudence: the Islamic position on IP protection, and the application of that position within the context of gene patenting and its impact on public interest.

The first issue to be determined is how compatible gene-related activities are with Islamic biomedical ethics and principles.[63] As seen, the issue of gene patenting is a relatively recent development; thus the challenge of accommodating such new developments under an Islamic framework arises, as it did in the case of IP protection. Ghaly et al. refer to this type of issue as *Nawāzil*, which translates as predicaments, 'a technical term in Islamic jurisprudence that refers to the novel issues that likely have rarely been addressed'.[64]

To apply the principles discussed earlier in this chapter, the context and extent of gene-related activities will have to be considered. In other words, what are the objectives (*Maqasid*) which such an activity aims to preserve, and is it in the public interest (*Maslaha*) of society to allow such activities? If it is, what are the limitations and conditions placed within this context?

There is no doubt that the use of modern technology and innovative medicines and methods to treat illness is compatible with Islamic principles based on preservation of life – as an objective of *Sharia* – and public interest considerations. Although this seems like an obvious conclusion, in practical terms the outcome may be different. For example, if the use of gene technology were for curing a certain disease, there would be no doubt that this should be encouraged. However, if

61 See American Medical Association (AMA), 'Gene Patenting', https://www.ama-assn.org/delivering-care/gene-patenting.

62 Naomi Hawkins, 'A Red Herring – Invalidity of Human Gene Sequence Patents', 38(2) *European Intellectual Property Review* (2016): 83–91.

63 As Athar explains, these principles include 'preservation of life, to promote and restore health, to alleviate human suffering, to respect patient's autonomy, to perform medical justice, to tell the truth, and to do no harm'. See Shahid Athar, 'Enhancement Technologies and the Person: An Islamic View', 36(1) *Journal of Law, Medicine and Ethics* (2008): 59–64, at 61.

64 As explained by Ghaly et al., '*Nawāzil* is a recurrent phenomenon throughout Islamic history, because each generation of Muslim jurists have had their own novel issues that their predecessors did not address, perhaps because they never thought such issues would exist'. Ghaly, *supra* n.45, at 27.

gene technology were used to alter the nature of humans by aiming towards directing or influencing their behaviour into a certain path, then the technology may not be compatible with the objectives of *Sharia*.[65] In light of this, one should keep in mind that the debate surrounding these types of activities within Islam revolves around the role played by humans in impacting God's creations.[66]

According to Islamic teaching, God created the world and entrusted humans with the responsibility of cultivating it for the common benefit in a balanced manner.[67] Within this context, the human body, like the soul, is a 'gift' from God, and therefore there is no absolute ownership over it.[68] As such, the outcome of gene-related activities should not be absolutely owned by those undertaking such activities. Rather, these activities are permissible under a divine 'licence' permitting them to take place within the objectives of the *Sharia* and for the common good. Consequently, determining the identity of the 'owner' within the framework of gene patenting raises some controversial questions from an Islamic perspective.

Nevertheless, this debate remains in its early stages in the majority of Muslim countries, since most of the innovative activities shaping this area of science remain largely concentrated in developed countries. Indeed, this is why such activities remain largely controversial today in the Muslim world. As Abd-Al-Haqq Guiderodni explains, 'To present the fundamental principles of ethics in Islam requires raising the question of Islamic anthropology which is entirely defined by man's relationship with God.'[69] Save for a few cases, the majority of Muslim countries have yet to tackle this issue through any institutional framework. For instance, in the Arab world today, only a limited number of countries, such as Qatar and the Kingdom of Saudi Arabia, have embarked on creating national genome programs.[70]

In line with this, there is a wide gap on gene-related issues in the legislative framework in the majority of Arab and Islamic states. For instance, a 2011 UNESCO Report on the Arab world found that there is a serious lack of legislation on bioethics in the region. More specifically, the Report found that[71]

65 Raquib comments on this type of scenario, stating that 'this argument is similar to the Islamic principle of blocking the rightful means to harmful ends. The ethico-legal principle of blocking the means to unlawful ends restricts certain otherwise lawful actions on the anticipation that they could play a role in the fruition of unlawful actions.' Raquib, *supra* n. 19, at 319.

66 The Quran in *Surat* Al Qamar (verse 49) explains: 'Indeed, all things did We create [according to] a precise measure'.

67 'Verily, we did offer the trust to the heavens, and the earth, and the mountain; but they refused to bear it because they were afraid of it. Yet man took it up.' *Surat* 33 (Al-Ahzab): verse 72. Also, 'He Who created the seven heavens one above another: No want of proportion wilt thou see in the Creation of Allah Most Gracious. So turn thy vision again, do you see any flaw?' *Surat* (67 (Al-Mulk): verse 3.

68 See generally Sachedina, *supra* n.46.

69 Quoted in UNESCO, *Ethics and Law in Biomedicine and Genetics: An Overview of National Regulations in the Arab States* (UNESCO, Cairo Office, 2011), at 4.

70 Ghaly et al., *supra* n.45.

71 UNESCO, *supra* n.69.

- Among seventeen Arab countries, none permits human reproductive cloning, and some countries explicitly or implicitly prohibit such practices by national guidelines or by virtue of an international declaration.
- The issue of genetic testing is insufficiently addressed in regulation in the Arab world. Only Lebanon has a written law specifically formulated to regulate it (No. 625 of 2004).
- No specific laws exist for human genome and gene analysis in the Arab world.

In dealing with the provided gaps in law, scholars and researchers have turned to religious institutions to seek guidance and justify these activities. As Chamsi-Pasha and Albar explain, 'In the absence of an organized "church" and ordained "clergy" in Islam, the determination of valid religious practice, and hence the resolution of bioethical issues, is left to qualified scholars of religious law, who are called upon to provide rulings on whether a proposed action is forbidden, discouraged, neutral, recommended or obligatory.'[72]

More specifically and with relation to gene therapy and genetic engineering, a number of Islamic edifies (*Fatwas*) and guidelines were issued with the aim of regulating practices in this area. In this regard, the Islamic Jurisprudence Council of the Organization of Islamic countries in its tenth session (28 June–3 July 1997) declared the following, under Decree number 100/2/D10:

- Human cloning is forbidden in any method that leads to human reproduction.
- It is permissible to use genetic engineering and cloning in the fields of germs, microorganisms, plants, and animals, following legitimate rules which lead to benefits and prevent harm.[73]

Building upon this, the Islamic Jurisprudence Council of the Islamic World League (Organization of Islamic countries), in its fifteenth session (31 October 1998), introduced the following guidelines:

- To use genetic engineering for disease prevention, treatment, or amelioration on the condition that such uses do not cause further damage;
- To forbid the use of genetic engineering for evil and criminal uses or what is forbidden religiously;
- To forbid using genetic engineering and its tools to change human personality and responsibility, or interfering with genes to improve the human race;
- To forbid doing any research or therapy of human genes except in extreme need, after critical evaluation of its benefits and dangers and after an official consent of the concerned, respecting the extreme confidentiality of the information and human rights and dignity as dictated by Islamic *Sharia*;

72 See Chamsi-Pasha and Albar, *supra* n.46.

73 Quoted in *Contemporary Bioethics: Islamic Perspective*, ch. 12: 'Ethical Issues in Genetics' (published online May 28, 2015), https://www.ncbi.nlm.nih.gov/books/NBK500187/.

- To allow the use of bio-engineering in the field of agriculture and animals, on the condition that precautions are taken not to inflict harm (even in the long term) on humans, animals or vegetation.[74]

Consequently, it is evident that there is nothing objectionable about gene-related innovation from an Islamic perspective, so long as (1) it is conducted for the prevention of harm and the promotion of health benefits, and (2) it is conducted on animals and plants and does not result in any harm to human beings in accordance with the provided guidelines.

As seen, Islamic jurists have taken a cautious and restrictive approach by placing strict conditions on gene-related activities. Interestingly, the restrictive approach taken by Islamic jurists is mirrored by other international declarations on the subject. For instance, the 2003 United Nations Educational, Scientific and Cultural Organization (UNESCO) Declaration on Human Genetic Data adopts a restrictive approach with relation to commercial exploitation of genetic resources and the requirement to share benefits. Moreover, the Universal Declaration on Bioethics and Human Rights adopted by UNESCO members in October 2003 commits member states to respect the principles of bioethics by stating explicitly under Article 15 that '[b]enefits resulting from any scientific research and its applications should be shared with society as a whole and within the international community, in particular with developing countries'.[75]

### C. *Ownership and Property Dimensions*

The next question in this context is who should own the information and research outcomes related to genes. How could we combine both issues: the permissibility of gene patenting and the permissibility, subject to the above guidelines, of gene-related activities where a conflict arises?

As explained, Islamic bioethics values take a clear line on this by assuming that all life – in its various forms and attributes – belongs to Allah. Consequently, no individual or entity could claim ownership and monopolization over any form of life, including genes. Based on this, it would be the duty of the state to take all necessary measures to ensure that such activities are undertaken within an environment that protects and serves the public interest. Moreover, if access to treatment was inaccessible due to high prices resulting from IP protection, then the Islamic approach would be to restrict IP protection and allow for wider use of the technologies in order to treat patients. This would meet the objectives of preserving life and serving the public interest of the society accordingly.

74 Quoted in ibid.

75 See UNESCO, 'Universal Declaration On Bioethics and Human Rights', 19 October 2005, at http://portal.unesco.org/en/ev.php-URL_ID=31058&URL_DO=DO_TOPIC&URL_SECTION=201.html.

Under the principles of *Maslaha* and the objectives of *Sharia*, the conflict which arises between accessibility to gene innovations and restrictions on private ownership would be settled in favour of the former. It would therefore be in the public interest to allow patients to access these technologies and innovations to cure their diseases. In fact, it would be the duty and obligation of the state to allow access, particularly since we are dealing with the need to provide public healthcare service, which is considered a public good from an Islamic perspective. Likewise, from the standpoint of the *Sharia* objectives, the right to life supersedes other considerations, including the right to private ownership.

## IV. NEED FOR REFORM AND ADAPTATION: RECOMMENDATIONS FOR ISLAMIC STATES

The purpose of technological innovations should be to enhance the lives of individuals and the progress of society. The driving force behind such a process should be the public interest, not the monopolization of such technologies for personal gain. This guiding principle represents the cornerstone of the Islamic framework of bioethics.

Several legislations prohibit the granting of patents on inventions which are contrary to morality or 'ordre public'.[76] The main challenge today is that, as already highlighted, many Muslim states do not have adequate patent legislation in place dealing with gene patenting and other related activities.[77] However, it could be argued that current laws may have some flexibility if they are implemented

[76] For instance, Article 53 of the European Patent Convention states that:

> European patents shall not be granted in respect of:
> (a) inventions the commercial exploitation of which would be contrary to 'ordre public' or morality; such exploitation shall not be deemed to be so contrary merely because it is prohibited by law or regulation in some or all of the Contracting States.

Based on this, the European Patent Office does not grant patents to any part of the body or the embryo, or a naturally occurring sequence or partial sequence of a gene. In line with this, the UK's Manual of Patent Practice (MOPP) explains, in Sec. 76A.02, that the following are not patentable inventions:

> (a) the human body, at the various stages of its formation and development, and the simple discovery of one of its elements, including the sequence or partial sequence of a gene;
> (b) processes for cloning human beings;
> (c) processes for modifying the germ line genetic identity of human beings;
> (d) uses of human embryos for industrial or commercial purposes;
> (e) processes for modifying the genetic identity of animals which are likely to cause them suffering without any substantial medical benefit to man or animal, and also animals resulting from such processes;
> (f) any variety of animal or plant or any essentially biological process for the production of animals or plants, not being a micro-biological or other technical process or the product of such a process.

[77] See UNESCO, *supra* n.69.

adequately. For instance, the Unified Gulf Cooperation Council Patent Law (the GCC Patent Law) states, under article 2, that

> [t]o qualify for a patent according to the provisions of this Regulation and its Bylaws, an invention shall be new, [shall] involve[e] an inventive step, and [shall be] industrially applicable. It shall not conflict with the laws of Islamic *Sharia*, or public rules of conduct observed in the Cooperation Council States, whether that was pertaining to new products, industrial processes, or to manufacturing methods.

Reference to Islamic *Sharia* under this article is of high relevance to our discussion. The TRIPS Agreement provides that a patent may be granted provided that it meets the patentability criteria under Article 27.1 of the Agreement: novelty, inventiveness, and industrial applicability. In addition, the Agreement further explicitly excludes inventions dangerous to human, animal, or plant life or health or seriously prejudicial to the environment. Consistent with the provided analysis, Article 2 of the GCC Patent Law may be invoked to prevent the monopolization through patenting of genes in member states of the Gulf Cooperation Council.

Moreover, some Muslim countries have already taken a more explicit approach in legislation. For instance, Article 2(5) of Egyptian Law No. 82 of 2002 on the Protection of Intellectual Property Rights states that 'organs, tissues, live cells, natural biological substances, nuclear acid and genome' shall be barred from patentability.

Nevertheless, Muslim countries should take a more proactive approach in this field. They should develop national policies and should focus on raising awareness, increasing investment, and amending current laws to deal with this matter adequately in order to preserve the public interest. The basis for such a framework is well established and developed under the principles of *Sharia*.

# 9

# Christian Libertarianism and the Curious Lack of Religious Objections to the Patenting of Life Forms in the United States

*Paul J. Heald*[*]

American Protestantism, especially in its more evangelical forms, is often portrayed as being anti-science or at least skeptical of science.[1] For example, many Christian groups continue to condemn the teaching of evolution in the classroom,[2] and evangelical leaders are prominent among climate change skeptics.[3] Hostility to scientific advances in birth control, stem cell research, and human cloning is long-standing.[4] No one was surprised, therefore, when 185 signatories from 80 different religious groups promulgated a "Joint Appeal Against Human and Animal Patenting" in 1995 in response to the US Patent Office's willingness to patent genes and animal life forms:

> We, the undersigned religious leaders, oppose the patenting of human and animal life forms. We are disturbed by the U.S. Patent Office's recent decision to patent human body parts and several genetically engineered animals. We believe that

* The author would like to thank Lou Hensler, Tom Berg, and attendees at the St. Edmund's College "Patents on Life" conference. Peter Legrand provided fine research assistance.

1 *See generally* Karl W. Giberson and Randall J. Stephens, *The Evangelical Rejection of Reason*, N.Y. TIMES (October 17, 2011), http://www.nytimes.com/2011/10/18/opinion/the-evangelical-rejection-of-reason.html?_r=0; Josh Rosenau, *Will Befriending Science Help Churches Grow?*, NATIONAL CENTER FOR SCIENCE EDUCATION (May 20, 2015), https://ncse.com/blog/2015/05/will-befriending-science-help-churches-grow-0016375; and Andrew Aghapour, *Are Evangelicals Really That "Science-Friendly"?*, RELIGION DISPATCHES (March 19, 2015), http://religiondispatches.org/are-evangelicals-really-that-science-friendly/.

2 *See generally* Michael Zimmerman, *Yet Another Call to Ban the Teaching of Evolution: Bad for Science, Worse for Religion and Legally Embarrassing*, HUFF. POST (October 27, 2013), http://www.huffingtonpost.com/michael-zimmerman/yet-another-call-to-ban-t_b_3810327.html.

3 *See generally* Chris Mooney, *New Study Reaffirms the Link between Conservative Religious Faith and Climate Change Doubt*, WASH. POST (May 29, 2015), https://www.washingtonpost.com/news/energy-environment/wp/2015/05/29/this-fascinating-chart-on-faith-and-climate-change-denial-has-been-reinforced-by-new-research/?utm_term=.9cd4d9480d5a.

4 *See* David A Skeel Jr. and William J. Stuntz, *Christianity and the (Modest) Rule of Law*, 8 U. PA. J. CONST. L. 809, 825 (2006); *see generally* Thomas C. Dienes, *Moral Beliefs and Legal Norms: Perspectives on Birth Control*, 11 ST. LOUIS U. L.J. 536 (1967); Shawn Markus Crincoli, *Religious Sex Status and the Implications for Transgender Nonconforming People*, 11 FIU L. REV. 137 (2015); Pew Research Center, *Religious Groups' Official Position on Stem Cell Research*, http://www.pewforum.org/2008/07/17/religious-groups-official-positions-on-stem-cell-research/.

humans and animals are creations of God, not humans, and as such should not be patented as human inventions.[5]

Such a response to the US Patent Office's embrace of gene patenting seemed entirely predictable. Human beings and animals, including their genetic makeup, are ostensibly creations of God, and any government recognition of a private property right in the building blocks of divine creation could be seen as an unwanted governmental intrusion into the sacred sphere. In 1995, America seemed poised on the brink of another major battle in the culture wars.

In reality, the Joint Appeal Against Human and Animal Patenting quickly fizzled. Audrey Chapman noted that "[a]n accompanying press release indicated that the signatories planned a nationwide educational campaign, [but] such a campaign never developed."[6] As of late 2017, a Google search of the text of the appeal returns only six online sources reproducing it in its entirety. Indeed, no coordinated religious attack on the patenting of life forms ever materialized, despite sporadic anti-patenting statements issued by some sectarian institutions. The absence of serious criticism of the patenting of life forms by American religious organizations is curious, and this chapter seeks to provide one partial explanation for the interesting silence on the issue.

First, this chapter will examine the occasional statements issued by American Christian institutions concerning the patenting of life forms. Although the evidence of opposition is underwhelming, the reasons offered by various religious groups suggest why no anti-patenting movement ever coalesced. Second, this chapter discusses the rise of "Christian libertarianism" in the latter half of the twentieth century, as described by Kevin Kruse in his book *One Nation Under God: How Corporate America Invented Christian America*.[7] The deliberate and successful incorporation of private property values into mainstream Christian theology may have made it problematic for many religious groups to mount a coherent critique of the granting of property rights in the fruits of scientific research. In other words, the sociology of twentieth-century American Christianity may explain the intriguing absence of religious voices in the debate over human gene, plant, and animal patenting.

## I. THE DIM VOICE OF OPPOSITION

One of the earliest statements in opposition to gene patenting came from the General Conference of the Methodist Church in 1984; it asserted that human genes were part of the common heritage of all peoples and should not be subject

5 Joint Appeal against Human and Animal Patenting, issued on May 18, 1995, at a Washington, DC, press conference, available at http://www.counterbalance.org/genetics/hugo-frame.html.

6 Audrey Chapman, *Religious Contributions to the Debate on the Patenting of Human Genes*, 10 St. Thomas L.J. 650, 663 (2013).

7 Basic Books, 2015.

to private ownership.[8] In a clarifying statement in 1988, the Conference (which was the only institutional signatory of the 1995 Joint Appeal) resolved that

> [t]he position of the church taken in 1984 is consistent with our understanding of the sanctity of God's creation and God's ownership of life. Therefore, exclusive ownership rights of genes as a means of making genetic technologies accessible raises serious theological concerns. While patents on organisms themselves are opposed, process patents wherein the method for engineering a new organism is patented provide a means of economic return on investment while avoiding exclusive ownership of the organism and can be supported.[9]

The statement is interesting for several reasons, one being the murkiness of the theological objection. Although God is the creator of life in Christian theology, the phrase "God's ownership of life" is bizarrely ill-fashioned. Ownership is a legal concept, a wholly human construct. A human might be responsible to his or her creator, or perhaps ultimately judged by divine authority, but Christian doctrine does not typically posit God's ownership of human beings in the way humans own cars or boats.

The notion of God "owning life" seems to be crafted specifically to respond to the claim that gene patents somehow give patentees ownership or control over life, but this misconceives patent law. A patent is a negative right allowing a patentee to enjoin a third party from making or using the invention.[10] A patent provides a patentee no right to use or exploit the invention.[11] Indeed, the practicing of an invention by its inventor may even be prohibited by law.[12] Moreover, patents on genes do not even provide a patentee control, much less ownership, over the gene's function. Even in the heyday of gene patenting before *Myriad*,[13] a patent on a gene was only allowed if it claimed the isolated form of the gene as it existed outside the human body. In other words, genes as they are present and functioning in a human body could never be properly claimed, and patents on genetic isolates provided patentees no control over the normal human use of our own DNA. Gene patents never provided anything like "ownership" over life.

8 *New Developments in Genetic Science*, from THE BOOK OF RESOLUTIONS OF THE UNITED METHODIST CHURCH – 2012, 1, 4 (United Methodist Publishing House 2012), http://www.umc.org/what-we-believe/new-developments-in-genetic-science.

9 *Id.*

10 Richard Benway, *Patents, Copyrights and Trademarks Compared and Distinguished*, 3 PORTIA L.J., Issue 1, 17, 17–18 (1967–68).

11 *Id.* at 17.

12 See, for example, a patent on the medicinal use of marijuana that would be illegal under federal law if practiced by the patentee: U.S. Patent No. 6,132,762 (filed May 5, 1998), http://patft.uspto.gov/netacgi/nph-Parser?Sect1=PTO2&Sect2=HITOFF&p=1&u=%2Fnetahtml%2FPTO%2Fsearch-bool.html&r=1&f=G&l=50&co1=AND&d=PTXT&s1=marijuana.TI.&OS=TTL/marijuana&RS=TTL/marijuana.

13 Association for Molecular Pathology v. Myriad Genetics, Inc., 569 U.S. 576 (2013) (holding that genes as found in the human body cannot be isolated and then patented).

Even more curious is the endorsement of process patents in the second part of the Methodist resolution, which fails to make an economic objection to gene patenting that could easily be framed as a theological objection. Imagine a process patent on a gene therapy for childhood leukemia. Although the gene is not claimed per se, the patentee would have the legal right to enjoin all others from using the therapy to save young lives. This monopoly power would allow the patentee to charge a high price for the therapy and would inevitably reduce its availability for treatment. The effect of patents (especially drug patents) on price and availability has frequently been demonstrated and is often used as an argument to exempt economically undeveloped countries from enforcing patents.[14] One can easily imagine a religious objection to process patents on the grounds they reduce the availability of medical treatment to those that need it. Yet, instead of seizing on the potential negative economic impact of gene patenting as a primary objection, the Methodist statement endorses gene process patents, which arguably have the same potentially negative effect on human welfare.[15]

An explanation for the Methodist statement's failure to frame the primary economic critique of patenting as a moral or theological objection might be found in the language guaranteeing researchers a "means of economic return on investment." Prior to the advent of Christian libertarianism described in Part II of this chapter, it might have been unusual to include concern for investors in what is essentially a theological policy statement. Yet hard work and investment, as protected by the institution of private property, are now routinely touted as Christian virtues built into the fabric of American democracy.[16] God may "own" genes, but God apparently does not own genetic processes discovered through significant investment. The Methodist Church reiterated its muddled position in 2014 in a declaration of support for the prohibition of gene patents advocated by the petitioners in *Myriad*.[17]

The Southern Baptist Convention has at times also voiced opposition to the patenting of human and animal life forms. Led almost single-handedly by Pastor Richard Land, the Convention passed a resolution in 1995 stating that "God alone is creator and owner of all . . . [W]e encourage . . . an immediate moratorium on the patenting of animal and human tissues and genetic sequences until a full and complete discussion has occurred."[18] Again, one can see the curious emphasis on God as "owner," which

[14] *See* Christopher Garrison, *Exceptions to Patent Rights in Developing Countries*, ICTSD-UNCTAD No. 17 (August 2006), http://unctad.org/en/docs/iteipc200612_en.pdf.

[15] *See* Chapman, *supra* note 6, at 681 ("The religious community did not discuss the detrimental impacts of genetic patents on scientific and medical research[or how] patents restricted the availability of needed genetic testing and obstructed patient care . . .. [T]he one amicus brief [in *Myriad*] filed on behalf of a religious agency did not raise broader ethical issues, such as the impact of genetic patents on human welfare and medical treatment.").

[16] *Id.* at 668.

[17] *BRCA – Statement of Support: United Methodist Board of Church and Society*, https://www.aclu.org/brca-statement-support-united-methodist-board-church-and-society.

[18] Southern Baptist Convention, *Resolution on the Patenting of Animal and Human Genes* (1995), http://www.sbc.net/resolutions/570/resolution-on-the-patenting-of-animal-and-human-genes.

becomes even less coherent when the resolution adds that "[humans] legitimately may own individual [*sic*] or groups of animals of a given species, but not an entire species or its progeny"[19] (which presumably is owned by God).

Land's advocacy on behalf of Baptists seems driven by concerns that go beyond the realm of the possible effects of patent law. He has stated that "marketing human life is a form of genetic slavery" and that patents hide "the hideous potential of increasing the number of abortions through prenatal genetic screening."[20] The powerful rhetoric of slavery and abortion suggest that Land's objection is not really to patent law, but rather to the science of genetic engineering itself. He explains, "We see altering life forms, creating new life forms, as a revolt against the sovereignty of God and an attempt to be God." This objection may or may not be theologically well-taken, but it is rather disconnected from the work of patent examiners. His decision to target the US Patent Office when the pertinent quarrel seems to be with National Institutes of Health may explain why his attempt to lead a mass Baptist revolt got nowhere.

Curiously missing from both the Methodist and Baptist resolutions is any concern for the patenting of living plants, seeds, and other forms of germplasm. Plants would seem to be as much God's living creation as humans and animals, and patenting seeds has raised serious concerns for food security worldwide.[21] Neither the "ownership" nor the "creator" concerns expressed in the Methodist and Baptist resolutions are applied to plants. Of course, the social and economic concerns over the availability and pricing of patented goods are equally applicable to patents in plants and seeds. In fact, many non-religious groups have voiced significant opposition to plant patents based on the patents' effect on poor farmers;[22] yet this concern remains unexpressed in either the Methodist or Baptist resolutions. Despite the fact that Baptists and other evangelical groups might be more science-skeptical than the typical Methodist, both of their denominations ended in the same place in the gene patenting debate.

The World Council of Churches has been more circumspect in its criticism of the patenting of life forms. After the Supreme Court in *Chakrabarty* (1980) endorsed the patenting of human-made forms of petroleum-eating bacteria, the Council issued a report worrying that the decision sought "to remove any distinction between living and nonliving matter [which] allows a shift in accepted ideas as to what may be done to living things."[23] Similarly, in a letter to President Carter, members of the National

19 *Id.*

20 John F. Peppin et al., eds., RELIGIOUS PERSPECTIVES ON BIOETHICS 104 (Taylor & Francis 2013).

21 *See* Philip McMichael, Food Security and Social Reproduction: Issues and Contradictions, in Isabella Bakker and Stephen Gill, eds., POWER, PRODUCTION AND SOCIAL REPRODUCTION 169–89 (Palgrave Macmillan 2003).

22 *See* Alan Bjerga and David McLaughlin, *Farmers Head to D.C. to Protest Agribusiness Consolidation*, BLOOMBERG MARKETS (September 9, 2016), http://www.bloomberg.com/news/articles/2016-09-09/farmers-head-to-d-c-to-protest-agribusiness-consolidation-wave.

23 Baruch A. Brody, *On Patenting Transgenic Animals*, THE AG BIOETHICS FORUM (February 17, 1997), http://www.biotech.iastate.edu/publications/bioethics_outreach/forum/brody.html.

Council of Churches, the Synagogue Council of America, and the US Catholic Conference asserted that *Chakrabarty* raised serious "moral, ethical, and religious questions [that] deal with the fundamental nature of human life and the dignity and worth of the individual human being."[24] The letter did not condemn the Patent Office but merely asked for a "reexamination" of patent laws. Testimony by several religious leaders to Congress in the subsequent 1987 Transgenic Animal Hearings was similarly restrained, calling for nothing more than a moratorium on patenting life forms.[25]

More comprehensive and coherent positions on the patenting of life forms have been taken by the Evangelical Lutheran Church of America and the Roman Catholic Church. Although neither institution attempted to lead a coordinated charge against gene or animal patents, their statements are illuminating. For example, in a 2001 booklet entitled *Genetics! Where Do We Stand as Christians?*, the ELCA engages in a highly informed discussion of the technicalities of patent law and criticizes the 1995 Joint Appeal as incoherent and wrong-headed. Instead, the publication emphasizes the relevance of a social gospel[26] and worries about "the monopoly of power over products gained by holding patents." It argues that "[g]enetic products are primarily medical in nature and beneficial to human health [and] Christian voices support this challenge to current patenting practice by pointing to the biblical message of God's special concern for the poor." Here is a glimpse of what might be seen as the most pointed theological argument against the patenting of life forms: that a Christian concern for the poor demands that the price and availability effects of patent law be mitigated.

Similarly, the United States Conference of Catholic Bishops (USCCB), in a publication entitled "A Call to Solidarity with Africa," has asked that "[t]he negotiation and implementation of trade agreements, including those involving intellectual property rights and patenting of life forms (plants, seeds), should be undertaken in consultation with African governments and civil society to protect the rights and dignity of all parties."[27] This position is elucidated in another publication

24 LeRoy Walters and Julie Gage Palmer, The Ethics of Human Gene Therapy 145 (Oxford University Press 1997).

25 Wesley Granberg-Michaelson of the National Council of Churches, representing thirty-two different denominations, stated, "We have a responsibility to preserve the integrity of creation . . . Creation is, in essence, held in trust." He added, "Reverence for all life created by God may be eroded by subtle economic pressures to view animal life as if it were an industrial product invented and manufactured by humans." Rabbi Michael Berenbaum supported a moratorium on the basis that "the religious community . . . has not done the type of homework" needed to evaluate the issue. Hearings before the Subcomm. on Courts, Civil Liberties, and Administration of Justice of the House Comm. on the Judiciary, *Patents and the Constitution: Transgenic Animals* 403 (1987), at https://catalog.hathitrust.org/Record/002774360.

26 This chapter will use the term "social gospel" to refer to theological positions that assert Christianity's primary responsibility is to care for the poor.

27 USCCB, A *Call to Solidarity with Africa* 1, 9 (November 14, 2001), http://www.usccb.org/beliefs-and-teachings/what-we-believe/catholic-social-teaching/upload/statement-a-call-to-solidarity-with-africa-2001-11-14.pdf.

entitled "For I Was Hungry and You Gave Me Food," which expresses the fear "that farmers will become dependent on seeds patented by a few companies, which could provide returns for investors at the expense of producers."[28] The point is made in an even more straightforward manner by the Vatican, which has warned that patented seeds could become "weapons of infliction of poverty and hunger."[29] The Roman Catholic publications are remarkable for two reasons. First, they show the plausibility of framing the economic critique of patents in terms of a social gospel message. Second, they consider plant patents as particularly pernicious, a possibility ignored in Methodist and Baptist communiques.

Given the size of the American Christian community and its high level of political activism, its barely visible engagement with the issue of patenting life forms is quite striking. Opposition to the patenting of life forms was never coordinated or vigorous, nor were congregations ever engaged in grass roots movements. Opposition has ranged from non-existent to sporadic and weak. When voices were raised, the emphasis tended to be on issues of God's "ownership" and sovereignty over his creation, rather than more coherent and compelling social concerns. This odd emphasis appears especially clearly in the almost complete failure by religious institutions to raise any significant objection to the patenting of two important life forms: plants and seeds.

## II. THE RELEVANCE OF THE RISE OF CHRISTIAN LIBERTARIANISM

Several factors may explain the failure of religious groups in the United States to mount a forceful attack on the patenting of life forms. First, the cost of understanding the precise nature of the rights granted by patent law is quite high. The law is not particularly intuitive, so both anti-patent activists and their potential audience may not be willing to invest the time necessary for effective communication. One should note, however, that the complexity of other intellectual property issues (for example, net neutrality or proposed anti-piracy copyright legislation like SOPA or PIPA[30]) did not prevent effective protest movements from coalescing. Second, objections based on God's "ownership" of life or "sovereignty" over life may not resonate with congregations, or those in church hierarchy, who are likely to be interested in policy making. The concepts may simply be too abstract for effective sloganeering. Third, attention and energy in religious institutions is limited, and

[28] USCCB, *For I Was Hungry and You Gave Me Food: A Catholic Agenda For Action – Pursuing a More Just Agricultural System – Part 4* 1, 12 (2003), http://www.usccb.org/issues-and-action/human-life-and-dignity/agriculture-nutrition-rural-issues/for-i-was-hungry-catholic-agenda-for-action-part-4.cfm.

[29] Carol Glatz, *Vatican Official Cautions against Genetically Modified Organisms*, Catholic News Service (March 10, 2010), http://www.catholicnews.com/services/englishnews/2010/vatican-official-cautions-against-genetically-modified-organisms.cfm.

[30] Stop Online Piracy Act, H.R. 3261 (2011); Protect IP Act of 2011, S.968 (2011). *See* Wikipedia, *Protests Against SOPA and PIPA*, https://en.wikipedia.org/wiki/Protests_against_SOPA_and_PIPA.

activists may choose to focus on other science-related issues like birth control, abortion, or surrogacy.

These factors, however, may not tell the whole story of the failure of religious institutions to successfully organize and effectively oppose the issuance of patents on life forms. Moreover, they do not adequately explain the general absence of a social-gospel basis for the objections that were voiced, nor the absence of any expression of concern over seed patents.

A further explanation may be found in the successful incorporation of libertarian notions of property rights into mainstream American Christianity in the latter half of the twentieth century. In *One Nation under God: How Corporate America Invented Christian America*, Princeton historian Kevin M. Kruse chronicles the deliberate crafting of a business-friendly gospel that gained impetus when the National Association of Manufacturers successfully co-opted influential evangelical preachers into its war on Franklin D. Roosevelt's economic reforms embodied in the New Deal. His research suggests that mainstream American Christianity might encounter serious internal conflicts when articulating an attack on a property right, like a patent, that is earned through hard work and risky investment.

Kruse's book contains many powerful images, but one of the most compelling is the "Credo of the American Way of Life" originally designed for the pro-business Freedoms Foundation.[31] The popular poster depicts the rock-solid basis of the "American Way of Life" as the "Fundamental Belief in God." All other American virtues are literally seen to rest on that mighty plinth. On the poster's left side, one sees a list of familiar constitutional freedoms, like the right to worship and the right to speak freely, while on the right, one encounters the libertarian agenda that the Freedoms Foundation linked to Christian theology: "The right to own private property"; "The right to engage in a business, compete, and make a profit"; "The right to contract"; and "The right to freedom from arbitrary government control." According to Kruse, the linking of democratic values, pro-business and anti-regulation libertarian values, and Christian theology was a deliberate attempt to counter FDR's socialist message that threatened the bottom line of many corporations. One need not spend too much time watching evangelical Christian television, nor look too deeply into the American political landscape, to find evidence of the success of this strategy.

One common narrative explaining the US government's open embrace of Christianity in the 1950s[32] argues that the linkage of capitalism and Christianity occurred as a response to the rise of atheistic communistic regimes in the Soviet Union and China.[33] Kruse, however, traces its origin much earlier, to the 1930s and

[31] Kruse, One Nation Under God, *supra* note 7, at 71.

[32] "Under God" was added to Pledge of Allegiance in 1954; "In God We Trust" was added to postage and paper money in 1954; "In God We Trust" became nation's official motto in 1956; the first National Prayer Breakfast was held in 1953; and in that same year, Nixon, Eisenhower, and prominent cabinet members declared that US government is based on biblical principles.

[33] Dianne Kirby, *Divinely Sanctioned: The Anglo-American Cold War Alliance and the Defence of Western Civilization and Christianity, 1945–48*, 35.3 J. of Contemp. Hist. 385 (2000).

the National Association of Manufacturers' (NAM) despair over Roosevelt's socialist tendencies. Members of NAM initially mounted open public attacks on Roosevelt's policies, but the attacks failed because the public saw the businesses' arguments as self-serving, concerned only with their own profit margins. According to Kruse, NAM and other organizations like it began to adopt a new strategy after "H.W. Prentis electrified the business world with a speech to the US Chamber of Commerce that called for the recruitment of religion in the public relations war against the New Deal" in December 1940.[34] Prentis, the president of NAM, enlisted prominent preachers like James Fitfield, pastor of the largest Congregational church in the world, who "convinced industrialists that clergymen could be the means of regaining the upper hand in their war on Roosevelt" and urged that "[e]very Christian should oppose the totalitarian trends of the New Deal."[35]

One key phrase in the campaign was "Freedom under God," and corporate leaders actively recruited and funded preachers who embraced the notion that America was a Christian nation founded on fundamental notions of human freedom and liberty.[36] The right to do business free of governmental regulation and to make a profit were not only consistent with Christian theology, but embedded within it. High taxes, the violation of property rights, and the redistribution of wealth were not only dangerously left-leaning, but anti-Christian, at least according to preachers like Abraham Verheide and Billy Graham. According to Kruse, "With ample funding from major corporations, prominent industrialists, and business lobbies [such as NAM and the US Chamber of Commerce], these new evangelists for free enterprise promoted a vision best characterized as 'Christian libertarianism.'"[37]

Once effective vehicle for the co-opting of religious leaders was the "Businessman's Advisory Committee for Spiritual Mobilization," which engaged in a successful campaign to recruit clergy.[38] Despite his own agnosticism, Friedrich Hayek's classic defense of libertarian economic principles, *The Road to Serfdom*, was distributed en masse to clergy and came to be a key text in the movement.[39] In addition to the distribution of texts, popular radio programs like "The Freedom Story" and magazines like "Faith and Freedom" were employed in the campaign. One executive, Howard Pew, captured the essence of the effort: "A large percentage of ministers in this country are completely ignorant of economic matters and have used their pulpits for the purpose of disseminating socialistic and totalitarian doctrines ... Much has been accomplished in the education of these ministers."[40]

34 Kruse, *supra* note 7, at 6.
35 *Id.* at 6, 12.
36 *Id.* at 26–27.
37 *Id.* at xiv.
38 *Id.* at 19.
39 *Id.*
40 *Id.* at 20.

Perhaps the most visible banner-carrier for the new Christian libertarianism was the massively popular evangelical preacher Billy Graham. His willingness to embrace corporate values and his access to the highest levels of the US government (he served as a spiritual advisor to several US presidents), combined with the popularity of his widely televised revivals, made him a fearsome advocate. For Graham, Christian paradise was a place with "no union dues, no labor leaders, no snakes, no disease."[41] He made his accommodation of corporate values and Christian plain for all to hear. "We have the suggestion from Scripture itself that faith and business, properly blended, can be a happy, wholesome, and even profitable mixture ... Thousands of businessmen have discovered the satisfaction of having God as a working partner."

This chapter can do no more than give a flavor of Kruse's painstakingly documented story of the materialistic turn taken by American Christianity in the last half of the twentieth century, when "Americans underwent an incredible transformation in how they understood the role of religion in public life," and where "Christianity and capitalism were inextricably intertwined."[42] Kruse does not claim that religious rhetoric is new to American politics. He points out the Roosevelt himself frequently resorted to scripture and biblical metaphor in his speeches. But parables like the rich man who has trouble entering heaven, or the disciple told to sell all he had and give to the poor, emphasized a social gospel that inadequately respected the property rights of the NAM and the US Chamber of Commerce (to their minds).[43] Kruse explains how American business successfully employed evangelical clergy to fight back. On a smaller (but still substantial) scale, Bethany Moreton's *To Serve God and Wal-Mart: The Making of Christian Free Enterprise* describes the use of Christian libertarian rhetoric to convince workers at Wal-Mart to tolerate low wages and to persuade shoppers to disregard the negative effect the mega-store might be having on their main street shops.[44]

The connection between the libertarian turn in American Christianity and the sputtering, insignificant religious opposition to the patenting of life forms may lay in the common conception of patents as "settled property rights."[45] If property rights are sacred in the rhetoric of Christian libertarianism, then it's difficult on religious grounds to oppose rules designed to protect the time and effort necessary to earn those rights. A consistent, and perhaps especially Protestant, impulse is to reward hard work with a property right. In the eyes of someone like Billy Graham (had he considered the issue), it would be anti-Christian for the government to interfere with

41 *Id.* at 37.

42 *Id.* at xiv, 8.

43 *Id.* at 39.

44 *See* Bethany Moreton, To Serve God and Wal-Mart 1 (Harvard Univ. Press 2009) ("the head of the powerful Christian Coalition understood the link between value shoppers and value voters," and "[j]ust as frequent church-going has proved a reliable predictor of support for the Republican Party since the 1980s, frequent Wal-Mart shopping correlates closely with conservative voting").

45 Parker v. Flook, 437 U.S. 584, 596 (1978).

and destroy the well-earned reward of a patent. Criticizing patents means opposing investment that can be seen as the natural outgrowth of Christian freedom and American democracy. In other words, criticizing the proprietization of life forms simply creates too much cognitive dissonance for the Christian libertarian speaker. To be anti-patent is to be anti-property, and to be anti-property is to be anti-Christian (and un-American).

In the 1990s, influential Roman Catholic thinker Michael Novak connected Abraham Lincoln's enthusiasm for patents[46] to Pope John Paul II's views on political economy, and argued, "As more and more nations take halting steps on the path of democracy and free markets, they will increasingly need the fire of invention, the fuel of interest."[47] Patents are natural rights, Novak said, and inventors should not be required to "bear the costs in time, effort, financial investment, and personal sacrifice necessary to produce their creations, while all others would be freeloaders."[48] Novak, however, managed to combine a rights-based intuition with concern for the common good, concluding that the poor, despite temporary monopoly pricing, are likely to be the net beneficiaries of an efficiently functioning patent system.

The failure to connect the dots between property rights and the common good, as Novak attempts to do, may explain why some religious institutions rely, incoherently, on God's "ownership" of life instead of expressing concern for patenting's potential effect on the poor. A critique of property rights as unfairly burdening the poor springs straight from Roosevelt's social-gospel playbook. In many American Christian fora, such rhetoric is no longer effective or even available. The hesitancy to employ a social-gospel rhetoric may also explain the failure of religious leaders (outside of the Catholic Church) to include plants in the patenting debate. The whole point of the work of the Businessman's Advisory Committee for Spiritual Mobilization was to make the marketplace safe for companies like Monsanto and Pfizer whose wealth is built on proprietizing new agricultural life forms. Criticizing intellectual property protection for plants would strike at the very heart of American agribusiness and the productivity of the American farmer.

Kruse, of course, does not claim that all American Christian institutions have been co-opted by what he describes as Christian libertarianism, and this chapter does not claim that the tepid religious debate over the patenting of life forms is entirely explained by his thesis. A liberal social gospel is still preached from many pulpits in the United States, as is seen in the growing acceptance of homosexual clergy and same-sex marriage in some mainstream churches. Then why haven't

46 *See* Abraham Lincoln, *Lecture on Discoveries and Inventions* (1858), http://www.abrahamlincolnonline.org/lincoln/speeches/discoveries.htm (describing patent laws' "great efficiency in facilitating all other inventions and discoveries" by "add[ing] the fuel of *interest* to the *fire* of genius") (emphases in original).

47 Michael Novak, The Fire of Invention, The Fuel of Interest: On Intellectual Property 10 (AEI Press 1996) (quoting Lincoln, *supra* note 46).

48 *Id.* at 26.

adherents to liberal strains of American Christianity risen up to oppose the patenting of life forms? One can offer three explanations:

First, Kruse suggests that Christian libertarians gradually won an important rhetorical battle: that in the political sphere, the use of social-gospel metaphor and imagery is less effective now than it was in the 1930s. In other words, social-gospel liberals who might be tempted to attack property rights in life forms as capitalism-gone-out-of-control understood the diminished impact of their religious arguments. More importantly, many Christians (myself included) may well harbor strongly held utilitarian views that lead them to be quite sanguine about the so-called patent threat. The traditional economic story of patents holds that all humanity benefits when we provide incentives for new inventions, especially medical advances and increased agricultural productivity. During the finite term of the patent, prices will rise and availability may diminish, but at the end of the patent term, the technology is free for all to use. Moreover, the invention may never have been commercialized in the absence of the incentive provided by patent law. Far from being threatened by patents on genes, animals, and plants, liberal Christian utilitarians may see patents as necessary to the eventual elimination of poverty. Interestingly, even Pope Francis's recent encyclical on the environment, *Laudato Si*, while generally very critical of global capitalism, takes a cautiously pragmatic view on agricultural advances that is theoretically reconcilable with the utilitarian view. It makes no condemnation of patent rights in genetically modified seeds; although it condemns the "dependency" of small farmers on "oligopolies for the production of cereals," it also admits that (patented) crops have "brought about economic growth which has helped to resolve [agricultural] problems."[49]

Finally, in *Seduced by Science: How American Religion Lost Its Way*, Steven Goldberg suggests that some religious institutions in the United States have accepted science in a way that crowds out traditional notions of spirituality. He looks at the religious opposition to gene patenting and is appalled by the assertion made by some religious leaders that "patenting should be banned because genes define who we are."[50] What happened to the soul, he wonders? Doesn't religion traditionally let God decide who we are or at least who we start out as? Why concede that genetic science holds the key to human identity? He criticizes the 1995 Joint Appeal as abandoning "any distinctive religious perspective in favor of precisely the materialistic point of view characteristic of science."[51] If he is correct that American Christianity has in part been partially co-opted by science, then one can discern an

[49] Pope Francis, *Laudato Si: On Care for Our Common Home*, 134 (2015), http://w2.vatican.va/content/francesco/en/encyclicals/documents/papa-francesco_20150524_enciclica-laudato-si.html; see *id.*, 135 (calling for a "balanced and prudent judgement [that] takes into account all the pertinent variables"). *See also* Steven Goldberg, Seduced by Science: How American Religion Lost its Way 93 (NYU Press 2000) ("The Christian tradition has never recognized the right to private property as absolute or inviolable, and has stressed the social purpose of all forms of private property.").

[50] *Id.* at 5.

[51] *Id.* at 9.

additional reason why even traditionally liberal churches might be slow to condemn life-form patenting as the handmaiden of scientific progress.

## CONCLUSION

Undoubtedly many factors could explain the absence of forceful religious voices in the debate over the patenting of life forms in the United States. If Kruse is correct in his history of American Christianity in the twentieth century, then one partial explanation may lie with the successful attempt of those he calls Christian libertarians to integrate strong protection for property rights into mainstream theology. The muddy and misguided arguments made in the 1995 Joint Appeal Against Human and Animal Patenting, and in other documents, support this theory. The absence of a general appeal to social-gospel rhetoric and imagery is also telling, as is the absence of seed patenting from the debate. In theory, religious voices on both the political right and the left in the United States should have good reasons to criticize the patenting of various sorts of life forms. Their general silence on the matter tells us little about science or law, but may help us understand the sociology of modern American Christianity.

# 10

# From *Chakrabarty* to *Myriad* and Beyond: Catholic Contributions to the Gene-Patenting Debate

*Paul J. Wojda*

The human animal, human society, flourishes, not to the extent that it possesses riches, but to the extent that we give life to each other, to the extent that we imitate the creativity of God.

–Herbert McCabe, "Poverty and God"[1]

## 1. RELIGIOUS CONTRIBUTIONS TO THE GENE-PATENTING DEBATE: AN OVERVIEW

In the now iconic photograph taken in their Cavendish laboratory, circa April 1953, a youthful James Watson and Francis Crick gaze up admiringly at the machine-shop model of the DNA molecule whose double-helical structure they had just successfully deduced.[2] Many things, doubtless, were going through their minds at that moment. Who *owned* the molecule, it seems safe to say, was not one of them.[3] After all, a good deal more experimental work, much of it conducted by Watson and Crick themselves, would still be necessary before the precise nature of their postulated "copying mechanism" of DNA would be fully understood.[4] More experimental work would be necessary before effective manipulation of the genome ("recombinant DNA" or "gene splicing") was possible.[5] However, by the early 1970s this more basic

[1] *God, Christ, and Us*, ed. Brian Davies (London: Continuum, 2005), p. 56.

[2] Reproduced in James D. Watson, *The Double Helix: A Personal Account of the Discovery of the Structure of DNA*, Norton Critical Edition, ed. Gunther Stent (New York: W. W. Norton & Company), p. 125.

[3] Watson would eventually weigh in on ownership of DNA. See Brief of James D. Watson, PhD, as Amicus Curiae in Support of Neither Party, in Assn. for Molecular Pathology *v.* Myriad Genetics, Inc., 569 U.S. 576 (2013), https://www.americanbar.org/content/dam/aba/publications/supreme_court_preview/briefs-v2/12-398_neither_amcu_watson.authcheckdam.pdf.

[4] This would be accomplished by Marshall Nirenberg and Heinrich Matthaei, who were the first to "crack the genetic code" in 1961. Their landmark experiment confirmed the existence of mRNA and its role in synthesizing polypeptide chains, the building-blocks of proteins. Cf. M. W. Nirenberg and H. J. Matthaei, "The Dependence Of Cell-Free Protein Synthesis in E. coli upon Naturally Occurring or Synthetic Polyribonucleotides," *Proceedings of the National Academy of Sciences of the United States of America* 47(10) (October 1961): pp. 1588–1602.

[5] D. Jackson, R. Symons, and P. Berg, "Biochemical Method for Inserting New Genetic Information into DNA of Simian Virus 40: Circular SV40 DNA Molecules Containing Lambda Phage Genes and the Galactose Operon of Escherichia coli," *Proceedings of the National Academy of Sciences of the*

and applied research in molecular biology was essentially complete. At that point the issue of ownership did in fact arise, and in a rather startling way.

In 1972, Ananda Chakrabarty, a microbiologist working for the General Electric Company, filed an application with the United States Patent Office (PTO) for a genetically modified bacterium, *Pseudomonas aeruginosa*, whose properties (breaking down crude oil) made it far more efficient in cleaning up oil spills than its non-modified precursors.[6] The PTO accepted Chakrabarty's application for a patent on the *process* for producing the bacterium, but rejected his application for a patent on the bacterium itself, arguing that the microorganism was a "product of nature," a "living thing," which was a category that had traditionally been excluded from patent eligibility under US patent law.[7] The case eventually reached the Supreme Court, which ruled in 1980 that Chakrabarty's bacterium was indeed patent eligible. Chief Justice Warren Burger, writing for a five-to-four majority, argued that the fact the bacterium was a "living thing" was irrelevant. All that mattered, at least as far as US patent law was concerned, was that Chakrabarty's bacterium, in addition to being useful, was "the result of human ingenuity and research."[8] Ownership of life – albeit the limited sort of ownership conferred by patent protection – was now legally permissible.

Public reaction to *Chakrabarty* was swift, in particular from the nation's major religious bodies.[9] Within a week of the Court's decision, the General Secretaries of the National Council of Churches, the Synagogue Council of America, and the United States Catholic Conference sent a letter to President Jimmy Carter expressing their concerns about the dangers of genetic engineering in general and the patenting of novel life forms in particular. They opened their letter with a series of urgent questions and provocative claims:

> Who shall determine how human good is best served when new life forms are being engineered? Who shall control genetic experimentation and its results which could have untold implications for human survival? Who will benefit and who will bear any adverse consequences, directly or indirectly? These are not ordinary questions.

*United States of America* 69(10) (1972), pp. 2904–9. See also Stanley Cohen, "The Manipulation of Genes," *Scientific American* 233 (July 1975), pp. 25–33.

6 See Diamond *v.* Chakrabarty, 447 U.S. 303 (1980).

7 35 U.S.C. § 101: "Whoever invents or discovers any new and useful process, machine, manufacture, or composition of matter, or any new and useful improvement thereof, may obtain a patent therefor, subject to the conditions and requirements of this title." Courts have interpreted these categories to *exclude* "laws of nature, natural phenomena, and abstract ideas." The only exception has been for certain asexually reproduced plants, over which Congress granted patent protection in its 1930 Plant Patent Act.

8 *Chakrabarty*, 447 U.S. at 310 ("[T]he patentee has produced a new bacterium with markedly different characteristics from any found in nature and one having the potential for significant utility. His discovery is not nature's handiwork, but his own; accordingly it is patentable subject matter under § 101.").

9 For another seminal response, see Leon Kass, "Patenting Life," *Commentary* 72(6) (1981), pp. 45–57.

> These are moral, ethical, and religious questions. They deal with the fundamental nature of human life and the dignity and worth of the individual human being.[10]

Using explicitly religious language – "Those who would play God will be tempted as never before" – the letter raises the specter of eugenic threats, a particularly sensitive issue given the ugly history of state-sanctioned eugenic policies in the United States, Nazi Germany, and many other countries earlier in the twentieth century.[11]

The letter concludes with, among other things, a call for Congressional action aimed at revising US patent laws:

> We know from experience that it would be naïve and unfair to ask private corporations to suddenly abandon the profit motive when it comes to genetic engineering . . . Yet when the products are new life forms, with all the risks entailed, shouldn't there be broader criteria than profit for determining their use and distribution?[12]

What these broader criteria might be, the Secretaries do not say, but in closing they pledge their respective religious communities' participation in more "urgent and organized" public debate on these "fundamental questions."

The letter had an immediate impact, far more powerful than its authors perhaps imagined: it provoked a major governmental report, *Splicing Life: The Social and Ethical Issues of Genetic Engineering with Human Beings*, by the recently impaneled "President's Commission for the Study of Ethical Problems in Medicine and Biomedical and Behavioral Research."[13] *Splicing Life* addressed both the narrower question about patents and the broader religious concerns raised by the Secretaries' letter – that is, those embodied by the phrase "playing God." On both counts, however, *Splicing Life* came to conclusions far more devastating than the Secretaries perhaps imagined, conclusions that would influence all subsequent efforts to contribute to this debate from a distinctively religious perspective.

10 See President's Commission for the Study of Ethical Problems in Medicine and Biomedical and Behavioral Research, *Splicing Life: The Social and Ethical Issues of Genetic Engineering with Human Beings* (Washington, DC: United States Government Printing Office, 1982), Appendix B, p. 96.

11 Ibid. The term "playing God" had been given prominence by Princeton theologian and first-generation "bioethicist" Paul Ramsey in his seminal study *Fabricated Man: The Ethics of Genetic Control* (New Haven: Yale University Press, 1970), p. 143, inter alia. The phrase was later adopted by biotech watchdogs Ted Howard and Jeremy Rifkin as the title of their book *Who Should Play God? The Artificial Creation of Life and What It Means for the Future of the Human Race* (New York: Dell, 1977).

12 *Splicing Life*, p. 96.

13 US Government Printing Office, November, 1982. A PDF version of this report is widely available online. *Splicing Life* was one of twelve reports produced by the President's Commission between 1980 and 1983. Albert Jonsen, who also served on a precursor Commission (the National Commission for the Protection of Human Subjects of Biomedical and Behavioral Research, 1974–78) considers *Splicing Life* to be the best report of either commission. See Albert Jonsen, *The Birth of Bioethics* (New York: Oxford University Press, 1998), pp. 90–122.

On the first issue, the report suggested that while "patents and trade secrets" were important for the future of genetic engineering, they were not, strictly speaking, "bioethical" issues at all; or at least that Commission members lacked the expertise for making informed judgments about them.[14] As it turned out, events would soon return "patents and trade secrets" to the table of mainstream bioethical debate (more on this as follows), but the issue has never garnered the attention in public bioethics accorded its more "glamorous" relatives such as gene therapy or enhancement. Among theological bioethicists, the issue receives even less attention.

Second, and more consequential, *Splicing Life* implied that religious contributions to public bioethical debate – and not just on genetic engineering – were largely superfluous. The phrase "playing God," it argued, was but another way of expressing concerns that could be better articulated in the language of secular moral theory:

> While religious leaders present theological bases for their concerns, essentially the same concerns have been raised – sometimes in slightly different words – by many thoughtful secular observers of contemporary science and technology. Concerns over unintended effects, over the morality of genetic manipulation in all its forms, and over the social and political consequences of new technologies are shared by religious and secular commentators. The examination of the various specific concerns need not be limited, therefore, to the religious format in which some of the issues have been raised.[15]

This view, that religious perspectives were marginal (at best) to public debate over issues of biomedical law and policy, would be one increasingly shared by mainstream bioethicists in the coming decades.[16]

In sum, *Splicing Life* left bioethically minded theologians gasping for air, struggling to articulate why religious perspectives deserved any hearing in the burgeoning sphere of public bioethical debate, and what, in any case, such perspectives ought to look like. Not much has changed in the almost four decades since *Splicing Life*. If anything, the questions have only grown sharper. Perhaps not surprisingly, some of the sharpest disagreement over these questions is intramural, as members of various religious communions contend with each other over how best to bring their own

14 *Splicing Life*, p. 88.

15 *Splicing Life*, p. 54; see also *The Birth of Bioethics*, p. 186.

16 On the marginalization of theology in mainstream bioethics, see John H. Evans, *Playing God? Human Genetic Engineering and the Rationalization of Public Bioethical Debate* (Chicago: University of Chicago Press, 2002); and, more recently, also by Evans, *The History and Future of Bioethics: A Sociological View* (New York: Oxford University Press, 2012). As important a moment as it was, *Splicing Life* was not the first time questions had been raised about the value of theological contributions to medical ethics and bioethics. The issue had been brewing steadily throughout the 1970s. In 1979 Alasdair MacIntyre famously challenged theologians to "explain why we [i.e., non-theologians] should not treat their discipline as we do astrology or phrenology. The distinctiveness and importance of what they have to say, if it is true, make this an urgent responsibility." "Theology, Ethics, and the Ethics of Medicine and Health Care: Comments on Papers by Novak, Mouw, Roach, Cahill, and Hartt," *The Journal of Medicine and Philosophy* 4 (1979), p. 443.

traditions to bear on the ever-growing array of bioethical issues, including, of course, the patenting of life.

Mindful of this history, and of these questions, the following chapter seeks to contribute to the contemporary public debate over human gene patenting from a distinctively Catholic theological perspective. It begins by taking up Audrey Chapman's critical overview of religious contributions to the post-*Chakrabarty* debate over human gene patenting. According to Chapman, such contributions have been intermittent, fragmented, and in general less interested in shaping actual patent law and policy, especially in the United States, than in engaging in "prophetic" critique of that law and policy, and in particular at what are taken to be its expansive claims about the extent of human power over nature. Religious contributions have reflected a largely "symbolic" approach, claims Chapman, which has come at the expense of "more careful theological and ethical analysis."[17]

In light of these criticisms, the next section of the chapter examines specifically Catholic contributions to the debate over gene patenting. Chapman does not devote a great deal of her attention to Catholic authors, and in fact she overlooks two important contributions to the debate by well-known Catholic theologians Lisa Sowle Cahill and Cathleen Kaveny. However, her criticisms are largely on-target when it comes to Catholic contributions overall. There simply isn't much to speak of. On the other hand, both Cahill's and Kaveny's work contains elements of a more robust Catholic theological contribution – or such, anyway, is what I will propose in the final section of this chapter.

A distinctively Catholic theological perspective on the issue of gene patenting, I argue, will focus principally on the questions of whether and under what conditions *ownership* of life – *all* life, but human life especially – is warranted. Religious critics of laws that permit the patenting of novel organisms and genes have argued that such laws are unjust (and possibly blasphemous), because they arrogate to human beings (creatures) the right to ownership of life that belongs properly only to God (the Creator). In response to these critics, other theologians have contended that God's ownership of life is to be construed not in an exclusive or absolute, but rather in a *regulative*, sense. That is, the *aim* of God's ownership, and not the bare fact of ownership itself, is what provides the model for a proper understanding of the nature and limits of human ownership. In other words, if the aims of human ownership are consistent with the aims of divine ownership, then there ought to be no principled objection to human ownership even of such things as living organisms and genes.

This line of argument will be familiar to those acquainted with the long tradition of Catholic teaching on the limited nature of all property rights. In the terms of that tradition, because God has destined the goods of creation to be shared by all ("the

17 Audrey R. Chapman, "Religious Contributions to the Debate on the Patenting of Human Genes," *University of St. Thomas Law Journal* 10(3) (Spring 2013): pp. 650, 681.

universal destination of goods"), any claim to private ownership of those goods must necessarily be weighed against the needs of the broader community – that is, the common good. As St. John Chrysostom famously put it, "not to share our own wealth with the poor is theft from the poor and deprivation of their means of life; we do not possess our own wealth but theirs."[18]

As time-honored as this argument is within the Catholic tradition, I will argue that it is less than adequate theologically, for by treating God as essentially a super-generous landlord, it misses the far more radical point, with its far more radical ethical (and political) implications, that God literally owns nothing at all. As the late Dominican theologian Herbert McCabe puts it, God "is literally poor, because he simply and literally has no possessions."[19] Recapturing this point, I conclude, would allow the formulation of a less fragmented and more ethically robust *theological* contribution to public bioethical debate over gene patenting.

## II. TOO MUCH PROPHECY, TOO LITTLE ANALYSIS? CRITICIZING RELIGIOUS CONTRIBUTIONS TO THE GENE-PATENTING DEBATE

Despite the conclusion of *Splicing Life* that "patents and trade secrets" on living organisms and genes were not properly bioethical concerns, the issue nevertheless returned to the edges of mainstream bioethical debate well before the end of the 1980s, where it has remained ever since.[20] At least five major developments are responsible for putting and keeping it there. First, while *Chakrabarty* itself had the obvious effect of increasing the number of patent applications for human-engineered biological organisms, even more influential was the Bayh-Dole Act, passed by Congress that same year, which for the first time permitted the private commercialization of federally funded scientific research.[21]

Second, in 1987, the US Patent Office decided to extend patent protection to "nonnaturally occurring nonhuman multicellular living organisms, including animals," and the following year granted a patent to Harvard University for the "oncomouse," a mouse genetically modified for the purpose of studying cancer.[22]

18 *On Wealth and Poverty*, trans. Catharine P. Roth (Crestwood: St. Vladimir's Seminary Press, 1984), p. 55.

19 Herbert McCabe, "Poverty and God," in *God, Christ, and Us*, ed. Brian Davies (London: Continuum, 2005), p. 55.

20 As early as 1985, Thomas Murray, one of the co-founders of the Hastings Center, identified "ownership of life forms" as one subset of concerns related to "the privatization and commercialization of genetic engineering." See "Ethical Issues in Genetic Engineering," *Social Research* 52/3 (1985), p. 472.

21 See 37 C.F.R. 401. For early bioethical commentary and critique on the commercialization of scientific research, see the collection of essays in *Science, Technology, & Human Values* 12(1) (Winter 1987).

22 US Patent 4,736,866

Third, in 1988 the Human Genome Project (HGP) was officially announced. The HGP, which formally began in the fall of 1990 and ended, ahead of schedule, in 2003, was a massive international (and publicly funded) effort to map the approximately thirty thousand genes of the human genome.[23] Five percent of the HGP budget was allocated to the study of the "Ethical, Legal, and Social Implications" of the project, assuring that a wide range of issues – including gene patenting – would receive bioethical attention.[24]

Fourth, in 1991 and continuously over the next twenty years, applications for patents on gene fragments isolated by researchers working for the HGP (as well as for private biotechnology firms) skyrocketed. In fact, in its formal announcement of the conclusion of the HGP, the National Institutes of Health, which was among the first to apply for such patents, could not say how many such applications had actually been granted, although some have estimated that by 2013 patents had been awarded for genes across the entire human genome.[25]

Fifth and finally, in the mid-1990s the biotechnology company Myriad Genetics was granted patent protection for two genes it had isolated, BRCA1 and BRCA2, mutations of which it identified as being linked to significantly higher risk of breast cancer. On the basis of its research, Myriad developed a proprietary genetic test for breast cancer, which earned it significant profits – until 2013, that is. That year, the US Supreme Court decided unanimously that *Myriad*'s patents on the BRCA1 and BRCA2 genes were invalid, arguing that a "naturally occurring gene segment" (such as BRCA1 and BRCA2) was not patent eligible merely because it had been isolated in the laboratory. On the other hand, "complementary DNA" (cDNA), a gene fragment synthesized in the laboratory and not "naturally occurring," could be patented.[26]

Mainstream bioethical response to these developments has clustered around two major issues. The first concerns the nature and purpose of US patent law itself. Here the debate has largely been between those who defend current law and policy as necessary to protect and promote the advancement of genetic science and medicine, and those who argue that patent law and policy should be guided by non-instrumental goods. The latter is more the case in Europe, where claims to patent protection can be limited by considerations of the public good ("ordre public").[27]

23 The HGP officially ended on April 14, 2003, almost fifty years to the day of Watson and Crick's original paper in *Nature*.

24 See "Ethical, Legal, and Social Issues Bibliography," in *Human Genome Project Information Archive 1990–2003*, http://web.ornl.gov/sci/techresources/Human_Genome/elsi/elsibiblio.pdf.

25 See National Human Genome Research Institute, "The Human Genome Project Completion: Frequently Asked Questions," https://www.genome.gov/11006943/human-genome-project-completion-frequently-asked-questions/; Chapman, "Religious Contributions," p. 663.

26 *Myriad*, 569 U.S. at 590–95, 133 S. Ct. at 2116–19.

27 See European Patent Convention, article 53(a), http://www.epo.org/law-practice/legal-texts/html/epc/2016/e/ar53.html.

The second issue is what such non-instrumental goods might be (e.g., what appeals to "public good" might mean), especially in the US context. On this much broader issue, the debate is not as easily summarized. However, it has tended to focus on questions that the "commodification" of living organisms and genes raises about the dignity of human life; about the meaning of embodiment; about the moral and metaphysical "status" of DNA, and the relation of genes to personal identity; and finally, about the requirements of justice in the allocation of the benefits of genetic research.[28]

On at least some of these issues and questions, one might reasonably expect that religious traditions would be able to offer clarifying insights if not actual guidance. Unfortunately, according to Audrey Chapman, that has not been the case. From *Chakrabarty* to *Myriad*, she argues, religious contributions have been both intermittent and fragmented (i.e., internally divided). Public statements, moreover, have tended to adopt a "prophetic" mode of discourse – for example, denouncing gene patenting as an unethical and even blasphemous arrogation to human beings of a right of ownership belonging properly only to God. Since prophetic discourse is not conducive to the sort of close, critical dialogue required in the analysis and clarification of policy and moral argument, the religious community, concludes Chapman, has not been able to make the sort of contribution to the gene-patenting debate that it otherwise might have.[29]

Much of Chapman's evidence for these claims comes from her experience as the founding director of the "Program of Dialogue on Science, Ethics, and Religion" of the AAAS, whose very first dialogue, in 1995, was dedicated to the topic of genetic patenting.[30] Like the events that provoked *Splicing Life* fifteen years earlier, Chapman's decision to focus her Program's first dialogue on the issue of gene patenting was also stimulated by a public statement from religious leaders. In this case, it was not a full-length letter, but a short declaration, entitled "A Joint Appeal against Human and Animal Patenting":

> We, the undersigned religious leaders, oppose the patenting of human and animal life forms. We are disturbed by the U.S. Patent Office's recent decision to patent body parts and several genetically engineered animals. We believe that humans and animals are creations of God, not humans, and as such should not be patented as human inventions.[31]

[28] Almost any contemporary anthology in bioethics will cover most of these issues, but see especially Justine Burley and John Harris, eds., *A Companion to Genethics* (Malden, MA: Blackwell, 2004).

[29] Chapman's implicit distinction between "prophetic," "ethical," and "policy" discourse relies on the influential typology first developed by James F. Gustafson, "Moral Discourse about Medicine: A Variety of Forms," *The Journal of Medicine and Philosophy* 15(2) (1990), pp. 125–42.

[30] Audrey Chapman, ed., *Perspectives on Genetic Patenting: Religion, Science, and Industry in Dialogue* (Washington, DC: American Association for the Advancement of Science, 1999). See also "The Ethics of Patenting Human Embryonic Stem Cells," *Kennedy Institute of Ethics Journal* 19(3) (September 2009), pp. 261–88; and Chapman, "Religious Contributions," pp. 650–82.

[31] Reprinted in Chapman, "Background and Overview," in Chapman, ed., *Perspectives on Genetic Patenting*, p. 19.

Leaders of more than 80 religious traditions, 180 persons in all, signed the statement, including 91 Catholic bishops (although not the US Catholic Conference as a whole). Much of the lead-up to the statement and its public announcement was organized by well-known anti-genetic engineering author and activist Jeremy Rifkin.

Unlike the authors of *Splicing Life*, of course, the AAAS dialogue group did not dismiss gene patenting as unworthy of bioethical attention. Neither did it suggest that the religiously framed concerns expressed by the Joint Appeal's authors – concerns that centered around the distinction between divine and human ownership – could just as well be translated without remainder into a presumably more widely shared "secular" frame (e.g., one focused on the concept of "commodification"). On the other hand, as both Chapman and other members of the dialogue group observed, the theological arguments put forward by defenders of the Joint Appeal, in particular those by Baptists C. Ben Mitchell and Richard Land, were underdeveloped. As a result, many of their assumptions were easily defeated, including the crucial assumption that divine and human ownership of life were necessarily mutually exclusive.[32] One theologian-member of the dialogue group, Mark J. Hanson, identified what was perhaps the fundamental problem with the religious critics' arguments: the very concept of divine "ownership" itself. "Religious claims about God's ownership," he noted, "are puzzling even to other theologians."[33]

Such claims were all the more puzzling to representatives in the dialogue group from science and industry. And in the end, it is this "failure to connect" that may well account for the paucity of religious contributions to the debate over gene patenting ever since. The inability of religious critics to translate their "ownership" objections into an intelligible secular idiom meant that, at certain crucial points, the dialogue broke down, taking the form of what Chapman describes as "a set of monologues putting forward varying and incompatible perspectives." Some supporters of patents, perhaps out of frustration, simply denied that the issue had anything to do with morality or religion at all, and was simply a legal matter.[34]

Ironically, then, when it came to the utility of religious contributions to the debate, the result of the AAAS dialogue on patents had the same practical effect as did the work of the President's Commission and *Splicing Life*, whatever their other differences. *Splicing Life* had concluded that religious perspectives were superfluous because they were so *easily* translated into secular perspectives. ("Playing God," the report said, was simply another, less precise way of expressing concern about unintended consequences.) In contrast, the AAAS dialogue concluded that religious

[32] See especially Ronald Cole-Turner, "Theological Perspectives on the Status of DNA: A Contribution to the Debate Over Genetic Patenting," in Chapman, *Perspectives on Genetic Patenting*, pp. 140–66.

[33] Mark Hanson, "Lessons from a Religious Objection to Genetic Patenting," in Chapman, ed., *Perspectives on Genetic Patenting*, 231.

[34] Chapman, "Background," p. 30.

perspectives were, if not exactly superfluous, at least not very helpful, because they *could not be* translated into less "religious" terms.

In something of a *coda* to her narrative, and as further evidence of the inadequacy of religious contributions to the debate over gene patenting, Chapman notes that only one, "poorly written," *amicus* brief from religious contributors was filed in the *Myriad* case, on behalf of the Southern Baptist Convention.[35]

## III. PROPHECY *AND* ANALYSIS: TWO CATHOLIC THEOLOGICAL VOICES

And Catholics? What contributions, if any, have they made to the debate over gene patenting since *Chakrabarty*, and more recently since *Myriad*? Do Chapman's criticisms – that religious contributions have been sporadic, fragmented, and less interested in shaping patent policy than in prophetically criticizing it – also apply to Catholic contributions?

In general, the answer to this question is yes, but a qualified yes. On the one hand, there has been no official magisterial statement on the issue of gene patenting as such. Official Catholic responses to *Chakrabarty* and *Myriad* were negligible. And while gene patenting has attracted the attention of some Catholic theologians, as will be noted as follows, that attention has not been sustained. In fact, the paucity of constructive theological responses to the issue makes it difficult to say whether or not there exists a consensus among Catholic theologians on many of the questions raised by the issue.[36] "Theology in process" is how Chapman once described the character of most religious contributions to the gene-patenting debate. The same can safely be said of more specifically Catholic contributions.

On the other hand, Chapman's narrative, which tends to focus on evangelical and mainline Protestant voices, overlooks two important contributions by well-known Catholic theologians: Lisa Sowle Cahill and Cathleen Kaveny.[37] Neither Cahill's nor Kaveny's work constitutes a fully developed theological contribution to the gene-patenting debate; however, they are more than simply "prophetic critiques." In fact, as I will argue in the final section of the chapter, their conclusions regarding patent law and policy are both ethically rigorous and deeply consonant with the most

[35] Chapman, "Religious Contributions," pp. 664, 682 (citing Brief of the Ethics and Religious Liberty Commission of the Southern Baptist Convention and Prof. D. Brian Scarnecchia as Amici Curiae in Support of Petitioners, in *Myriad*, https://www.americanbar.org/content/dam/aba/publications/supreme_court_preview/briefs-v2/12-398_pet_amcu_southernbaptist.authcheckdam.pdf).

[36] Interestingly, the "dean" of American Catholic medical ethics, Richard McCormick, entirely neglected the issue in his influential annual essay "Notes on Moral Theology" in the pages of *Theological Studies*.

[37] David Byers, then Director of the Committee on Science and Human Values of the US Catholic Conference, was a member of the AAAS dialogue program, but his contributions are more about the relationship between "science and religion" than about gene patenting. They are also not theological in any robust sense of that term.

fundamental Catholic claims regarding the most important theological *topos* of the debate (i.e., ownership, both divine and human).

Cahill engages the debate over gene patenting in the context of her constructive efforts to articulate a Christian theological bioethics centered on the idea of "justice in access to health care resources" and, on that basis and with that end in mind, dedicated to a wide-ranging, "participatory" engagement in the transformation of contemporary law, policy, and social structures.[38] Cahill's project is something of a *cri de coeur* to her fellow theologians, Catholics and others. She calls on her colleagues to challenge the reigning powers of mainstream bioethics, by questioning the adequacy of its underlying theories of the human person and human sociality (excessively "thin" and individualistic, respectively), as well as by taking a broader view of the relevant field of bioethical action. Influencing public law and policy, while important, should not be seen as the only or ultimate goal of the theologian engaged in public bioethical debate. For a "participatory" bioethics, finding common cause and working with other like-minded advocates for social change on behalf of the poor will be as important, often more so, than jockeying for a place at the tables of policy-forming commissions and work-groups (or lamenting one's exclusion from them). "The role of a theological worldview," she proposes, "is at least as much to challenge a social ethos as it is to specify policy outcomes that are required by the worldview."

For Cahill, the debate over gene patenting provides an illustrative case for "participatory bioethics" so conceived. There is nothing inherently wrong with patents, even patents on life, states Cahill. However, she also cites Pope John Paul II's use of the metaphor of a "social mortgage" on property, including both "intellectual property" and "knowledge," to "advance the ethos of solidarity in both ethical and socially prophetic terms." The "social mortgage" metaphor does not "specify exactly what policy is necessary to guarantee that the mortgage will be paid," but it does serve as a reminder that the "use of legal patent rights is morally problematic when it becomes a means to protect disproportionate financial gain for some while preventing others from gaining access to lifesaving medicines or treatments."[39]

The 1995 "Joint Appeal," argues Cahill, also constitutes an instance of "participatory" bioethics, "in that it gathered interreligious participants in a common action and received significant public attention in the press and from scientists and churches." In this respect, Cahill's judgment about the Joint Appeal differs from those of Chapman and many other critics. On Cahill's terms, it is not problematic that the Joint Appeal is more of a prophetic statement, one that moves from "religious symbols to a policy conclusion" without much mediating ethical analysis. "Sometimes it is better to express moral concerns and the religious values behind

[38] Lisa Sowle Cahill, *Theological Bioethics: Participation, Justice, Change* (Washington, DC: Georgetown University Press, 2005), p. 1.

[39] Cahill, *Theological Bioethics*, p. 240.

them, then join with others in seeking to establish better alternative practices that address those concerns, even if some moral issues remain unresolved."[40]

Cahill does not elaborate on what such alternative practices might involve when it comes to the issue of patent law. Rather, she emphasizes the growing recognition that such laws typically serve to enrich multinational biotechnology firms, not those who would most benefit from production of the medicines the patent laws are designed to promote. She highlights, in this regard, the publication of a study in 2000 by an international organization of fifteen Catholic NGOs. The study encouraged "affected nations to resist patent law by making as wide a use as possible of the discretionary power allowed them under the World Trade Organization regulations and to seek the moral high ground with Catholic social teaching, especially the pope's declaration that there is a 'social mortgage on property' requiring the rich to fulfill obligations to the developing world."[41]

On the other hand, there is nothing in Cahill's account of "participatory bioethics" that would preclude more constructive proposals – for example, the establishment of agreements between biotech companies, NGOs, and developing nations over the costs and access to drugs and other therapies. Indeed, socially conscious biotech firms looking for a principled justification for such agreements (and other policies) should find Cahill's proposals congenial.

Echoing many of Cahill's concerns, Cathleen Kaveny, who is both a legal scholar and a theologian, considers the debate over gene patenting in the context of arguments about the larger purposes of law in liberal society.[42] For Kaveny, the issue at the heart of the debate over the patenting of genetic material is in fact the "symbolic" meaning of that material itself.[43] She agrees with the argument of Mark Hanson (and many others) that the "commodification" of life forms, which is at least implied in the practice of granting patents on them, threatens to alter our understanding of life. "Will we begin to value human genes, and the traits they code for, in terms of their ability to meet the desires of the marketplace? Will not the market not only satisfy purchaser's desires with respect to improvements in human geneotypes [*sic*], but also create and shape those desires?" For Kaveny, "that is the specter lurking in the inmost recesses of the controversy over the patenting of living materials."[44]

Confronting this specter, Kaveny argues, will require a rethinking, not simply of patent law, but of "the fundamental normative assumptions undergirding the way

40 Ibid., p. 241.

41 Ibid., p. 244 and n.80 (citing CIDSE [International Cooperation for Development and Solidarity], *Biopatenting and the Threat to Food Security* [2000], retrieved from www.cidse.org/pubs).

42 M. Cathleen Kaveny, "Jurisprudence and Genetics," *Theological Studies* 60 (1999), pp. 135–47. More generally, see Chapter 5, "Genetic Information and Razian Autonomy," in her *Law's Virtues: Fostering Autonomy and Solidarity in American Society* (Washington, DC: Georgetown University Press, 2012), pp. 111–40.

43 In this respect, Kaveny differs from Chapman, who, as we noted previously, considers the "symbolic" questions as secondary to the ethical. For Kaveny, the symbolic questions *are* ethical, and vice versa.

44 "Jurisprudence and Genetics," p. 146.

the law deals with the broader issues of human reproduction, medical confidentiality, and the relationship of medicine to the market." More specifically, "it will be necessary to counteract the multifaceted individualism that permeates the legal system in these areas, particularly in the U.S."[45] Here Kaveny emphasizes the resources available for this critical task within the Catholic tradition, in particular the tradition's long history of "uncompromising recognition of both the equal dignity and the essentially social nature of the human person." Given such resources, the "fundamental jurisprudential question" on these genetic issues, Kaveny concludes, should be something like the following: "What legal stance on cloning, genetic privacy, and gene patenting will give the virtue of solidarity the best chance of taking root and flourishing in the hearts of our neighbors and in our own hearts?"[46]

## IV. THE POVERTY OF GOD: RECOVERING THE RADICAL ROOTS OF CATHOLIC TEACHING ON PROPERTY

Human dignity, essential sociality, solidarity, equal justice, virtue, option for the poor, law as pedagogue, a "social mortgage" on property – taken together, these terms and concepts, all operative in both Cahill's and Kaveny's work, constitute the elements of an unmistakably Catholic approach to the issue of gene patenting. As such, their work stands as an exception to most, even if not all, of Chapman's criticisms of religious contributions to that debate in the period from *Chakrabarty* to *Myriad*. Theirs is a constructive, not merely reactive, approach; it does not sacrifice theological and ethical rigor for the sake of prophetic critique. Indeed, it explicitly insists on and engages the specifically ethical questions raised by patent law and policy while maintaining its prophetic edge. It even, with Cahill's emphasis on the "participatory," suggests where, other than the policy arena, theologians might look to effect real change.

As a contribution to the shaping of patent law and policy specifically, both Cahill and Kaveny, consistent with the long tradition of Catholic social teaching, criticize approaches that subsume moral considerations, especially about the dignity and sociality of the human person, to economic criteria alone. Not everything in the area of genetic research can or should be justified by appeal to the protection and promotion of invention and innovation, regardless of how promising such research is said to be for the health and well-being of future generations. It should not be assumed, in other words, that the benefits of a regulated genetic marketplace to humankind are simply automatic.[47]

45 Ibid., pp. 146–7.
46 Ibid., p. 147.
47 In a similar vein, but on much different grounds, J. Benjamin Hurlbut argues that in contemporary bioethical debate, appeals to the unquestioned benefits of scientific progress are themselves rhetorical

Cahill's and Kaveny's claims about the dignity and sociality of the human person are ultimately rooted in the theological anthropology articulated most succinctly by the Second Vatican Council's *Pastoral Constitution on the Church in the Modern World* (*Gaudium et spes*). In the mystery of the Incarnation the mystery of the human person comes fully to light.[48] Human persons are uniquely called to share in the divine life, which, because it is itself a communion of persons, requires of each the "sincere gift of self" to others.[49] Only through practices of solidarity, in other words, do we reach our full stature as persons.

The implications of this view for evaluating political, economic, and social systems and policies is clearly stated later in the same document:

> In the economic and social realms ... the dignity and complete vocation of the human person and the welfare of society as a whole are to be respected and promoted. For man is the source, the center, and the purpose of all economic and social life.[50]

As Charles Curran and others have observed, since the pontificate of John Paul II, the trajectory of official Catholic social teaching, and of Catholic social thought in general, has grown *more* explicitly theological on these points. Pope Francis is simply extending this trajectory when he emphasizes commitment to the poor as the decisive test for how seriously we take the Incarnation:

> For the Church, the option for the poor is primarily a theological category rather than a cultural, sociological, political or philosophical one. God shows the poor "his first mercy." This divine preference has consequences for the faith life of all Christians, since we are called to have "this mind ... which was in Jesus Christ" (Phil 2:5). Inspired by this, the Church has made an option for the poor which is understood as a "special form of primacy in the exercise of Christian charity, to which the whole tradition of the Church bears witness". This option – as Benedict XVI has taught – "is implicit in our Christian faith in a God who became poor for us, so as to enrich us with his poverty."[51]

Given this background, it is not at all surprising that both Cahill and Kaveny gesture toward revisions to patent law and policy that call for greater attention to human dignity, to social solidarity, and to an option for the poor. Neither fleshes out a proposal with any concreteness, though it is likely both would support moving US

exercises that tend to assume that "science" is either pre-political or immune from political challenge, much to the detriment of democratic modes of civic engagement. Views that cannot claim the mantle of "science" are systematically excluded from serious debate. See *Experiments in Democracy: Human Embryo Research and the Politics of Bioethics* (Columbia University Press, 2017).

48 *Gaudium et spes* (1965), 22, http://www.vatican.va/archive/hist_councils/ii_vatican_council/documents/vat-ii_const_19651207_gaudium-et-spes_en.html.

49 Ibid., 24.

50 Ibid., 63.

51 *Evangelii Gaudium* (2013), 198, https://w2.vatican.va/content/francesco/en/apost_exhortations/documents/papa-francesco_esortazione-ap_20131124_evangelii-gaudium.html.

patent law closer to the current European model, with its greater openness to limits based in public morality.[52]

More prominent in both Cahill's and Kaveny's approach is the appeal to the limited nature of property rights in the Catholic tradition. Cahill in particular refers explicitly to John Paul II's notion of a "social mortgage" on property, a fairly recent development of the Catholic doctrine of limited property rights. However, both Cahill and Kaveny tend to accept this teaching as a given. They do not explore the theological claims upon which it rests, perhaps because those claims are so familiar: only God, the creator, can be said to have absolute property rights over the goods of creation. In fact, property ownership, strictly speaking, belongs only to God.[53] A particularly generous owner, God intends that all, and not merely some, benefit from the good things of the earth. Hence, to speak in the terms usually used to describe the doctrine, there is a "universal destination of goods."[54]

In closing this chapter, I want to challenge this account of the Catholic tradition as theologically inadequate, not in order to call either Cahill's or Kaveny's conclusions into question (in fact, I would concur with all of them), but to expose what I take to be the more radical claims underlying traditional Catholic teaching on property and ownership – namely, that God owns nothing and that, correlatively, we creatures are likewise invited to "aim at poverty." These claims, I suggest, provide better, albeit more challenging, foundations for a Catholic contribution to the debate over gene patenting.

In order to make sense of such claims that "God owns nothing" and that we are to "aim at poverty," it is first necessary to recall the peculiar character of all theological discourse about God in the Abrahamic religious traditions. Affirmed (indeed praised) as transcendent creator, God is not and cannot be an object *in* the world, like computers, coffee mugs, or cats. Hence, language about God, at least in its predicative mode, cannot function the way it does when referring to such objects as "computers," "coffee mugs," and "cats." A certain wariness is required, therefore, when speaking about God, even when affirming God as "transcendent creator," if we are to avoid making basic theological mistakes, or worse, falling into an unwitting idolatry that substitutes the created for the creator. All of which is to say that, in the Abrahamic traditions, a heightened sensitivity to all language about the divine is warranted, with the primary task of theology, accordingly, being the search for ever less *inadequate* ways of speaking about God.

52 See text in n. 26 of this chapter.

53 Paul Griffiths puts the point nicely: "[God] brought all creatures into being *ex nihilo* – out of nothing – and without constraint by anything external to himself, can sequester any one of them, or all of them together, without impediment. Ownership, in its pure form, belongs to the LORD." "The Natural Right to Property and the Impossibility of Owning the Intangible: A Tension in Catholic Thought," *University of St. Thomas Law Journal* 10(3) (2013), p. 592.

54 Pontifical Council for Justice and Peace, *Compendium of the Social Teaching of the Church*, nn. 171–84.

On its face, then, there is nothing wrong with asserting that God *owns* the world and everything, even everyone, in it. Such expressions can be found throughout the Bible, for example, and on practically every page of Christian theology on the topic. The expression also serves a range of uses, one of which is simply to praise God, and invite others to praise God, as creator, as in Psalm 95:

> Let us come before him with a song of praise,
> joyfully sing out our psalms.
> For the LORD is the great God,
> the great king over all gods,
> Whose hand holds the depths of the earth;
> who owns the tops of the mountains.
> The sea and dry land belong to God,
> who made them, formed them by hand.
> Enter, let us bow down in worship;
> let us kneel before the LORD who made us.
> For he is our God,
> we are the people he shepherds,
> the sheep in his hands. (Ps. 95:2–7 NAB)

The most common moral use of the concept of divine ownership, as we have seen repeatedly, is to remind us of our essentially limited dominion over the world and its goods.

On the other hand, if Hanson is correct that theologians are divided about what divine "ownership" might mean, at least in the context of the debate over gene patenting,[55] then it may be less troublesome to think about God as *not owning* anything. In other words, despite its long theological pedigree, "ownership" may be a very misleading, and thus ethically inadequate, metaphor for God's relation to the world. This idea was explored in a characteristically provocative way by the late Dominican theologian Herbert McCabe, in a short homily titled "Poverty and God."[56]

The text for McCabe's homily is the story of the rich young man in the Gospel of Mark (10:17–22). Just as that story turns on the distinction between poverty and riches, so too does McCabe's homily. His primary point is to suggest "that there is something godlike about being able to live in poverty . . . and something less than human about needing to live with riches."

Both riches and poverty, McCabe proposes, are ideals. "Riches" represents the ideal of taking the world for your own use. "Poverty" represents the ideal of complete freedom from possessions. "Possessions," he continues, drawing another key distinction, are things that can be taken (or refused), even when offered as gifts. *Being* (or life), in contrast, is "sheer gift," and is not for the taking. "No one can take upon

[55] See n. 33 of this chapter.

[56] *God, Christ, and Us*, ed. Brian Davies (London: Continuum, 2005), pp. 53–57. As this is a very short text, I will not cite each quotation separately.

herself life; nothing can bring itself into existence. Always we receive being from another or from others."

To seek riches, McCabe continues, is to "aim at taking possession of things, even, perhaps taking possession of people." In contrast, to seek poverty "is to aim at the giving of life, and this comes from gratitude for receiving life ourselves. And giving life is a specially godlike activity."

McCabe acknowledges the ubiquity of language about divine ownership in the Christian tradition, but he argues nevertheless that "the riches of God" is not something we can speak about literally, because God "has not, and could not have, any possessions." However, we *can* speak literally about the poverty of God, for precisely the same reason.

> Of course, his poverty is not the same as our poverty, as his wisdom is not the same as ours. But he is literally poor because he simply and literally has no possessions. He takes nothing for his own use. He only has life and being. And, if you want to press the point, he does not even *have* life, as he does not *have* wisdom or *have* goodness. In God, being alive or being wise or being good are just simply being God and nothing more, nothing extra he *has*.

On this view, God's act of creation is itself an act of God's poverty, "for God gains nothing by it." Again, God does not stand over creation as owner to owned, as possessor to possessed. Creation has nothing to do with the logic of possession, which is taking. It is, rather, "that characteristic act of love which is the giving of life."

Thus, McCabe concludes, "to aim at poverty is for us to aim in the direction of the simplicity and poverty of God – a direction away from possessing to being." To aim at poverty is thus to become godlike. But it is also to become more human, because we can only flourish as human beings to the extent we give life to each other, which for McCabe (and Aquinas) requires above all the cultivation of those non-manipulative relationships of equal regard we call *friendship*.

That friendship is necessary for human flourishing is a platitude, according to McCabe. Nonetheless, by ending his homily with it, he draws out the political implications of his premises about God's poverty, and the distinction between possessions and being (or life). Personal friendship, he avers, is a far better and more useful model for human society than the word "market," whose current popularity as a metaphor for our life together he finds "bizarre." However good and even necessary markets may be, they are ultimately about possession, about taking, and not giving.[57]

[57] For McCabe, a society founded on the sacredness of possessions is one example of the "idolatrous society":

> The idolatrous society . . . presents two faces; on the one hand it is a religious society with great respect for traditional ways; it will be a society in which patriotism is highly valued and in which there is much concern for the country's "heritage." On the other hand it will also be a society of institutionalised violence in which brutality and injustice is either hidden or given a mask of legality. It is important to see that any society may become idolatrous in this way, that in fact

In emphasizing the poverty of God, therefore, McCabe invites us to think afresh about traditional Catholic teaching on property and ownership, and thus on the possible contributions of that tradition to the contemporary debate over gene patenting. On the specific questions of patent law and policy, McCabe's claims would support, or at least not seriously undermine, the sorts of practical conclusions to which both Cahill and Kaveny (and others) point – for example, supporting greater participation by all stakeholders (including local communities, NGOs, developing countries, and other international bodies) in crafting and applying patent law sensitive (and accountable to) "ordre public." However, in his emphasis on God's poverty, our call to seek it, and the political implications of both, McCabe more effectively exposes the radical foundations of these practical proposals, and does so in four distinct but interrelated ways.

First, the notion of God's poverty, resting as it does on the distinction between *possessions* and *being*, brings forward, and does so with far greater clarity than the metaphor of God as cosmic landlord, the deep incompatibility between ownership and life – not just human life, but all life. The landlord metaphor tempts us to think of ourselves as competing with God over the ownership of real estate, "creation." Moreover, because God seems to be something of an "absentee" landlord, then we, at least in practice, tend to claim effective ownership for ourselves. However, if we do not conceive of God as a landlord to begin with, then we can better avoid such temptations. In short, speaking about God's "poverty" (much like John Paul II's metaphor of "social mortgage") brings out in a more challenging way the fundamentally limited nature of property "rights."

Second, emphasizing creation as an act of divine poverty in the way McCabe proposes also emphasizes the kenotic dimensions of divine action (God a "ruined millionaire"[58]) and the correlative ethical demands on those infused with the grace of participation in that action. All creation – including all artistic, technological, and scientific work – is *for* others. And only in turning to others in our work, and by turning our work to others, do we come into our full stature as human persons.[59] Emphasis on divine poverty, in sum, "tightens" the connection between ownership and the demands of charity. The necessary control "of" or "over" resources hailed by defenders of property rights (on the "divine ownership" model) is first and foremost control "for" the other.

Third, the language of divine "poverty" also allows a closer connection between our economic activity and the claims on us of those who are, in fact, literally poor.

> every society betrays a built-in tendency to such a worship of the work of man's hands. In any society men are liable to find their identity simply in what they themselves have achieved; the rejection of this is the beginning of the discovery of Yahweh.

*Law, Love, and Language* (1968; London: Continuum, 2011), pp. 122–23.

58 T. S. Eliot, "East Coker," IV, from *The Four Quartets*, http://oedipa.tripod.com/eliot-2.html.

59 Pope Saint John Paul II has made these themes central to Catholic social teaching on the dignity of labor. See especially *Laborem Exercens* (1981), http://w2.vatican.va/content/john-paul-ii/en/encyclicals/documents/hf_jp-ii_enc_14091981_laborem-exercens.html.

Even if patenting sequences of the human genome is not intrinsically wrong, it still takes place against the backdrop of ongoing inequities in the distribution of the goods necessary for human flourishing.

Fourth, the internal connection McCabe makes between our seeking divine poverty and the good of friendship is a reminder that the central question in the debate over gene patenting, indeed the central question of all bioethics, is political, in the broadest sense of that term, as having to do with the form of our life together.[60] As McCabe rightly says, Jesus does not tell us to have no possessions; he tells us "to *aim* at poverty, to move towards it, and certainly not to aim at *riches*."

To aim at riches, "by building ever more secure fences of possessions around ourselves," is ultimately corrosive of both friendship and human society. It is for this reason, at least in the long Catholic tradition on property, that the sequestration of created things – ownership – constitutes one of the enduring obstacles to our ability to live together, peaceably, in friendship. Those who support gene patenting, the sequestration of the human genome, are obliged to explain why such a practice is anything other than the most recent of such obstacles.

60 One of the two architectonic questions of bioethics, according to one of its founders, Daniel Callahan, is "[w]hat kind of a society ought we to want in order that the life sciences will be encouraged and helped to make their best contribution to human welfare?" *Encyclopedia of Bioethics*, revised edition, ed. W. Reich (New York: MacMillan, 1995), volume 1, p. 255.

# III

# Social Justice and Political Aspects

# 11

# "Thou Shalt Not Steal": The Morality of Compulsory Licensing of Pharmaceutical Patents[1]

*Margo A. Bagley*[2]

## I. INTRODUCTION

*The story is told of a woman in Europe who was near death from a rare form of cancer. There was one drug that her doctor thought might save her. It was a type of radium that a druggist in the same town had recently discovered. The drug was expensive to make, but the druggist was charging four thousand dollars, which was ten times the cost of manufacture. The sick woman's husband, Heinz, went to everyone he knew to borrow money, but he could only get together about two thousand dollars, half of the drug's price. He asked the druggist to sell it cheaper or let him pay later. But the druggist said, "No, I discovered the drug and I'm going to make money from it."*

*So, having tried every legal means, Heinz, in desperation, considers breaking into the druggist's store to steal the drug for his wife. Should Heinz steal the drug?*

Lawrence Kohlberg used the way study participants responded to this and other stories to assess their stage of moral development.[3] The moral dilemma arises because allowing someone to die when you could save their life is morally wrong. But stealing is also morally wrong, and for Heinz to take the drug without permission would be considered stealing, even if it were ultimately deemed excusable.

On Kohlberg's scale, the more a person felt bound by the dictates of law, the less sophisticated his or her moral development. The more he considered and valued the rights and attitudes of all parties involved, the more advanced his moral thinking was

[1] Much of this chapter was first published as *The Morality of Compulsory Licensing as an Access to Medicines Tool*, 100 MINN. L. REV. 101 (2018).

[2] Special thanks to Tom Berg, Anne Coughlin, Cynthia Ho, Justin Hughes, Joe Miller, Caleb Nelson, Ruth Okediji, and participants at the Patents on Life Conference (St. Edmund's College, Cambridge University), at my Hardy Cross Dillard Chair lecture at the University of Virginia School of Law, and the Georgia IP Scholars workshop, and the Loyola Los Angeles IP Colloquium. Thanks also to Joseph Babitz, Samuel Bagley II, Shawn Gannon, and Eric Pettis for superb research assistance, to the editors of the *Minnesota Law Review* for their gracious flexibility, and to the librarians at the University of Virginia Law Library and the Emory McMillan Law Library for stellar research support. I owe a debt of gratitude to Professor (and Rabbi) Michael Broyde for directing me to the Heinz scenario, and for his helpful insights into the Levitical gleaning laws and the concept of *pe'ah*.

[3] Lawrence Kohlberg, THE PSYCHOLOGY OF MORAL DEVELOPMENT: THE NATURE AND VALIDITY OF MORAL STAGES 640 (San Francisco: Harper and Row, 1984).

deemed to be. The highest level of morality would recognize universal ethical principles, human rights, which might conflict with, and supersede, a society's statutory laws.

But is it possible for Heinz taking the drug without permission to not be viewed as theft at all? Might the druggist instead be the thief if he tried to keep the drug from Heinz? If so, what theory would explain such a conclusion? Such philosophical questions are surprisingly relevant to an important aspect of the access to medicines discussion today: the right and ability of countries to grant compulsory licenses on patented pharmaceutical products, including biologic drugs produced in living organisms.[4]

This chapter contemplates the validity of theft rhetoric in relation to the right of countries to grant compulsory licenses from an unconventional perspective[5] – that of biblical teachings on what it means to steal. Part II describes the use of theft rhetoric in relation to intellectual property infringement broadly and drug patent compulsory licenses in particular. Part III challenges the contention, implicit in theft rhetoric, that compulsory licenses are morally wrong as a form of stealing. It does this by exploring the meaning of theft in the context of its Judeo-Christian origins. Part IV then considers and critiques the assertion that the issuance of compulsory licenses in developing countries destroys pharmaceutical company innovation incentives. Part V concludes that expanding the definition of theft to include, as the Bible does, the possibility that a property owner may be stealing from the poor, can help us to properly evaluate the morality of drug patent compulsory licenses.

[4] According to the US Food and Drug Administration (FDA):

> Biological products [biologics] include a wide range of products such as vaccines, blood and blood components, allergenics, somatic cells, gene therapy, tissues, and recombinant therapeutic proteins. Biologics can be composed of sugars, proteins, or nucleic acids or complex combinations of these substances, or may be living entities such as cells and tissues. Biologics are isolated from a variety of natural sources – human, animal, or microorganism – and may be produced by biotechnology methods and other cutting-edge technologies. Gene-based and cellular biologics, for example, often are at the forefront of biomedical research, and may be used to treat a variety of medical conditions for which no other treatments are available.

US Food and Drug Administration, *What Are 'Biologics'? Questions and Answers*, https://www.fda.gov/AboutFDA/CentersOffices/OfficeofMedicalProductsandTobacco/CBER/ucm133077.htm.

[5] While the use of biblical insights by legal scholars is not common, other legal scholars have applied them to legal questions as diverse as environmental ethics, capital punishment, dispute resolution, and professionalism. I too have previously used insights from biblical parables to illuminate issues relating to patents and self-replicating technologies. *See* Margo A. Bagley, *"Grant Me Justice Against My Adversary": What Parables Can Teach Us About* Organic Seed Growers & Trade Assoc. v. Monsanto, in Irene Calboli and Srividyha Ragavan eds., Diversity in Intellectual Property: Identities, Interests, and Intersections (Cambridge University Press 2015); Margo A. Bagley, *The Wheat and the (GM) Tares: Lessons for Plant Patent Litigation from the Parables of Christ*, 10 U. St. Thomas L.J. 683 (2013).

## II. PATENTS, DRUGS, AND THE LANGUAGE OF EXPROPRIATION

Patents and drug products seem to go hand in hand.[6] Through the patent mechanism, a government grants an inventor the right to exclude others from making using, selling, offering to sell, or importing a patented invention into a country for the patent term – typically twenty years from the filing date.[7] Patent rights are territorial, so inventors must apply in each country or region where they desire protection, and the resulting patent is only valid in that country.[8] Moreover, because patent laws differ by country, an inventor may obtain a patent in one country and not in another on the same invention.

The right to exclude granted by a patent is widely considered essential to pharmaceutical investment and development. Moreover, the deadweight losses that patents create are generally considered justifiable for pharmaceuticals, as companies need to price above marginal cost to recoup significant fixed development expenses.[9]

However, these exclusionary rights translate to soaring drug prices in both rich and poor countries,[10] despite global calls for access to affordable medicines.[11] Moreover, the problem is exacerbated by the fact that biologics, which can be

[6] *See, e.g.*, Rebecca S. Eisenberg and W. Nicholson Price, II, *Promoting Healthcare Innovation on the Demand Side*, 4 J.L. & Biosci. 3, 3 (2017) ("Policy mechanisms to promote biopharmaceutical innovation often focus on fortifying incentives for firms to develop new products. Biopharmaceutical firms favor exclusionary rights [such as patents] that defer competition, allowing them to profit by charging higher prices prior to generic entry.").

[7] TRIPS: Agreement on Trade-Related Aspects of Intellectual Property Rights, April 15, 1994, Marrakesh Agreement Establishing the World Trade Organization, Annex 1C, 1869 U.N.T.S. 299, 332–34, Articles 28 and 33 (1994) (hereinafter TRIPS).

[8] Paris Convention for the Protection of Industrial Property, art. 4, March 20, 1883, 828 U.N.T.S. 305 (1979).

[9] *See* Joseph A. DiMasi, *Innovation in the Pharmaceutical Industry: New Estimates of R&D Costs*, 47 J. Health Econ. 20, 21 (May 2016). *See infra* note 66.

[10] *See, e.g.*, Sharon Begley, *Cancer Drugs Are Unaffordable in the Developing World*, STAT, June 6, 2016, https://www.statnews.com/2016/06/06/cancer-drug-prices-developing-world/; Joanna M. Shepherd, *Bio logic Drugs, Biosimilars, and Barriers to Entry*, 25 Health Matrix 139, 159 (2015), https://scholarlycom mons.law.case.edu/healthmatrix/vol25/iss1/8; IRIN, *Essential New AIDS Drugs Unaffordable for Developing Countries*, http://www.irinnews.org/news/2007/07/25/essential-new-aids-drugs-unaffordable-developing-countries.

[11] For example, Goal 3.B of the United Nations Sustainable Development Goals 2030 includes the targets of supporting "the research and development of vaccines and medicine for the communicable and non-communicable diseases that primarily affect developing countries" and providing "access to affordable essential medicines and vaccines, in accordance with the Doha Declaration on the TRIPs Agreement and Public Health." UN Sustainable Development Goals 2030 (SDGs), http://www .un.org/sustainabledevelopment/health/. *See also* Report of the United Nations Secretary-General's High-Level Panel on Access to Medicines 8 (September 2016), http://www .unsgaccessmeds.org/final-report/ ("Over the last few decades, medical innovation has dramatically improved the lives of millions of people across the globe . . . Despite this noteworthy progress, millions of people continue to suffer and die from treatable conditions because of a lack of access to health technologies.").

covered by a web of "life" patents,[12] and cost an average of twenty-two times as much as a small molecule drug for treating the same condition,[13] comprise an increasing proportion of new drugs in development.[14] Yet efforts by governments to reduce costs, using mechanisms like compulsory licenses, routinely meet with censure at the hubris of even considering harm to the patent "goose" that lays the golden eggs of new medical breakthroughs.[15]

Through compulsory licensing, a nation allows a third party to practice a patented invention without the patent owner's permission, and requires that third party to pay a government-specified royalty to the patent owner.[16] A compulsory license does not require the patent owner to *do* anything except sit back and receive the royalties from the third party. But it does prevent the patent owner from stopping that third party from practicing the patented invention.

Viewed from a property perspective, criticizing compulsory licenses might seem quite justified, as the right to exclude is the paradigmatic feature of private property. However, patents and other forms of intellectual property (IP) lack some attributes of property, despite sharing others. For example, patented subject matter is nonrivalrous, lacks clear boundaries, provides minimal notice, and includes only a right to exclude, not a right to use.[17] Thus it should not be surprising that governments would choose to employ, via a compulsory license, a liability rule

[12] *See, e.g.*, Cynthia Koons, *This Shield of Patents Protects the World's Best-Selling Drug*, BLOOMBERG NEWS, September 7, 2017, https://www.bloomberg.com/news/articles/2017-09-07/this-shield-of-patents-protects-the-world-s-best-selling-drug (describing the web of more than one hundred patents covering AbbVie's $16 billion/year blockbuster biologic Humira); Jayne O'Donnell, *Biologic Drugmakers Use Patents, Suits, to Thwart Generic Competition*, USA TODAY, August 4, 2016 ("Biologic makers have been 'asserting many, many patents to protect their franchise'"), https://www.usatoday.com/story/news/politics/2016/08/04/biologic-drugmakers-patents-suits-generic-competition/879157347.

[13] Anthony D. So & Samuel L. Katz, *Biologics Boondoggle*, N.Y. TIMES, March 7, 2010, at A23.

[14] Ameet Sarpatwari, Jerry Avorn, and Aaron S. Kesselheim, *Progress and Hurdles for Follow-On Biologics*, 372 N. ENGL. J. MED. 2380 ("Though [biologics] account for less than 1% of all prescriptions dispensed in the United States, expenditures on them amount to 28% of prescription drug spending, and both their use and cost are forecast to grow sharply.").

[15] *See, e.g.*, William W. Fisher III and Talha Syed, INFECTION: THE HEALTH CRISIS IN THE DEVELOPING WORLD AND WHAT WE SHOULD DO ABOUT IT, ch. 6 (Stanford University Press, forthcoming 2018) (noting several reasons for the infrequent use of compulsory licenses despite the strong need for cheaper drugs, including that "the pharmaceutical firms disadvantaged by compulsory licenses and the governments of the countries in which those firms are based sometimes retaliate [or threaten to retaliate] against the countries that use them."), https://cyber.harvard.edu/people/tfisher/Infection_Regulations.pdf; Muhammad Zaheer Abbas, *Pros and Cons of Compulsory Licensing: An Analysis of Arguments*, 3 INT'L J. SOC. SCI. & HUMAN 254, 254–55 (2013) (cataloging arguments against compulsory licenses).

[16] Compulsory Licenses Under TRIPS: An Introduction, in Cynthia M. Ho, ACCESS TO MEDICINE IN THE GLOBAL ECONOMY: INTERNATIONAL AGREEMENTS ON PATENTS AND RELATED RIGHTS 127 (Oxford University Press 2011).

[17] *Id.* at 6, 8, 54. *See also* Mark A. Lemley, Peter S. Menell, and Robert P. Merges, INTELLECTUAL PROPERTY IN THE NEW TECHNOLOGICAL AGE, I-2 (Clause 8 Publ. 2016) ("The fact that the possession and use of ideas is largely 'nonrivalrous' is critical to intellectual property theory because it means that the traditional economic justification for tangible property does not fit intellectual property. In the

(the patent owner is entitled to compensation for use of the invention) and not a property rule (the patent holder can preclude use of the invention) for patent infringement in certain circumstances.[18] Patent rights are not absolute.

The legitimacy of compulsory licenses under international law is not truly in question. The United States and several European countries have issued numerous explicit and de facto compulsory licenses in various technological areas over the years.[19] Moreover, the 1995 World Trade Organization Agreement on Trade Related Aspects of Intellectual Property (TRIPS) explicitly allows compulsory licenses under certain conditions.[20] Nevertheless, the moral "rightness" of countries in the Global South issuing compulsory licenses for pharmaceuticals seems very much in question, with such tools often being labeled as "theft" and otherwise mischaracterized as expropriation.[21]

Theft rhetoric in patent law is not new, but it has a particularly pernicious effect in this context. Theft rhetoric tends to constrain policy choices and government actions, overly extending the boundaries of the patent grant beyond the social bargain for products that can mean life or death to millions of individuals, especially those in low- and middle-income countries (LMICs).

When a third party makes, uses, sells, or offers to sell an invention in a country, or imports a patented invention into that country, without the patent owner's permission, that third party has engaged in patent infringement. Garden-variety patent infringement is a strict liability offense; no particular mental state is required to

state of nature, there is no danger of overusing or overdistributing an idea, and no danger of fighting over who gets to use it. Everyone can use the idea without diminishing its value.").

18 *See, e.g.*, Guido Calabresi and A. Douglas Melamed, *Property Rules, Liability Rules, and Inalienability: One View of the Cathedral*, 85 Harv. L. Rev. 1089, 1092 (1972).

19 *See, e.g.*, *Recent Examples of Compulsory Licensing of Patents*, KEI RN 2007:2, available at https://www.keionline.org/book/publications-and-research-notes/kei-rn-2007–2-recent-examples-of-compulsory-licensing-of-patents/ (describing the issuance or threat of issuance of compulsory licenses by a variety of countries). *See also* James Love, Knowledge Ecology International (KEI), written comments and notice of intent to testify at the Special 301 Public Hearing Monday, February 24, 2014, at the offices of USTR (Feb. 7, 2014):

> The United States itself is [a] major user of compulsory licenses ... [This includes] the extensive non-voluntary uses of patents and copyrights under 28 USC 1498, such as the compulsory licenses benefiting contractors for NASA ..., and the growing number of compulsory licenses granted by federal judges, when they forgo the granting of injunctions in patent infringement cases, in favor of forward looking royalties as a remedy for infringement [per] *eBay Inc. v. MercExchange*. These include a number of medical inventions.

20 TRIPS Art. 31.

21 *See* Letter from Livia Leu, Swiss Ambassador, to Dr. Carolina Gomez, Adviser, Colombian Ministry of Health and Social Protection (May 26, 2015), https://www.minsalud.gov.co/sites/rid/Lists/BibliotecaDigital/RIDE/VS/MET/patent-of-Imatinib-glive-closing-arguments.pdf (criticizing Colombia's plan to issue a compulsory license on the cancer drug Gleevic). *See also* Letter from Dr. Deborah Gleeson et al. to Juan Manuel Santos, President of the Republic of Colombia (May 16, 2016), https://www.keionline.org/wp-content/uploads/5–16-2016-letter-to-colombia-santos-imatinib-license.pdf (letter from 122 health, intellectual property, and trade experts defending Colombia's right to issue a compulsory license).

violate a patent owner's rights. Nevertheless, enhanced damages may be awarded for willful infringement, and, controversially, some countries even provide criminal remedies.[22] Thus, it is perhaps unsurprising that patent infringement is sometimes characterized as "theft." Consider the following examples:

- "[W]hat makes *patent theft* so attractive is that *infringement* is not a criminal act and those found guilty face no jail time."[23]
- "It is estimated that patent theft costs the US economy billions each year."[24]
- "Licensing is simply a legitimate cost of doing business. Not paying that cost is *theft*."[25]

The use of theft framing is not new to intellectual property (IP).[26] Trade secret misappropriation is often labeled as theft,[27] and in copyright and trademark law, criminal penalties are available for counterfeiting, which also involves a close link with the notion of theft.[28] The ubiquitous FBI notices that appear at the beginning of copyright-protected movies have long linked copyright infringement and piracy, a form of theft.[29] On some level, infringement does involve the unpermitted use of something to which a government-granted right attaches, and certainly some of the activities being characterized in this way are more egregious than others.

The notion that theft is wrong, that it is immoral even, derives, at least in part, from the Judeo-Christian roots of the Ten Commandments and their injunction against stealing.[30] The Ten Commandments, or Decalogue, is known as "the moral

22 *See* J. W. Baxter et al., World Patent Law And Practice, App. 2A.00 (Matthew Bender 2008) (providing a list of countries with criminal penalties for patent infringement).

23 Pat Choate, *Patent Theft as a Business Strategy*, Huffington Post, May 23, 2010, https://www.huffingtonpost.com/pat-choate/patent-theft-as-a-busines_b_508780.html (emphasis added).

24 Russ Genet, *What's the Cost of Allowing Patent Theft? Don't Wait to Find Out*, Intellectual Property Watch (January 9, 2019), http://www.ip-watch.org/2019/01/09/whats-cost-allowing-patent-theft-dont-wait-find/. *See also* Richard Stim, *Intellectual Property Crimes*, http://www.intellectualpropertylawfirms.com/intellectual-property/ip-crimes.htm. ("The theft [or 'infringement'] of a patent, ... is typically handled as a civil matter.")

25 John Wiley, *Patent Infringement Is Theft, Plain and Simple*, Wash. Post, November 17, 2015, https://www.washingtonpost.com/news/in-theory/wp/2015/11/17/patent-infringement-is-theft-plain-and-simple. Moreover, President Donald Trump recently tweeted: "The U.S. is acting swiftly on Intellectual Property theft. We cannot allow this to happen as it has for many years!" President Donald J. Trump (@realDonaldTrump), Twitter (Mar. 7, 2018, 10:38 AM), https://twitter.com/realDonaldTrump/status/971409845453762560.

26 See Justin Hughes, *Copyright and Incomplete Historiographies: Of Piracy, Propertization, and Thomas Jefferson*, 79 S. Cal. L. Rev. 993, 1008-10 (2006) (highlighting pre-Statute of Anne [1710] uses of piracy and other theft-related terms to describe copyright infringement).

27 There is even a law entitled the "Theft of Trade Secrets Clarification Act of 2012," 18 U.S.C. §§ 1831–39.

28 18 U.S.C. § 2320(b)(1) ("Whoever commits an offense under subsection (a) if an individual, shall be fined not more than $2,000,000 or imprisoned not more than 10 years, or both").

29 *See* FBI Anti-Piracy Warning Seal, https://www.fbi.gov/investigate/white-collar-crime/piracy-ip-theft/fbi-anti-piracy-warning-seal.

30 "Thou shalt not steal." Exodus 20:15 (KJV).

law,"[31] and violations of many of its tenets, such as those against adultery, lying, and stealing, are widely deemed immoral in the United States even today.

One striking example of linking IP infringement to violation of a biblical commandment is found in *Grand Upright Music v. Warner Bros.*,[32] where the defendant was accused of sampling a small portion from one of the plaintiff's recordings. The opinion begins as follows:

> "*Thou shalt not steal*" has been an admonition followed since the dawn of civilization. Unfortunately, in the modern world of business this admonition is not always followed ... The conduct of the defendants herein, however, violates not only the *Seventh Commandment*, but also the copyright laws of this country.[33]

Importantly, the judge explicitly classified copyright infringement as a violation of the biblical commandment against stealing, a moral wrong. This is not an isolated occurrence; references to the Ten Commandments, or Decalogue, have appeared in at least 846 cases since the early 1800s. As John Eidsmoe notes, "These usages demonstrate how thoroughly ingrained into our culture the Ten Commandments have become, to the point that the usage of this terminology lends instant recognition and moral authority to the injunctions they describe."[34]

A surprising number of theft of property cases, dealing with tangible and intangible property, mention the seventh commandment (or eighth, depending on which version of the Bible is used).[35] While many of these cases are old, some are rather new. And while the courts are not explicitly relying on the commandment as the basis for the decision, invoking the biblical admonishment is doing some work for the judge; a link is being made between the moral law and theft. In *Edgenet, Inc. v. The Home Depot USA, Inc.*, the court also alluded to modern IP laws as being derived from the seventh commandment:

> The ancient admonition "thou shalt not steal" expresses a value that is basic to and underlies the law of property: one cannot freely appropriate to him or herself what

[31] *See* John A. Eidsmoe, *The Use of the Ten Commandments in American Courts*, 3 Liberty U. L. Rev. 15, 18, 19 (2009), http://digitalcommons.liberty.edu/cgi/viewcontent.cgi?article=1050&context=lu_law_review. Professor Eidsmoe identified 515 cases mentioning the Ten Commandments and a further 331 referencing the Decalogue. He notes that "these case citations cover a time span from the early 1800s to the present time, demonstrating an unbroken tradition of looking to the Ten Commandments as the moral foundation of [US] law." *Id.* at 19.

[32] 780 F. Supp. 182 (S.D.N.Y. 1991).

[33] *Id.* at 183 (emphases added).

[34] Eidsmoe, *supra* note 31, at 43–44.

[35] *See, e.g.*, Rochon v. Iberia Parish Sch. Bd., 601 So.2d 808, 809–10 (La. App. 3d Cir. 1992) ("One of the commandments is 'Thou shalt not steal.' Stealing and theft are synonymous and are criminal acts ... So, this Court holds that theft (especially of these funds) is an immoral act."); Fountain v. State, 109 So. 463, 464 (Fla. 1926) (noting that the meaning of the word steal "has been pretty thoroughly understood since the eighth commandment was brought down from Sinai, or at least since it was translated into English"); *Grand Upright Music Ltd.*, 780 F. Supp. at 183; State v. Jim, 508 P.2d 462 (Or. Ct. App. 1973) ("Theft is a synoptic concept: the Eighth Commandment condemns theft without explaining every possible nuance and contrivance in its accomplishment.").

> has been produced by the labor, efforts, or capital of another. While the plaintiff did not formally allege that the defendants . . . contravened the seventh commandment, [the Plaintiff] has accused the defendants of violating nearly every modern derivation of the biblical edict.[36]

It is unlikely that courts and commentators are linking the moral law and IP infringement unconsciously or even thoughtlessly, especially in light of the increasing domestic and international efforts to criminalize IP infringement.[37] However, such rhetoric is particularly concerning when compulsory licensing of drug patents is involved. With a compulsory license, a third party may practice a patented invention without the permission of the patent owner, who is entitled to receive adequate compensation for such use.[38] Theft rhetoric in relation to compulsory licensing employs moral overtones, and it seems designed to counter what might otherwise appear to be a moral obligation to save life when it will not harm one to do so. Consider the following examples.

A few years ago, the *Wall Street Journal* published an op-ed castigating the Thai government's issuance of compulsory licenses on two HIV/AIDS drugs and a heart disease drug. According to Boston University School of Law Dean Emeritus Ronald Cass, Thailand had engaged in "theft":

> The EU's trade commissioner . . . recently joined the U.S. in protesting Thailand's effective theft of pharmaceutical companies' intellectual property . . . [T]here is a growing appreciation that trampling patents to allow a middle-income nation to cut its spending on drugs seriously threatens the world's system of protections for innovation.[39]

Though Thailand is considered a middle-income nation, its per capita gross national income was a mere $5,640 in 2016.[40] One of the op-ed's striking aspects was the characterization of Thailand as both ripping off drug companies *and* harming innovation, when it was in fact complying with its international obligations. The license was for the Thai public health system and thus constituted "public non-commercial use" within the meaning of TRIPS Article 31(b).[41] Also, as Jill Johnstone and James Love noted, the public health reasons for the licenses were quite legitimate:

36 Edgenet, Inc. v. The Home Depot USA, Inc., 2010 U.S. Dist. LEXIS 2276 (E.D. Wisc. 2010), *aff'd*, 658 F.3d 662 (7th Cir. 2011).

37 *See, e.g.*, Jonathan Masur and Christopher Buccafusco, *Innovation and Incarceration: An Economic Analysis of Criminal Intellectual Property Law*, 87 S. Cal. L. Rev. 275 (2014).

38 Because the government generally determines what comprises "adequate" or "reasonable" compensation, there may be disagreement regarding whether that standard is met in a given case.

39 Ronald A. Cass, *Patent Remedy*, Wall. St. J., August 28, 2007, https://www.wsj.com/articles/SB118824874547610202.

40 *Data for Upper Middle Income: Thailand*, World Bank, https://data.worldbank.org/?locations=XT-TH (last visited February 22, 2018).

41 TRIPS Article 31(b). See also *Compulsory Licenses Under TRIPS: An Introduction*; in Ho, Access To Medicine, *supra* note 16, at 134.

In 2006, Thailand had an average per capita income of $8.19 per day. For the bottom 80 percent of the income distribution, the average was $5.22 per day. To illustrate the Thai concerns regarding affordability of products, it is useful to note that the pre-compulsory licensing price for the heart disease drug Plavix was $2 per day, or nearly 40 percent of the average income of the bottom 80 percent of the population.[42]

Like Cass, other commentators have used theft language to describe compulsory licenses, including the CEO of pharmaceutical giant Bayer AG, characterizing a different compulsory license as "essentially theft."[43] The Indian government granted the compulsory license on a cancer drug that Bayer had priced so high in India that it was available to only 2 percent of the patients who needed it.[44] When using this theft rhetoric or framing, commentators generally send two related but distinct messages:

(1) compulsory licenses are morally wrong because stealing is morally wrong, and
(2) compulsory licenses will harm innovation and society will not get the new drugs it needs.[45]

They thus appeal both to moral outrage and to understandable self-interest. Each message will be considered in turn.

42 Letter from James Love, U.S. Co-Chair, and Jill Johnstone, European Co-Chair, Trans Atlantic Consumer Dialogue Working Group on Intellectual Property, to Susan C. Schwab, Ambassador and U.S. Trade Representative, and Peter Mandelson, European Commissioner for International Trade (March 17, 2008), http://test.tacd.org/wp-content/uploads/2013/09/TACD-IP-2008-Letter-To-S.-Schwab-and-P-Mandelson-on-Compulsory-Licensing.pdf. *See also* Inthira Yamabhai et. al., *Government Use Licenses in Thailand: An Assessment of the Health and Economic Impacts*, 7 Global Health 28 (2011), http://www.ncbi.nlm.nih.gov/pmc/articles/PMC3176479/ (concluding that Thailand's compulsory licenses resulted in increased access to the medicines, with no appreciable decline in exports or foreign direct investment).

43 Ketaki Gokhale, *Merck to Bristol-Myers Face More Threats on India Drug Patents*, Bloomberg News, January 28, 2014, http://www.bloomberg.com/businessweek/news/2014–01-21/merck-to-bristol-myers-face-more-threats-on-india-drug-patents/. *See also* Merrill Matthews, "Miracle Drugs Ripped Off," Geelong Advertiser (Australia), 2008 WLNR 23639900 ("In Thailand and Brazil, politicians have decided to reward inventors of medical miracles not with accolades but with compulsory licenses. This is a fancy name for simple theft."); Editorial, "Theft, Extortion, and AIDS," Chi. Trib., 2005 WLNR 233841123 ("No matter how you package it, Brazil's campaign against patents [threatening to issue compulsory licenses] held by U.S. pharmaceutical firms amounts to international thievery and extortion.")

44 *Id.*

45 *See also* Jason Urbach, *State Will Undercut Itself with New IP Policy*, Business Day (South Africa), 2014 WLNR 5917470 ("Compulsory licenses allow the government to break a patent and give a license to a local manufacturer to produce a drug. This amounts to *state-sanctioned theft of property* ... If we weaken drug patents, we will *discourage companies and investors from developing the next generation of medicines*." [emphasis added]); Phil Kerpen, *Time to Get Tough On India For Stealing All of Our Stuff*, Baxter Bulletin, 2014 WLNR 10679268 ("Quite simply, from *pharmaceuticals* to motion pictures ... India is *stealing our stuff* .... Strong protections for *IP rights of innovators are absolutely critical to raise the capital and justify investing it in developing new cures*. The *greatest threat to innovation globally* now comes from India's policies." [emphases added]).

## III. COMPULSORY LICENSES: MORALLY WRONG OR MORALLY RIGHT?

First, if compulsory licenses are morally wrong because it is morally wrong to steal, and the moral wrongness of theft derives in some measure from the Judeo-Christian Ten Commandments, then it may be instructive to take a closer look at the meaning of theft under that moral law.

Accusing someone of stealing is a serious matter. In the Bible theft is a sin, a violation of a command that is sacred in Judeo-Christian traditions. However, the Ten Commandments exist not in isolation, but rather as part of a larger, richer context that both prohibits actions and prescribes affirmative duties to humankind. As such, the Ten Commandments are far more than just a bunch of "thou shalt nots."

Probably the most illuminating way to think about their true meaning is through the explanation Jesus gave when, asked by an expert in the law which was the greatest commandment, He said: "'Love the Lord your God with all your heart, and with all your soul and with all your mind.' This is the first and greatest commandment. And the second is like it; 'Love your neighbor as yourself.' All the Law and the Prophets hang on these two commandments."[46] In other words, Jesus explains that the Ten Commandments, perhaps surprisingly, are about love. Love for God in the first four, and love for your neighbor in the last six.

We show love to God by not putting other gods before him, not bowing down to idols, not taking his name in vain, and choosing to spend his Sabbath with him. In terms of our love for, or duty to, our neighbor, dishonoring our parents is not loving; neither is murder, nor cheating on our spouses. Nor is lying on our neighbor, or coveting our neighbor's possessions. And stealing is definitely not loving.[47]

Perhaps because of that "love" orientation and the context of duties and obligations, there are explicit instructions in the Bible regarding a variety of actions that might normally be considered stealing.[48] And some of those actions, embodied in

[46] Matthew 22:37–40 (NIV). Jesus' answer to the lawyer is repeated in Mark 12:28–34, where it is footnoted as referring to Deuteronomy 6:5 and Leviticus 19:18. It also appears in the context of the lawyer answering Jesus and Jesus affirming his answer, in Luke 10:25–28, and serves as a prelude to the parable of the Good Samaritan. See also Romans 13:8–10, which makes the connection explicit:

> Let no debt remain outstanding, except the continuing debt to love one another, for whoever loves others has fulfilled the law. The commandments, "You shall not commit adultery," "You shall not murder," "You shall not steal," "You shall not covet," and whatever other command there may be, are summed up in this one command: "Love your neighbor as yourself." Love does no harm to a neighbor. Therefore love is the fulfillment of the law.
>
> Romans 13:8–10 (NIV)

[47] Exodus 20:2–17 (NIV).

[48] The eighth commandment is "You shall not steal." *Id.*, vs. 15. To be sure, the Bible does condemn ordinary theft, and even requires restitution when someone steals because they are hungry. *See, e.g.*, Exodus 21:16 (dealing with kidnapping, which is stealing someone's freedom); Exodus 22:1–4 (dealing with stealing livestock); Proverbs 6:30–31 (saying one should not despise a thief who steals because he

the Jewish concept of *pe'ah*,[49] provide an interesting and useful analogy that can inform the way we think about compulsory licenses on pharmaceutical patents. The descriptions of *pe'ah* can be found in Leviticus 19 and 23, and in Deuteronomy 24:

> When you reap the harvest of your land, . . . Do not go over your vineyard a second time or pick up the grapes that have fallen. Leave them for the poor and the foreigner. I am the Lord your God.[50]

The instructions apply not only to vineyards but to other crops:

> When you reap the harvest of your land, do not reap to the very edges of your field or gather the gleanings of your harvest. Leave them for the poor and for the foreigner residing among you. I am the LORD your God.[51]
>
> When you beat the olives from your trees, do not go over the branches a second time. Leave what remains for the foreigner, the fatherless and the widow.[52]
>
> When you are harvesting in your field and you overlook a sheaf, do not go back to get it. Leave it for the foreigner, the fatherless and the widow, so that the LORD your God may bless you in all the work of your hands.[53]

In these examples, the poor have the *right* to enter onto the land of a property owner and harvest the remnants of the crop. These are not leftovers after the workers have cleaned every scrap of salvageable fruit or grain. The property owner is instructed not to do too good a job in harvesting the crop, so that he leaves some portion of the crop for the poor and the foreigner. In fact, these examples illustrate that loving one's neighbor is so important that not only are the poor *not* stealing (nor even trespassing), the landowner is *enjoined* from harvesting to the edges of his field. *Or* going back to get the forgotten sheaf. *Or* making a second sweep to pick up missed grapes in his vineyard or missed olives in his grove. What is left does not belong to the landowner, *and it will not harm him to let the poor have it.* It seems God wanted people to trust him, not wealth, to supply their needs.

These broad concepts are also reflected in Christian thought and tradition. For example, the Catholic Catechism recognizes the "universal destination" of goods as being provided by God for the benefit of the *whole* human race, and while it affirms the importance of respect for private property, it also considers as theft "forcing up

is hungry, but that he should still be punished for his crime); Luke 10:30 (dealing with stealing money); Hosea 12:7 (referring specifically to stealing by tax collectors).

49 In Hebrew, *pe'ah* means corner or side. *See Strong's Concordance*, BIBLE HUB, http://biblehub.com/hebrew/6285.htm. As Jeffrey Spitzer explains: "The Bible's model of tzedakah (social justice and support) included a variety of agricultural gifts. Grain and produce that were left or forgotten during the harvest were available for the poor to glean. The corners of the fields (pe'ah) were also designated for the poor." Jeffrey Spitzer, *Pe'ah: The Corners of Our Fields*, MY JEWISH LEARNING, https://www.myjewishlearning.com/article/peah-the-corners-of-our-fields.

50 Leviticus 19:9–11 (NIV).

51 Leviticus 23:22 (NIV).

52 Deuteronomy 24:20 (NIV).

53 Deuteronomy 24:19 (NIV).

prices by taking advantage of . . . the hardship of another."[54] As Professor Tom Berg explains:

> Intellectual property, like other forms of property, serves important purposes related to human dignity, productivity, and especially creativity. Therefore, Catholic teaching affirms intellectual property. But for a variety of reasons, limits on intellectual property are equally important: in particular, the full extension of IP rights may harm the poor, and certain limits on those rights are important to benefiting and empowering the poor.[55]

Protestant perspectives are not to the contrary. As Baptist theologian Gary Gunderson notes: "At the core of every major faith tradition stands an explicit commitment to be with the sick the poor, the alienated, the marginal, the wounded, and the dying. The commitments are very old, but the implications are forever new and increasingly radical because of remarkable changes in health science."[56]

Now to be clear, neither the United States nor any other liberal democracy is a theocracy, and this chapter does not posit that the Ten Commandments or even some of them are the foundation of US law.[57] Nor is it suggesting the adoption of Levitical approaches to dealing with current-day legal issues. Rather, this chapter is

[54] Catechism of the Catholic Church, Pt. III, Sec. II, Ch. II, Art. 7, 2409, http://www.vatican.va/archive/ccc_css/archive/catechism/p3s2c2a7.htm. *See also* Pope John Paul II, *The Catechism of the Catholic Church* 578, 2408 (United States Conference of Catholic Bishops, 2nd ed. 2000) ("The seventh commandment forbids theft . . . But there is no theft if consent can be presumed or if refusal is contrary to reason and the universal destination of goods. This is the case in obvious and urgent necessity when the only way to provide for immediate, essential needs [food, shelter, clothing . . .] is to put at one's disposal and use the property of others."). The Catholic Church set forth this position even more forcefully in a 2001 submission to the World Intellectual Property Organization:

> All men and women of all nations are entitled to have whatever they need for their subsistence and personal advancement, taking it from all the resources available at any given time in history. The provisions protecting private property cannot therefore ever lose sight of the common destiny of all goods, so much so that it has to be said that all private property is subject to a social encumbrance. Consequently, should there be an institutional conflict between acquired private rights and overriding community demands, it is for the public authorities to set about resolving it with active involvement on the part of individuals and social groups.

Document Submitted by the Holy See on Intellectual Property and Genetic Resources, Traditional Knowledge and Folklore for the First Session of the Intergovernmental Committee on Intellectual Property and Genetic Resources, Traditional Knowledge and Folklore, WIPO/GRTKF/IC/1/7, at 4 (April 26, 2001), http://www.wipo.int/edocs/mdocs/tk/en/wipo_grtkf_ic_1/wipo_grtkf_ic_1_7.pdf.

[55] Thomas C. Berg, *Intellectual Property and the Preferential Option for the Poor*, 5 J. Cath. Soc. Thought 193, 199 (2008) (advocating a "degree of skepticism toward broad IP rights" particularly in relation to pharmaceutical patents, as necessary to effectuate the Catholic principle of the preferential option for benefitting or empowering the poor among us).

[56] Gary Gunderson, Deeply Woven Roots: Improving the Quality of Life in Your Community 5 (Fortress Press 1997).

[57] This is not to say that the Ten Commandments and other Biblical provisions, as well as the body of Jewish law that developed around the Torah, have not influenced the common law; the case references identified by Professor Eidsmoe attest to their influence. See *supra* notes 31–34. However, the influence has not always been explicit. *See* Michael J. Broyde, *The Hidden Influence*

- an invitation to consider another perspective on the meaning of theft, drawn from a text widely considered as sacred; and
- an argument that current uses of theft rhetoric in patent law, to the extent they aim to resonate with Judeo-Christian notions of morality, are incomplete, dangerous, and a misapplication of Biblical scripture.

Such uses are incomplete because they do not take account of the full definition of theft in that Judeo-Christian moral context, a definition that includes the possibility of the property owner stealing from the poor. These uses represent a misapplication of scripture, because when one invokes the power of the commandment against stealing, one must also import the underlying norms and values embodied in that commandment. Such uses are also dangerous, because they delegitimize a government's moral obligation to provide for the health and well-being of its citizens. When governments are threatened with economic sanctions for issuing compulsory licenses based on theft framing, they may be reluctant to issue such licenses to alleviate access to medicines problems, and some people may die as a result.

It is perhaps worthwhile to remember that patent rights are limited property rights, at best, and that their contours and scope are constantly being adjusted through judicial, legislative, and administrative action. Every court or United States Patent Office (USPTO) tribunal invalidation of a patent based on obviousness is a rejection of an absolute-rights theory. Governments grant patents to meet utilitarian societal goals, and governments decide the subject matter, scope, and duration of patents. Two inventors in the same country may independently work countless hours and develop the same invention; however, only one of them will receive a patent claiming that exact invention from that country's patent office.[58] In the United States, Article 1, Section 8, clause 8 of the Constitution authorizes, but does not require, Congress "to promote the progress of Science and the useful arts by securing to authors and inventors for *limited times*, the exclusive right to their respective writings and discoveries."[59] Moreover, an inventor does not have a natural right to a patent. As Thomas Jefferson eloquently stated:

> Stable ownership is the gift of social law, and is given late in the progress of society. It would be curious then, if an idea, the fugitive fermentation of an individual brain, could, of natural right, be claimed in exclusive and stable property ... Inventions then cannot, in nature, be a subject of property. Society may give an exclusive right to the profits arising from them, as an encouragement to men to pursue ideas which

*of Jewish Law on the Common Law: One Lost Example*, 57 EMORY L.J. 1403 (2008) (providing an example of the hidden influence of Jewish law on the common law of bailments).

58 In rare cases, a patent office may inadvertently issue two patents on the same invention, but one would be invalid. In the United States, 35 USC § 291 provides for a civil action to determine which inventor has priority and is entitled to the patent.

59 U.S. Const. art. I, § 8, cl. 8 (emphasis added).

may produce utility, but this may or may not be done, according to the will and convenience of the society, without claim or complaint from anybody.[60]

Thus, for example, when the US Supreme Court concluded in *eBay v. MercExchange* that having a right to exclude does not mean the remedy for the violation of that right is an injunction, the Court did not engage in theft, even though *eBay* has resulted in many de facto compulsory licenses where injunctive relief was denied and an adjudged infringer was able to continue practicing the invention by paying an ongoing royalty.[61] The fact that there may be important reasons for creating property rights to allow owners to internalize externalities does not mean that the owner should be able to internalize *all* of the positive externalities.[62] Allowing the public to absorb some beneficial spillovers without positively harming the property owner can be a virtuous policy choice.

Because governments create patent rights, governments can impose appropriate limitations on those rights without such limitations being theft. And making sure the poor have access[63] to the drugs they need in order to live, in a way that does not harm the patent holder, should be viewed as part of the social bargain inherent in the patent system and should be deemed morally right, not morally wrong.

## IV. THE FALLACY OF COMPULSORY LICENSES AS DESTROYERS OF INNOVATION

Regarding the second message, that compulsory licenses will harm innovation resulting in our not getting the drugs we need,[64] two points bear mentioning: (1) We already are not getting many of the drugs we need, and (2) the ones we get we often cannot afford. This is because drug companies are incentivized, but not in the ways one might think.

60 Thomas Jefferson to Isaac McPherson (August 13, 1813), in Andrew A. Lipscomb, ed., 13 The Writings of Thomas Jefferson 333 (Thomas Jefferson Memorial Association of the United States, 1904–05). *But see* Adam Mossoff, *Who Cares What Thomas Jefferson Thought about Patents? Reevaluating the Patent "Privilege" in Historical Context*, 92 Cornell L. Rev. 953 (2007) (arguing that natural rights philosophy undergirded the grant of patents in early America); Justin Hughes, *Copyright and Incomplete Historiographies: Of Piracy, Propertization, and Thomas Jefferson*, 79 S. Cal. L. Rev. 993, 1008-10 (2006) (noting the fluidity of Jefferson's views on patents and his lack of participation in the drafting of the IP Clause).

61 eBay Inc. v. MercExchange, LLC., 547 U.S. 388 (2006). *See also* Paice LLC v. Toyota Motor Co., 504 F.3d 1293 (2007) (Rader, C.J., concurring) ("calling a compulsory license an 'ongoing royalty' does not make it any less a compulsory license").

62 *See* Harold Demsetz, *Toward a Theory of Property Rights*, 57 Am. Econ. Rev. 347, 348 (1967).

63 Of course, having low-cost drugs produced does not guarantee patient access to those drugs. There may be several factors at play that impact whether drugs actually reach the people who need them. *See* Lucie White, *Getting Real about Essential Medicines: "The Last Kilometer,"* 31 Maryland J. Int'l. L. 79 (2016) (exploring the complex barriers that may exist to getting drugs the last mile, or last kilometer, to the patients who need them).

64 *See, e.g.*, Rachel E. Sachs and Austin B. Frakt, *Innovation-Innovation Tradeoffs in Drug Pricing*, 165 Ann. Intern. Med. 871 (2016) (describing pharmaceutical company arguments that reducing drug prices will reduce new drug development).

## A. *Pharmaceutical Company Incentives: Other Markets than the Global South*

It is instructive to consider big pharma's profitability and decisional inputs. The pharmaceutical industry has long been one of the top-five, often the top-two, most profitable industries in the world. The numbers in Figure 1 are astounding.

The ability to obtain such profits is, not surprisingly, what drives big pharma drug-development decisions, since their goal, as for-profit corporations, is to maximize shareholder value.[65] They are not charitable institutions and we should not expect them to be.

The costs of bringing a drug to market are high and do involve significant risk.[66] But companies sell drugs so far above marginal cost that the prices more than make up for the costs of drug development, as their annual profits clearly show.[67] As described in a World Health Organization submission, "launch prices as set by companies are not based on R&D investment or production costs, but on the outcomes of economic calculations

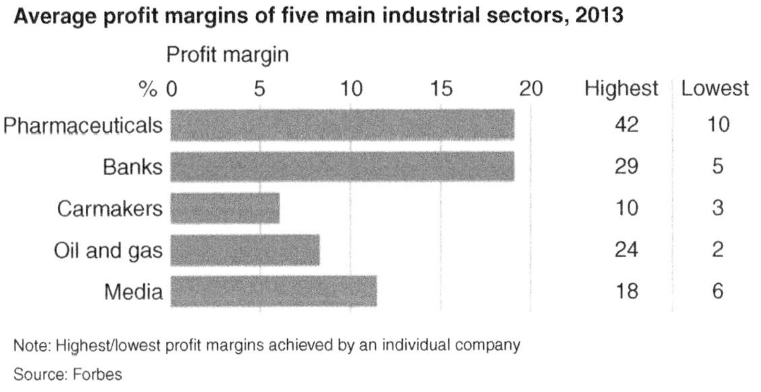


FIGURE 1: Pharmaceutical Profits, 2013[68]

65 *See* Jia Lynn Yang, *Maximizing Shareholder Value: The Goal That Changed Corporate America*, WASH. POST, August 26, 2013, https://www.washingtonpost.com/business/economy/maximizing-shareholder-value-the-goal-that-changed-corporate-america/2013/08/26/26e9ca8e-ed74-11e2-9008-61e94a7ea20d_story.html?utm_term=.493500d2be45.

66 *See, e.g.*, Rhona Finkel, *The 5 Most Profitable Medications Ever Produced*, May 24, 2014, http://www.drugsdb.com/blog/the-5-most-profitable-medications-ever-produced.html ("It is certainly true that – from the outside – it looks like drug companies live on easy street, . . . but the reality is that they invest billions of dollars every year into research and development. However, . . . only one out of every 5,000 to 10,000 compounds that the companies study in preclinical trials actually makes its way to market . . .. And of those, very, very few ever become blockbusters. You don't have to cry for them – but it is more complicated than it initially appears.").

67 *See* James Bessen and Michael J. Meurer, PATENT FAILURE 88 (Princeton University Press 2008) ("The higher prices that pharmaceutical firms charge while they are still 'on-patent' allow them to earn above-normal profits, or 'rents,' that more than recoup their development expenses.").

68 Richard Anderson, *Pharmaceutical Industry Gets High on Profits*, BBC NEWS, November 6, 2014, http://www.bbc.com/news/business-28212223.

that aim to identify the highest possible profit margin the market will tolerate."[69] A recent US Government Accountability Office study also notes that

> [a]bout 67 percent of all drug companies saw an increase in their annual average profit margins from 2006–2016. Among the largest 25 companies, the annual average profit margin fluctuated between 15 and 20 percent. For comparison, the annual average profit margin across non-drug companies among the largest 500 globally fluctuated between 4 and 9 percent.[70]

Consider, for example, Iminitib, also known as Gleevic®, a treatment for chronic myeloid leukemia or CML, a disease that used to be a death sentence but has been turned into a chronic condition. According to an article by one hundred CML experts, Iminitib was originally priced at thirty thousand dollars per year of treatment; in 2001, there were thirty thousand patients in the United States, which would generate nine hundred million dollars per year. At that rate, the developer would recoup the costs of development in its first few years on the market, and generate profits after that.[71] The drug was far more successful than anticipated at prolonging life for CML patients; the ten-year survival rate increased from 20 percent to 80 percent.[72] However, the price of Iminatib increased from thirty thousand dollars per year to ninety-two thousand dollars per year in 2012, even though all research costs were accounted for in the original price; new indications were developed, and the patient population increased dramatically.[73]

Despite having made tremendous profits from sales of Iminatib, Novartis, the owner of several patents on the drug, engineered the assertion of significant negative pressure on the Colombian government when it announced it was considering granting a compulsory license on the drug in 2015.[74]

High profits also encourage risk minimization. As a result, pharmaceutical companies have focused much of their marketing and development efforts either on drugs for rare indications for which exorbitant prices can be charged,[75] or on "me too" drugs with the same mechanism of action as a prior blockbuster drug and with the same active

[69] WHO Submission to the UN SG High Level Panel on Access to Medicines 5 (2016), https://static1.squarespace.com/static/562094dee4b0d00c1a3ef761/t/56e746279f7266a586c2b893/1457997352055/WHO_HLP_Submission_7Mar2016.pdf (citing Ed Schoonveld, The Price of Global Health (Routledge 2011)).

[70] US Gov't Accountability Office, Drug Industry Profits, Research and Development Spending, and Mergers and Acquisition Deals, GAO-18–40 (November 2017), https://www.gao.gov/products/GAO-18-40.

[71] Experts in Chronic Myeloid Leukemia, *The Price of Drugs for Chronic Myeloid Leukemia (CML) Is a Reflection of the Unsustainable Prices of Cancer Drugs: From the Perspective of a Large Group of CML Experts*, Blood J. (April 25, 2013), http://www.bloodjournal.org/content/121/22/4439?sso-checked=true.

[72] *Id.*

[73] *Id.*

[74] See *supra* note 21. Two sources of the pressure were the Swiss government (home to Novartis, the patent holder) and the US.

[75] Sara Jane Tribble and Sydney Lupkin. *High Prices for Orphan Drugs Strain Families and Insurers*, NPR, January 17, 2017, https://www.npr.org/sections/health-shots/2017/01/17/509507035/high-prices-for-orphan-drugs-strain-families-and-insurers.

ingredient, but perhaps a different form, dosing regime, and so forth. According to the US Government Accountability Office, in 2006, 60 percent of new FDA approvals were for "me too" drugs.[76]

"Neglected tropical diseases" such as malaria and tuberculosis affect more than one billion people each year.[77] Ten million people die from these diseases annually. Millions more are so incapacitated by disease that they cannot work, care for themselves, or care for their children. These diseases predominantly affect the poorest people in least developed countries.[78] Yet as the chart in Figure 2 from the European Patent Office shows, pharmaceutical R&D disproportionately focuses on chronic, non-communicable diseases, even though many more disability-adjusted life years are lost

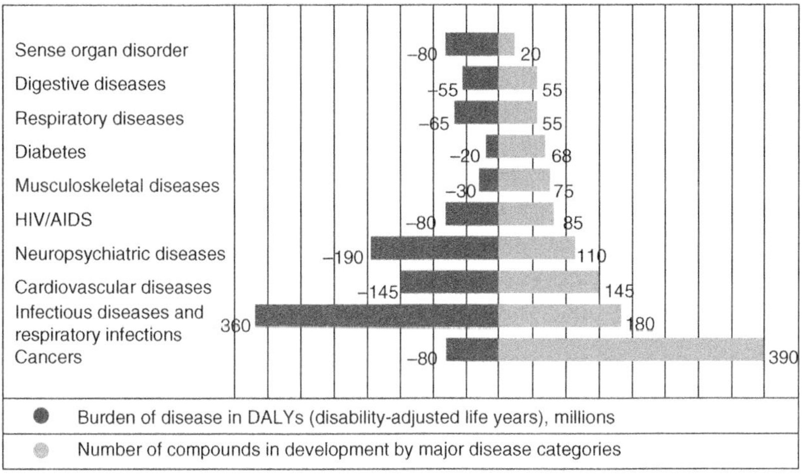


*Source: Courtesy of International Policy Network based on PhRMA (2005) and WHO (2004)*

**Above: R&D follows the money, not the disease** High profile diseases afflicting Western countries tend to attract the highest levels of research (as measured by the number of compounds in development). There is a particularly disproportionate level of research into cancers, diabetes and musculoskeletal diseases compared to their impact. Infectious diseases, which are paticularly prevalent in the developing world and cause by far the highest disease burden, ranks a distant second to cancers in terms of R&D.

FIGURE 2: The Global Disease Burden versus Number of Compounds in Development[79]

[76] US Gov't Accountability Office, GAO-07-49, New Drug Development: Science, Business, Regulatory, and Intellectual Property Issues Cited as Hampering Drug Development Efforts 17 (2006), https://www.gao.gov/new.items/d0749.pdf.

[77] Centers for Disease Control and Prevention, *Neglected Tropical Diseases*, https://www.cdc.gov/globalhealth/ntd/index.html.

[78] Suerie Moon, *Powerful Ideas for Global Access to Medicines*, N. Engl. J. of Medicine, February 9, 2017, http://www.nejm.org/doi/full/10.1056/NEJMp1613861 ("Whether low- and middle-income countries [LMICs] are struggling to treat millions of people living with HIV or to immunize refugee children against pneumonia, unaffordable prices mean that many people simply go without. Meanwhile, despite billions of public and private dollars invested in pharmaceutical research and development, urgent needs for new antibiotics and tools for other public health priorities go unmet.").

[79] European Patent Office, Scenarios for the Future 77 (2007), http://documents.epo.org/projects/babylon/eponet.nsf/0/63A726D28B589B5BC12572DB00597683/$File/EPO_scenarios_bookmarked.pdf.

for infectious diseases. So we as a global community are already not getting the best mix of drugs that we need.[80]

In 2014, a whopping 40 percent of new drugs approved by the FDA targeted rare diseases for which high prices and generous profit margins are common.[81] In 2012, such specialty drugs represented only one percent of prescriptions written nationwide, but accounted for 25 percent of the $263.3 billion spent on all prescription drugs.[82] With many of the other approvals being for "me too" drugs, we arguably are not getting enough of the kinds of drug innovations we need due to the current system's skewed incentives.[83]

But the adverse effects of the pharmaceutical industry's skewed incentives manifest in ways beyond a lack of cures for the Global South. Throughout the world, even in first-world countries, companies are incentivized to develop compounds that turn deadly diseases into chronic conditions so that people can live longer and keep taking the drug. Pharmaceutical companies have few incentives to develop cures for diseases.[84] This is not to say that they never develop cures – sometimes they do – but mostly they develop treatments because cures are problematic, generating a one-time payment, not a revenue stream. And if developers price cures too high, they may encounter significant resistance and censure in the marketplace. The controversial blockbuster hepatitis C cure Sovaldi® provides a timely example of this phenomenon.[85]

So what does this mean for compulsory licenses in developing countries? They are often both necessary and justifiable.[86] Pharmaceutical companies are not basing

80 *Id.* See also UN Report on Access to Medicines, *supra* note 11, at 13.

81 John Jenkins, *CDER Approved Many Innovative Drugs in 2014*, FDA Voice, Jan. 14, 2015, https://blogs.fda.gov/fdavoice/index.php/2015/01/cder-approved-many-innovative-drugs-in-2014/.

82 Laura Fegraus and Murray Ross, *Sovaldi, Harvoni, and Why It's Different This Time*, Health Affairs, November 21, 2014, https://www.healthaffairs.org/action/showDoPubSecure?doi=10.1377%2Fhblog20141121.042908&format=full.

83 *See* US Gov't Accountability Office, New Drug Development, *supra* note 76.

84 Finkel, *supra* note 66 (quoting Martin Kuehne, University of Vermont chemist) ("Pharmaceutical companies don't like cures. Really, they don't – that's the sad thing. They like treatment. Something for cholesterol or high blood pressure that you take for years and years, every day. That's where the profit is.").

85 See S. Prt. No. 114–20, Staff of S. Comm. on Finance, 114th Cong., *The Price of Sovaldi® and Its Impact on the U.S. Health Care System* (2015), https://www.finance.senate.gov/imo/media/doc/Srpt%20114-20.pdf (emphasis added). The Senate report detailed how Solvaldi® and its follow-on drug Harvoni® came to be priced so high, generating $26 billion for Gilead in less than two years on a product it acquired for $11 billion. The report concluded that "Gilead's marketing, pricing, and contracting strategies were focused on maximizing revenue – even as the company's analysis showed a lower price would allow more patients to be treated." *Id.* at 117. It is beyond the scope of this chapter to address the situation in the United States, where access to drugs is also becoming a very serious problem for reasons including, but not limited to, patents. Having compulsory licensing discussions in the United States adds a further complexity due to the fact that R&D decisions are based, in large part, on the US market.

86 As a Malaysian report justifying a compulsory license on Sovaldi® (Sofosbuvir) notes:

> Sofosbuvir ... is patented in Malaysia, which means it must be sold at the originator companies' proposed Malaysian price (US$12,000 for twelve weeks of treatment). Given that it is estimated that the drug can be produced at US$171 to US$360 for twelve weeks of treatment, at volume, pricing is likely to be based on maximisation of profit margins rather than any rational stratification based on country developmental levels (as claimed by pharmaceutical companies).

their R&D decisions on the possibility of generating profits in developing countries. They are basing them on the needs of the United States, Europe, and other highly developed markets where the bulk of their revenue originates.[87]

As such, whatever drug companies earn in poor countries is largely gravy. Bayer CEO Marijn Dekkers confirmed this when he noted, after calling the Nexavar® compulsory license "essentially theft":

> Is this going to have a big effect on our business model? No, because we did not develop this product for the Indian market, let's be honest. We developed this product for Western patients who can afford this product, quite honestly.[88]

This is not to suggest that the relatively small percentage of revenue is not a meaningful sum that companies would not wish to lose. However, this potential loss of revenue does not justify pressuring countries to forego compulsory licensing using theft rhetoric. In both the Thai and Indian compulsory license examples, the drugs were going to poor people who would not be able to buy the drug at the price the originator pharmaceutical company was charging. These were not lost sales; these sales *never would have been made*.

Moreover, in situations where the compulsory license is for public non-commercial use, the generic drugs would only be supplied to patients in the public health system; rich citizens would continue to buy the expensive branded drugs sold by the patent owner.[89] This is important because LMICs tend to have high income inequality, a factor that pharmaceutical firms count on in pricing their drugs. As Flynn et al. explain in an example involving South Africa, where the richest 10 percent of the population earn 58 percent of the income:

> The [pharmaceutical company] maximizes its sales in South Africa by selling at the price that only the top 10% can afford. At this price, the firm makes $814.6 million in total revenue. If the firm lowers its price to be able to make sales to 20% of the affected individuals (at $396 per patient), then it will sell twice as many medicines at a price less than half of the profit-maximizing price, earning substantially less ($435.6 million). As the monopolist continues to cut prices and raise production, revenues fall further at almost every level of output and corresponding price. In

Malaysian AIDS Council, At the Edge of a Miracle: The Hepatitis C Virus (HCV) Epidemic in Malaysia 5 (Fifa Rahman ed. 2017), http://www.mac.org.my/v3/wp-content/uploads/2017/07/HCV_Report_A4_FA_23052017-web.pdf.

87 According to Statista, in 2015, global pharmaceutical market revenues were $940 billion, up from $712 billion in 2008. Moreover, "[i]n 2016, the United States was still the largest single pharmaceutical market, generating almost 450 billion U.S. dollars of revenue. Europe was responsible for generating around 200 billion U.S. dollars." *Global Pharmaceutical Sales from 2015 to 2017, by Region*, https://www.statista.com/statistics/272181/world-pharmaceutical-sales-by-region/.

88 Gokhale, *supra* note 43.

89 *See* Ho, Access to Medicines, *supra* note 16, at 137 (defending Thailand's compulsory license on heart treatment Plavix with respect to "patients covered by government health insurance, who are typically the lowest-income citizens").

other words, the firm will maximize its profits by setting a price *unaffordable for at least 90% of people in need.*[90]

Because wealthy patients in LMICs would still purchase the originator pharmaceutical company's products, compulsory licenses are not likely to meaningfully affect even that further 20–25 percent of revenue in LMICs. This also shows that there is little incentive for pharmaceutical companies to lower their prices in LMICs despite the vast numbers of people in need of access to their medicines.

This precise scenario resulted in the catastrophic loss of life on the African continent due to the pricing of HIV/AIDS drugs out of the reach of millions of poor people in the late 1990s and 2000s. As vividly described in the movie *Fire in the Blood*, the entire continent of Africa accounted for only 1 percent of originator pharma HIV/AIDS drug sales.[91] Yet the companies took a hard line, keeping their prices at ten to fifteen thousand dollars per patient per year in Africa, just as in the United States, and fought against the introduction of generic drugs while millions of lives were lost.[92]

### B. The Moral "Rightness" of Compulsory Licensing

Thus it is worth asking: if the incentive to develop new drugs is not meaningfully[93] affected by developing country markets, should a country allow a drug patent (one most likely obtained by a foreign entity) that had little or nothing to do with incentivizing research on that drug, and on which the drug developer did not anticipate making a substantial profit, to restrain competition and result in lost lives? Should we watch people die when it costs nothing, or very little, to let them live?[94] Or, to put it in the context of *pe'ah*, shouldn't the poor have a right to needed drugs when such use of the patented invention will not harm the patent holder or future innovations?

[90] Sean Flynn, Aidan Hollis, and Mike Palmedo, *An Economic Justification for Open Access to Essential Medicine Patents in Developing Countries*, 37 J.L. Med. & Ethics 184, 189 (2009) (emphasis added).

[91] *See* www.fireintheblood.com.

[92] *Id.*

[93] This is not to say that there is no possible effect on innovation at all. Nevertheless, as Flynn et al. note (*supra* note 90, at 192):

> [F]or markets in ... most developing countries[,] the patent system will be ineffectual in delivering much innovation ... Ultimately, the problem of finding an adequate and equitable mechanism to fund research and development for medicines in developing countries must be found elsewhere.

[94] There is also a self-interest argument for compulsory licenses in the infectious disease context. We live in a world where we increasingly see global pandemics. SARS, H1N1, MERS, bird flu, swine flu, Ebola, Zika: the list keeps growing. These diseases cross borders and traverse oceans. As such, the health of the developing world affects us all. To the extent more people in the developing world are immune compromised or lack treatment, there may be a greater spread of such diseases, or the development of mutations that may affect the ability of HICs to fight these pandemics when they reach their shores.

So to return to the story at the beginning of this chapter, is it possible that Heinz has a right to the drug for his wife? Or, to put it into a patent context, that a government has a *duty* to allow a third party to provide a drug, via compulsory license or some other mechanism, to its citizens in a way that will not harm the patent owner? Certainly. When we consider the broader meaning of theft illuminated by the biblical concept of *pe'ah*, we should be more willing to conclude that an LMIC issuing a compulsory license is not theft and should not be characterized as such. We also should be able to recognize that the poor do have a right to free or low-cost life-saving drugs where providing such drugs does not unduly harm the drug developer.

Moreover, if there is virtue in thinking that Heinz's situation should not be viewed as theft, issuing a compulsory license in accordance with TRIPS should not be viewed as a morally culpable action at all; it is far removed from theft. It is not even defined as stealing under international law, and it involves compensation to the patent owner. In addition, the very legality of compulsory licenses channels people away from the kinds of "self-help" mechanisms contemplated by Heinz that can lead to interpersonal violence and anarchy. In a sense then, compulsory licenses can be seen as anti-mayhem devices in addition to their other life-saving virtues.[95] Yet they too often are characterized as theft in a way that appears to give pharmaceutical companies the moral high ground, allowing them to play the victim in terms of public relations and inciting governmental action against offending countries. In fact, it seems appropriate to turn the tables and label, from a moral perspective, the pharmaceutical companies trying to keep needed drugs from the poor as "thieves."

If, to some extent, our notions of the moral wrongness of stealing derive from the Ten Commandments, we should look at those commandments in context, not in isolation. As Professor Gnuse explains:

> Though the gleaning laws are not extensive, they provide us with significant insight into the command against theft. What the modern mind might call theft was not so defined in the Old Testament ... Human need had a right of access to the basic essentials of life ... For the poor to take food from another person's land was not theft, but it was wrong for the more affluent person to withhold it.[96]

Using the label of theft to describe a particular set of actions and simultaneously playing the innovation card effectively erects a barrier that may limit a government's ability to appropriately utilize compulsory licensing to meet pressing public health needs. So the hard questions regarding who really is stealing from whom do not get asked. This theft framing has no correlation to the most productive innovation, nor to providing optimal innovation; rather, it is employed to produce optimal profit.[97]

95 Special thanks to Joe Miller for illuminating this point.

96 Robert Gnuse, You Shall Not Steal: Community and Property in the Biblical Tradition 29 (Wipf and Stock 2011).

97 Ramsi Woodcock, *Property, Efficiency, the Commons, and Theft*, in Research Handbook on Political Economy and Law 561 (Ugo Mattei and John D. Haskell eds., Edward Elgar 2015).

In view of the access to medicines crisis occurring in countries around the globe, it is critically important for us to tear down that barrier.

## V. CONCLUSIONS

The misapplication of a Biblical precept in the context of access to medicines tools like compulsory licensing in the Global South is profoundly disturbing, especially in light of the deafening silence that prevails with respect to the use of compulsory licenses in the Global North. When we consider the biblical analogy of *pe'ah*, it allows us to begin to reframe the discussion and ask who is "stealing" from whom. Are the poor are being robbed of what they are due? These are not easy issues to address, and there are no risk-free solutions, no clear way to know exactly how to optimize access and innovation at the same time. But we may be better able to achieve win-win outcomes if we open our eyes to other ways of viewing competing interests, and allow our legal discussions to be enlightened by analogies from important traditions that have in the past informed our constructions of right and wrong, of morality itself.

The production of essential medicines is important and worthy of powerful incentives because they may improve the duration and quality of human life. But without proper controls, patents and other incentives may produce harms, which is why the patent system has always included various limits and safety valves on the scope of the patent grant. If we consider a just law to be one that improves the lives of human beings, then drug compulsory licenses issued in accordance with international law are not only legal, but they are also moral and just. Governments have a moral obligation to provide access to life-saving treatments to their citizens, and the use of incomplete and dangerous theft rhetoric to stigmatize and denigrate such efforts undermines that duty in profoundly important ways.

Both patent law and the Biblical commandment against stealing exist within larger systems of rights and obligations. Recognizing this allows us to put together, as Christ did, the two phrases "thou shalt not steal" and "thou shalt love thy neighbor as thyself." Blending these two phrasings of the commandment allows us to view theft in relation to pharmaceutical patents in its proper light. We can seek and develop appropriate exceptions to patent rights that do not eviscerate protection, but provide the balance society needs to enhance access to essential medicines. Perhaps then, in the pharmaceutical context at least, we will begin to see "justice run down like water, and righteousness like a mighty stream."[98]

[98] Amos 5:24 (NKJV).

12

# Genetic Resources and Patents: In Search of Ethical Solutions to Global IP Discord

*Roman Cholij*

This paper considers the question of patents and genetic resources, as currently being discussed in various international fora, in the light of a broad understanding of the function of intellectual property and of Catholic Social Teaching.[1] The particular issue concerns the proposal by developing countries, also supported by the Vatican (Holy See), to require, in an international legal instrument, mandatory disclosure of the origin of genetic resources and associated traditional knowledge (TK) in patent applications, with a view to realising in an effective way fair and equitable benefits sharing as required by the Convention on Biological Diversity (CBD).

## I. THE CONVENTION ON BIOLOGICAL DIVERSITY AND THE ISSUES SURROUNDING MANDATORY DISCLOSURE IN PATENT LAW

The CBD is an international legal instrument for the conservation and sustainable use of biological diversity, presented for signature at the Rio 'Earth Summit' in 1992.[2] It asserts the sovereign rights of nations over their national resources, and their right to determine access according to national legislation with the aim of facilitating the sustainable use of these resources, promoting access and their common use. It requires that access to genetic resources be on the basis of prior informed consent, and on mutually agreed terms that provide fair and equitable sharing of the results of R&D and the benefits of commercialisation and utilisation.[3] It also calls for the fair

[1] This chapter is drawn from the report by Roman Cholij, 'Genetic Resources, Patents, and Catholic Social Teaching', in *Patents on Genetic Resources?*, Caritas in Veritate Foundation, 2014, pp. 15–63, http://www.fciv.org/downloads/FCIV%20WP3%20Patents%20on%20Genetic%20Resources.pdf. Portions of the report discussing Catholic Social Teaching have been greatly condensed here; readers can find fuller discussion in the report and in the Neves, Colecchi, and Wojda chapters.

[2] The text of the Convention (1992) and the texts of its further developments – the Cartagena Protocol on Biosafety (2000) and the Nagoya Protocol on Access and Benefit Sharing ('ABS') – can be accessed at http://www.cbd.int/.

[3] 'Genetic resources' (GR) are defined in the CBD as 'genetic material of actual or potential value'. Genetic material in turn is defined as 'any material of plant, animal, microbial or other origin containing functional units of heredity'. CBD, *supra* note 2, Article 2.

and equitable sharing of the benefits derived from the use of traditional knowledge (TK). The custodians of genetic resources with associated TK are (the often marginalised) indigenous peoples and/or poor local communities located, often but not always, in developing or least developed countries.

In 2008, indigenous peoples from around the world gathered together before the 34th G8 Summit in Hokkaido, Japan. The *Nibutani Declaration* that followed called upon the G8 nations to 'stop the theft and piracy of our traditional indigenous knowledge, traditional cultural expressions . . . bio-genetic resources including our human genetic resources, by biotechnology corporations, cultural industry, and even by States and individual scientists and researchers'.[4]

### *Permanent Sovereignty and Conservation of Genetic Resources*

To anchor the rights of indigenous and local communities over the biological material in their custody, the CBD begins with the principle of *permanent sovereignty*, a fundamental feature of international law and of the nation state system. Independent states have permanent sovereignty over their territories, including their natural biological resources. This is explicitly upheld in the CBD, although, by way of balance, it also states that 'States are responsible for conserving their biological diversity and for using their biological resources in a sustainable manner', biological diversity being a common concern of mankind. The CBD underlines that biological life is a resource to which access can be given to another by a custodian/provider, rather than being considered part of a common heritage (except in a loose ethical sense).

The relation of indigenous people and local communities to the State with regard to property ownership, including land and its biological and natural resources, is dictated by national law subject to obligations of international treaties, including the CBD.[5] It is an internationally accepted principle that indigenous peoples should fully participate in decisions that affect their livelihood and development including, implicitly, arrangements of benefit sharing with regard to third-party access to the genetic resources under their custody.[6] The institutional proprietary nature of biological life as found on land within a territory, determined by international and

[4] *Nibutani Declaration of the 2008 Indigenous Peoples Summit in Ainu Mosir*, § 17, http://www.tebtebba.org/index.php/content/100-nibutani-declaration-of-2008.

[5] CBD, Article 8(j).

[6] The Indigenous and Tribal Peoples Convention (1989), Article 7.1, states:

> The peoples concerned shall have the right to decide their own priorities for the process of development as it affects their lives, beliefs, institutions and spiritual well-being and the lands they occupy or otherwise use, and to exercise control, to the extent possible, over their own economic, social and cultural development. In addition, they shall participate in the formulation, implementation and evaluation of plans and programmes for national and regional development which may affect them directly.

Available at https://www.ilo.org/dyn/normlex/en/f?p=NORMLEXPUB:12100:0::NO::P12100_ILO_CODE:C169.

national law, permits control of access to what are now regarded as 'resources' in the CBD. This control is not necessarily in conflict with the idea of biological or genetic resources being part of the 'common heritage' of mankind, in an ethical sense – since in controlling access, the authorities are also bound to act with a mind to the common or public international good. Without an appropriate balance, allowing free exchange of information based on reasonable access, situations could arise such as that of the assertion of 'viral sovereignty', where a nation prevents access for foreign medical researchers to new viruses that could potentially cross borders and cause pandemics affecting all peoples of the world.[7]

A related proprietary issue is the protection against misappropriation of TK as such, a knowledge that reflects historical and on-going communal shared-inventiveness regarding, inter alia, the uses of genetic resources, where such knowledge is used without consent.[8] A 2010 report of the United Nations Conference on Trade and Development (UNCTAD) expresses concern over

> a lack of balance in the current intellectual property rights (IPR) system ... One type of IP – that generally produced and owned by entities in developed countries – is well protected. That category of IP in which developing countries have comparative advantage, namely TK, is generally considered free for all takers.[9]

Among its objectives, the CBD (in Article 1) aims to ensure 'the fair and equitable sharing of the benefits arising out of the utilization of genetic resources, including by appropriate access to genetic resources and by appropriate transfer of relevant technologies'. This objective arises from concerns about 'biopiracy' or unauthorised bioprospecting.

### *Biopiracy or Bioprospecting?*

'Biopiracy' through the grant of a patent can arise in two different ways:

1. ***The granting of 'wrong' patents.*** These are patents granted for inventions that are either not novel or are not inventive having regard to traditional knowledge already in the public domain. Such patents may be granted due to oversights during the examination of the patent or simply because the patent examiner did not have access

[7] In Indonesia in 2008, for example, the Health Minister of the time asserted that deadly viruses are the sovereign property of individual nations – in the context of Indonesia having had, according to the WHO, fifty-three outbreaks of bird flu viruses appearing in both humans and chickens. Cf. Richard Holbrooke and Laurie Garrett, '"Sovereignty" That Risks Global Health', *Washington Post*, 10 August 2008, http://www.washingtonpost.com/wp-dyn/content/article/2008/08/08/AR2008080802919.html??noredirect=on.

[8] Called a 'silent theft of centuries of knowledge from some of the poorest communities in developing countries'. *Human Development Report (1999)*, United Nations Development Programme (UNDP), p. 6, http://hdr.undp.org/sites/default/files/reports/260/hdr_1999_en_nostats.pdf.

[9] UNCTAD, *Intellectual Property in the World Trade Organization: Turning It into Developing Countries' Real Property*, 2010, p. 42, http://unctad.org/en/docs/ditctncd20068_en.pdf.

to the knowledge. This may be because it is written down but not accessible using the tools available to the examiner, or because it is unwritten knowledge.

2. ***The granting of 'right' patents.*** These are patents that may have been correctly granted according to national law on inventions derived from a community's traditional knowledge or genetic resources. It can be argued this constitutes 'biopiracy' on the following grounds:

- First, patenting standards may be too low. Patents are allowed, for instance, for inventions which amount to little more than discoveries. Or the national patent regime may not recognise some forms of public disclosure of traditional knowledge as prior art defeating a patent (e.g., until recently, the United States refused to recognise oral disclosures overseas).
- Second, even if the patent represents a genuine invention, no arrangements may have been made to obtain the prior informed consent (PIC) of the communities providing the knowledge or resource, and for sharing the benefits of commercialisation with them in accordance with the CBD.

The term 'bioprospecting', on the other hand, covers the practice of collecting and screening plant and other biological material for scientific or industrial purposes. It may also involve collecting associated local knowledge, such as from healers. It has also been argued that bioprospecting plays a wider role in encouraging the preservation of biodiversity.[10]

A continuum of activities, ranging from the criminal to the legal but perhaps unethical, have been classed as biopiracy. Examples may be unauthorised and/or uncompensated use of common TK, the deceptive acquisition of TK, the commercial use of TK based on a literature search without compensating the source holders, claiming TK within a patent in the form it was acquired; claiming a refinement of TK, claiming an invention based on TK, the unauthorised and uncompensated extraction and use of widespread genetic resources, or a patent claim claiming the resource or purified version of the resource or a derivative.[11]

An oft-cited example of 'biopiracy' is that involving neem (*Azadirachta indica* or 'Indian Lilac'), a plant that is endemic to the Indian subcontinent and traditional medicine practitioners for its various therapeutic properties. The neem tree has been the subject of a considerable number of patents. At the time of a 2004 research project, there were more than 40 in the United States alone and at least 150 worldwide. Traditional knowledge was used as a starting point in each case. This appropriation (or misappropriation) of TK together with a GR has aroused considerable

10 Graham Dutfield, *Bioprospecting: Legitimate Research or Biopiracy?*, Policy Briefs SciDev Net, 26 May 2003.

11 See Graham Dutfield, *Intellectual Property, Biogenetic Resources and Traditional Knowledge* (Abingdon: Earthscan, 2004), 17–20.

controversy, especially in India.[12] One patent, a '[m]ethod for controlling fungi on plants by the aid of a hydrophobic extracted neem oil', owned by a US company, was revoked at the European Patent Office when challenged by activists, based on evidence supplied to show that there was a lack of novelty and inventive step.[13]

Another patent on a '[m]ethod of promoting healing of a wound by administering turmeric to a patient afflicted with the wound', owned by the University of Mississippi, involved turmeric, a herbaceous plant native to tropical South Asia, likewise known through traditional knowledge to have healing properties.[14] This was similarly challenged successfully, based on evidence of prior art – in this case, an ancient Sanskrit text – that was used by the Indian government to show lack of novelty.[15]

In the light of these and other celebrated examples, together with the Prior Informed Consent and Access and Benefit Sharing arrangements of the CBD, developing countries (in particular) have sought an effective international legal instrument that would embrace the objectives of the CBD in patent systems through a strict requirement of disclosure.

### *Incorporating CBD Obligations into Patent Law*

The most obvious and the most appropriate place to integrate the IP-related objectives of the CBD is the Trade-Related Aspects of Intellectual Property Rights (TRIPS) Agreement, as this is global in its extent and has a well-established and effective enforcement mechanism potentially involving trade sanctions. However, the developed countries – in particular the United States, the principal architect of the agreement – argue that since TRIPS says nothing about the CBD, nor indeed the CBD about TRIPS, there can be no conflict between the two agreements. As patenting often leads to commercialisation, which generates the benefits that are a prerequisite to any benefit-sharing arrangement, it is further argued that TRIPS actually supports the CBD in its objectives.

The developing countries, NGOs, academics, independent studies, and some sympathetic developed countries have to the contrary argued that, since patenting based on the use of genetic resources is allowed under TRIPS (subject to meeting

[12] Report on Disclosure of Origin in Patent Applications, Queen Mary Intellectual Property Research Institute for the European Commission, DG-Trade ('QM Report'), October 2004, p. 21, http://trade.ec.europa.eu/doclib/docs/2005/june/tradoc_123533.pdf. The search term 'neem', entered into the English-language database of Espacenet of the European Patent Office on 13 January 2019, produced 323 patents worldwide with 'neem' in the title.

[13] European Patent No. 0436257 B1, 1994. See *Thermo Trilogy Corporation, et al. v. Aelvoet, Magda, MEP, The Green Group in the European Parliament, et al.*, Boards of Appeal of the European Patent Office, T 0416/01 – 3.3.2, 8 March 2005.

[14] U.S. Patent No. 5, 401, 504, 1993.

[15] There are a number of other dubious applications on the patents' database for patent protection, based on turmeric, including from Korea to protect such 'inventions' as 'chocolate containing turmeric and method for preparing the same', 'turmeric vinegar' and similar.

patentability criteria), this does not support the objectives of the CBD because the criteria do not include prior informed consent or mutually agreed terms for benefit sharing. Nor is there any provision for the protection of traditional knowledge as such. The absence of these requirements and provisions, it is argued, in fact encourage biopiracy and inequitable behaviour towards poorer nations and the underprivileged.

In 2002, the Sixth Meeting of the Conference of the Parties (COP) to the CBD adopted the Bonn Guidelines to address access to genetic resources and fair and equitable benefit sharing arising from use of those resources.[16] In the Bonn Guidelines, the CBD COP invited Parties and governments to encourage disclosure of the country of origin of genetic resources and associated traditional knowledge in applications for intellectual property, where the subject matter of the application concerns or makes use of such knowledge in its development. The types of disclosure required within the patent applications would be principally: a) genetic resources utilised in the development of the claimed inventions; b) the country of origin of such genetic resources; c) associated traditional knowledge (ATK), innovations and practices utilised in the development of the claimed inventions; d) the source of ATK, innovations, and practices; and e) evidence of prior informed consent.

Disclosure is an already established prerequisite of patent law. A patent application may already need to disclose GR or TK and their source as the way to enable a person skilled in the art to carry out the invention or its best-known mode, or to allow for a proper examination of novelty and/or non-obviousness. These are all conventional disclosure requirements. The disclosure requirements relating to the CBD, however, go beyond these to include not only the link between the invention as such and GR or TK, but the legal relationship between the applicant/inventor and the access to GR or TK.

A 2004 WIPO technical study describes how the implementation of the CBD objectives could be achieved via the patent system, including an analysis of application procedures and requirements and the various options of the consequences of failure to fully disclose.[17] From the study, it is apparent that the two regimes could be integrated, but it does not offer any direct recommendations. In contrast, an UNCTAD study authored by Professors Joshua Sarnoff and Carlos Correa is very clear about the need for mandatory disclosure requirements and why the TRIPS agreement, ideally, should in their view be amended accordingly:

> An international system of mandatory disclosure of origin requirements is needed to prevent misappropriation of genetic resources and associated traditional knowledge, to promote compliance with CBD access and benefit sharing requirements,

[16] *Report of the Sixth Meeting of the Conference of the Parties to the Convention on Biological Diversity*, 2002, UNEP/CBD/COP/6/20.

[17] WIPO *Technical Study on Patent Disclosure Requirements Related to Genetic Resources and Traditional Knowledge*, WIPO, 2004, Study no. 3, https://www.wipo.int/edocs/pubdocs/en/tk/786/wipo_pub_786.pdf.

> and to prevent misuse of the intellectual property system . . . Objections raised to mandating adoption of disclosure of origin requirements through new international treaty provisions either do not stand up to analysis or do not outweigh the benefits to be obtained.[18]

Likewise, an independent international study commissioned by the British Government in 2002 also supports a mandatory disclosure of information of the geographical source of genetic resources with sanctions applied in cases where patentees fail to make this disclosure or deliberately mislead about the source. It also recommends a system whereby patent offices examining patent applications that identify the geographical source of genetic resources or traditional knowledge pass on that information, either to the country concerned or to WIPO, which may act as a depository for patent-related information on alleged biopiracy.[19]

Several countries in fact have implemented CBD obligations through their national laws, requiring applicants for intellectual property to disclose the source and country of origin of genetic resources and associated traditional knowledge, along with relevant documentary information regarding compliance with access and benefit sharing (ABS) requirements.[20] However, '[d]eveloping countries, in particular, do not have the resources to follow each and every patent issued outside their territories on the use of their resources'.[21] Similarly, indigenous and local communities typically lack the resources to effectively enforce patent and other rights relating to genetic resources and traditional knowledge. New international treaty provisions imposing mandatory disclosure of origin obligations would reduce uncertainties regarding recognition and enforcement of such national disclosure requirements in foreign intellectual property applications, and would thereby reduce the burdens and costs of preventing and remedying biopiracy and misappropriation.[22]

The goal of cross-border harmonisation on ABS regimes as obligated by the CBD has been significantly advanced by the Nagoya Protocol (2010), a side agreement of the CBD, ratified by 114 parties, including the UK and EU.[23] It provides for an international regime on access to genetic resources and benefit sharing, at the same time providing for greater legal certainty and transparency for both providers and users of genetic resources.

18 UNCTAD, *Analysis of Options for Implementing Disclosure of Origin Requirements in Intellectual Property Applications*, 2006, UNCTAD/DITC/TED/2004/14.

19 Commission on Intellectual Property Rights, *Integrating Intellectual Property Rights and Development Policy*, 2002, pp 86–7, http://iprcommission.org/papers/pdfs/final_report/ciprfullfinal.pdf.

20 Such as India, Brazil, China, the Andean community. See also responses to a questionnaire on this topic: WIPO/GRTKF/IC/Q.3, Part 1, Annex; Michael Blakeney, *Proposals for the Disclosure of Origin of Genetic Resources in Patent Applications*, Queen Mary Intellectual Property Research Institute. WIPO/IP/GR/05/01, Annex.

21 Cf. Submission to WTO, 24 June 2002, § 12, WTO/IP/C/W/356.

22 UNCTAD, *Analysis of Options*, *supra* note 18, pp. 21–2.

23 Nagoya Protocol, https://www.cbd.int/abs/about/default.shtml.

WIPO, however, is today the main centre for discussions and negotiations on the CBD compliance disclosure requirement. The differences, broadly speaking, between North and South, or developed and developing countries, are manifest in the manner in which the 'Consolidated Document Relating to Intellectual Property and Genetic Resources',[24] the negotiation text of the IGC, is divided into two sets of diametrically opposed propositions. In the latest text, the first proposition found in Articles 2–6 is under the heading '[MANDATORY] DISCLOSURE'. There then follows a section with the second opposed proposition entitled '**ALTERNATIVES TO ARTICLES 2–6, NO NEW DISCLOSURE REQUIREMENT**'. The position today, as before, at the TRIPS Council is that delegates from the developing countries, such as China, the African Group, Brazil, India and Peru from the cross-regional 'Like-Minded Countries' and the Development Agenda Group (DAG)[25] are for mandatory disclosure, while the developed countries, in particular the United States – backed by a small handful of other countries (such as Japan, Canada and South Korea) – maintain a 'no change to the patent system' position (except in regard to facilitating better examination procedures to eliminate 'bad' patents through uses of TK databases, as proposed particularly by Japan). The EU has a softer position than the United States and in principle is in favour of a disclosure requirement of some sort which is uniform globally, providing transparency, but which does not act as a de facto or de jure formal or substantial patentability criterion. Legal consequences of non-compliance, it is suggested, would generally lie outside the ambit of patent law.[26]

The United States maintains its position that patent laws are not the appropriate means to address the misappropriation of genetic resources and traditional knowledge. While claiming to be supportive of the objectives of the CBD in principle,[27] it believes the best way of achieving these ends is through contractual obligations which might include disclosure of origin within a patent application, the absence of which would become a breach of contract but would not invalidate the patent.[28] The

[24] *Consolidated Document Relating to Intellectual Property and Genetic Resources (Rev. 2)*, 10 April 2018, https://www.wipo.int/meetings/en/doc_details.jsp?doc_id=404077. Further work on the text has since been done. See Catherine Saez, 'WIPO IP and Genetic Resources Committee Makes Progress Despite Block at End,' IP Watch, 2 July 2018, http://www.ip-watch.org/2018/07/02/wipo-ip-genetic-resources-committee-makes-progress-despite-block-end/.

[25] The members of the 'Like-Minded' group of countries include Algeria, Angola, Bangladesh, Botswana, Brazil, Ecuador, Egypt, Guatemala, India, Indonesia, Iran, Malaysia, Mexico, Nepal, Oman, Peru, the Philippines, Senegal, South Africa, Sri Lanka, Thailand, Yemen, Zambia and Zimbabwe The DAG is a group of developing countries committed to mainstreaming the development dimension into all areas of WIPO's work. Members also overlap with those of the Like-Minded group. Proposals from the developing countries can be seen at WIPO/GRTKF/IC/20/INF/12 (African Group); WIPO/GRTKF/IC/20/20/6 (Like-Minded Countries). We might note, also, that some developing countries (e.g. Bolivia and Venezuela), and the indigenous communities, are opposed in principle to the patenting of living organisms, even if modified by human intervention.

[26] IP/C/W/383, § 53. QM Report, *supra* note 12, p. 38.

[27] Although the US has not in fact accepted or ratified it.

[28] QM Report, p. 43.

United States and Northern countries are prolific filers of patent applications, and the greatest users of the PCT System for patent filings administered by WIPO, whereas Southern countries account for a small fraction of applications. In 2017 the whole African continent accounted for just 475, or 0.2%, of the 243,500 total applications using the PCT system, these coming mostly from South Africa.[29] Not only is it therefore not in the particular interest of the United States or Northern countries to address these issues through the patent system, but it is argued new degrees of legal uncertainty would be introduced, as well as administrative and fiscal burdens that would not be justifiable if there are alternative mechanisms available to achieve the same end. Prior informed consent and recording of interests in inventions that arise from access to or use of genetic resources are not relevant to traditional patentability criteria unless they impact on prior art and/or enabling disclosure to 'work' the invention. There is no vision of evolving the system to include new pro-development objectives.

As discussed previously, however, these arguments do not appear to be either insurmountable or particularly compelling given the reasons which argue in favour of amending the global patent regime in view of the common good of the international community and the benefits to be achieved (as discussed later). Mandatory disclosure has been seen to be workable in various countries, including Switzerland.[30] The Vatican (or more properly the 'Holy See')[31] has weighed in on this debate too in favour of the position of developing countries. It has done so in its capacity as permanent observer at the UN and voting delegate of the WIPO IGC in

29 *PCT Yearly Review, 2018*, WIPO, pp. 24–5, www.wipo.int/edocs/pubdocs/en/wipo_pub_901_2018.pdf. Statistics for 2017 show China only second to the United States in PCT filings (48,882) – closely followed by Japan. US-based applicants have filed the largest number of PCT applications every year since the PCT system began operating in 1978.

30 Cf. the Swiss intervention at (IWG 3): WIPO/GRTKF/IC/20/5, § 83.
For further reading on this topic of patents and mandatory disclosure, see, for example, C .A. Conde Gutiérrez, *Access to Genetic Resources and Intellectual Property Rights: Opportunities and Perils for Developing Countries Rich in Biodiversity and the Pharmaceutical Industry*, M. Phil Thesis, University of Sheffield School of Law, May 2015, etheses.whiterose.ac.uk/11796/1/SUPER%20FINAL%20VERSION.docx; Donna O. Perdue, 'Patent Disclosure Requirements Related to Genetic Resources: The Right Tool for the Job?', in *Biotechnology Law Report* 36(6):285–6 (December 2017); Wallace Feng, 'Appropriation without Benefit-Sharing: Origin-of-Resource Disclosure Requirements and Enforcement under TRIPS and the Nagoya Protocol', *Chicago Journal of International Law*, vol. 18: no. 1, Article 7 (2017), http://chicagounbound.uchicago.edu/cjil/vol18/iss1/7.

31 The Holy See is the universal government of the Catholic Church, headquartered in the Vatican city-state, the head being the bishop of Rome, the Pope. Before the UN it has permanent observer status with all the rights of a full member, except for a vote in the UN General Assembly. Its work in the field of intellectual property is described in Silvano M. Tomasi (former Apostolic nuncio before the UN in Geneva), *The Vatican in the Family of Nations: Diplomatic Actions of the Holy See at the UN and Other International Organizations in Geneva* (Cambridge: Cambridge University Press, 2017), III,3: 'Intellectual Property: Production and Inventive Activity for the Sake of the Common Good', pp. 415–49.

view of its understanding of the social function of intellectual property and its tradition of Catholic Social Teaching.

## II. THE WIDER CONTEXT OF INTELLECTUAL PROPERTY: ITS SOCIAL FUNCTION AND TRIPS

In broader debates over the international IP regime, one side sees IP as a mere economic tool to drive prosperity and, in the international context, as a tool of promotion of trade that in practice most benefits rights holders of developed countries. This traditional emphasis, born of the economic theories predominant in Northern countries, contrasts with the newer developments of viewing IP much more as having a more rounded social purpose in society, including international society. These different emphases on the nature and purpose of IP, which we might call the 'old' versus the 'new', are at the heart of the divisions between North and South at both WIPO and the WTO on IP matters that impact on the societies of the developing world. This includes matters on the patenting of living forms, where there are deep divisions, and the willingness to adapt the IP system to accommodate the newer thinking on the purpose of IP in relation to the objectives of the CBD.

Traditionally, and in most countries, IP has been thought of exclusively in its economic rationale: 'Intellectual property protects applications of ideas and information that are of commercial value.'[32] The grant of a patent monopoly is given in order to stimulate innovation through the reward to an innovator of a legal property right that facilitates market advantage through a time-limited monopoly. This is given in return for sharing publicly details of inventiveness which are socially useful, and which benefit society through public disclosure of new and useful information that might stimulate yet further innovation by others and allow advancement through technology transfer, usually by licensing. This arrangement, which is in essence a legal 'property' fiction – namely a bundle of rights that can be bought and sold and transferred like real property – is a long-established 'quid pro quo' arrangement with society that, although creating a temporary market inefficiency, has generally been seen, especially with its own appropriate checks and balances, to facilitate economic progress. It is there to protect legitimate economic interests, while also being an encouragement to benefit, with the approval of society, from one's own intellectual labours.

A prominent issue with the traditional utilitarian approach to understanding intellectual property – incentivising the creation of more 'knowledge goods' within the conceptual framework of markets and the economy – is that it does not embrace 'non-utility' concerns, such as rights, freedoms and the ability to participate in the process of knowledge creation. Utilitarianism does not ask the question who makes

[32] W. Cornish and D. Llewelyn, *Intellectual Property: Patents, Copyright, Trade Marks and Allied Rights*, 6th ed. (Sweet & Maxwell, 2007), p. 6.

the goods or whether they are fairly distributed in society to all who need them. A broader understanding is needed to see IP as an end and a means to integral human development.[33]

## *The International IP Regime of TRIPS and Its Impact on Developing Countries*

The international aspects of IP, especially in relation to developing countries, contextualise our previous discussions on genetic resources, IP and disclosure. All members of the World Trade Organization (WTO) are at the same time signatories to the TRIPS agreement that emerged from the Uruguay Round of the General Agreement on Tariffs and Trade (GATT) negotiations which took place between 1986 and 1994. It represented a victory for industry groups and multinational companies who had lobbied hard for raised international IP standards and increased protection of their intellectual asset interests in developing countries.

The drive for TRIPS was fuelled by major social and economic interests, including the estimated US$650 billion per year global pharmaceutical industry, the estimated $21 billion per year commercial seed industry and the estimated $800 billion per year global software and entertainment industries. These industries – where often just a handful of companies monopolised the markets – were thoroughly dependent on securing IP protection. These corporations, together with other research centres and individuals from developed countries, held more than 80 per cent of the world's IP rights, making the negotiating table an uneven playing field.[34]

At the negotiating table, the proponents of stronger IP protection for developing nations argued that this would encourage foreign direct investment (FDI), innovation, and technology transfer, and spur the development of national cultural and creative industries. Given the growing trade in counterfeit medicines and other products, stronger IP protection was presented as a way to help protect public health and safety.

On the other hand, critics of this approach warned that while stronger IP protection *might* foster such outcomes in some cases, it would require the right conditions, carefully tailored policies and laws, and a range of complementary measures. In fact, the critics warned, industrial development could actually be slowed by constraining opportunities to copy and adapt technologies. Many developing countries have sought to employ the same strategies of copying and reverse-engineering that had served developed countries at similar stages of development.

[33] Cf. Tzen Wong, 'Intellectual Property through the Lens of Human Development,' in *Intellectual Property and Human Development: Current Trends and Future Scenarios*, eds. Tzen Wong and Graham Dutfield (Cambridge: Cambridge University Press, 2011), p. 31.

[34] Carolyn Deere, *The Implementation Game: The TRIPS Agreement and the Global Politics of Intellectual Property Reform in Developing Countries* (Oxford: Oxford University Press, 2009).

IP law making has always been a political process involving disputes, policy shifts, and revisions. The powerful industrialised countries of today have themselves in the past – in fact for centuries – adopted IP policies quite fluidly to advance their own industrial policies and trade interests without regard to any broader implications affecting other countries and their interests. As detailed in a 2010 UNCTAD report:

> Numerous countries had – at times – exempted various kinds of invention in certain sectors of industry from patent protection. Major industrial countries such as Italy, Japan, and Switzerland serve as examples: they adopted pharmaceuticals patent protection only when their per capita income had reached about $20,000. . . . For some countries, notably, those at a very low level of technological and economic development, the protection of IPRs . . . may not generate any positive implications. These differences suggest that there is need for flexibility allowing each country to design the IP system that best suits its particular developmental needs, and to carefully weigh the costs and benefits of IPRs in each specific circumstance.[35]

At the time of the negotiations that led to TRIPS, for the majority of developing countries for whose national IP agendas were potentially being dictated by third parties, the technical expertise necessary to handle the negotiations was absent, and their approach was more reactive than proactive. Their participation was far from optimal, leading to an outcome that was one sided. For example, even within the flexibilities of the agreement allowing staged implementation, the criteria set – namely specific deadlines, considered by many to be completely arbitrary – left no consideration of either the actual level of development reached by that deadline for the countries concerned or whether in the round implementation would be for the overall good of the country as determined by the country itself.

### *Addressing the Imbalances of Intellectual Property Regimes*

The 'imbalance' in the global IP regime – what has been called the 'TRIPS development deficit' – has led to important initiatives in recent years. To take just one example, a report in August 2000 on intellectual property and human rights by the United Nations High Commissioner for Human Rights stressed the primacy of human rights obligations over economic policies and agreements, while noting what might be called the potential 'human rights deficits' of TRIPS. It calls upon governments and intergovernmental organisations to integrate within their laws and/or policies 'provisions, in accordance with international human rights obligations and principles, that protect the social function of intellectual property'.[36]

The WTO Doha Ministerial Declaration on the TRIPS Agreement and Public Health in 2001, as well as the modified amendment in 2005 of the TRIPS Agreement

35 UNCTAD, *Intellectual Property in the World Trade Organization*, *supra* note 9, p. 1.

36 Subcommission on the Promotion and Protection of Human Rights, Res. 2000/7, §§ 5 & 6, https://www.aaas.org/sites/default/files/SRHRL/PDF/IHRDArticle15/E-CN_4-SUB_2-RES-2000-7_Eng.pdf.

in Hong Kong, reaffirmed the flexibilities of the agreement in favour of development policy. It was a major statement to remind the developed countries of the WTO membership of the Objectives and Principles of the TRIPS Agreement articulated in Articles 7 and 8, which until that time had not been operationalised in any meaningful way, but which in theory provided balance in the direction of policy space in the text of the agreement.[37]

A development in the direction of this growing 'paradigm shift' in the thinking of the purpose of IP, as embracing policy objectives of matters of genuine public interest, as well as the traditional equation of 'reward for innovation', has been the adoption of a new Development Agenda at WIPO, the fruit of an initiative in September 2004 of a group of fourteen developing countries. Formally establishing the agenda three years later by all member states in 2007 with the adoption of forty-five Development Agenda recommendations, WIPO has committed itself to ensuring that all countries are able to benefit from the use of IP not just for economic purposes, but also for social and cultural development.[38]

## III. CATHOLIC SOCIAL TEACHING AND INTELLECTUAL PROPERTY

The Catholic Church, as has been stated previously and elsewhere in this volume, has in recent times taken an active interest in the intellectual property regime, principally in relation to its impact on developing nations as a consequence of TRIPS. Its interest, concerns and interventions flow from its own tradition of biblical reflections and Catholic Social Teaching (CST). The insights from this body of thought feed into a religious articulation of important non-utilitarian, social aspects of IP. Given the increasing number of IP interventions coming from the Holy See in particular, the following brief explanation of some CST principles should assist in a better understanding of the Vatican mindset.

[37] UNCTAD, *Intellectual Property in the World Trade Organization*, *supra* note 9, pp. 27–8. TRIPS Article 7 states:

> Objectives: The protection and enforcement of intellectual property rights should contribute to the promotion of technological innovation and to the transfer and dissemination of technology, to the mutual advantage of producers and users of technological knowledge and in a manner conducive to social and economic welfare, and to the balance of rights and obligations.

Article 8 states:

> Principles: Members may in formulating or amending their laws and regulations, adopt measures necessary to protect public health and nutrition, and to promote the public interest in sectors of vital importance to their socio-economic and technological development, provided that such measures are consistent with the provisions of this agreement. Appropriate measures, provided that they are consistent with the provisions of this agreement, may be needed to prevent the abuse of intellectual property rights by right holders or the resort to practices which unreasonably restrain trade or adversely affect the international transfer of technology.

[38] See https://www.wipo.int/ip-development/en/agenda/.

## *Common Good*

The principle of the common good stems from the dignity, unity and equality of all people. According to its primary and broadly accepted sense, the common good indicates 'the sum total of social conditions which allow people, either as groups or as individuals, to reach their fulfilment more fully and more easily'.[39] Its centrality stems from the social nature of man (and ultimately, in the Christian view, the fact that man was created in the image of God, who is Trinity and who has an inner communitarian life). 'The human person cannot find fulfilment in himself, that is, apart from the fact that he exists "with" others and "for" others.'[40] Distinctive, if not unique, to Catholic teaching is the extension of the idea of the common good internationally, 'challenging mentalities of national self-interest'.[41]

## *Solidarity*

Solidarity highlights in a particular way the intrinsic social nature of the human person. 'In the presence of the phenomenon of interdependence and its constant expansion, however, there persist in every part of the world stark inequalities between developed and developing countries. The acceleration of interdependence between persons and peoples [must] be accompanied by equally intense efforts on the ethical-social plane.'[42]

## *Property Ownership, Universal Destination of All Goods and Preferential Option for the Poor*

Pope John Paul II, along with his predecessors, affirms a quasi-natural right to property: 'It is through work that man, using his intelligence and exercising his freedom, succeeds in dominating the earth and making it a fitting home. In this way, he makes part of the earth his own, precisely the part which he has acquired through work; this is the origin of individual property.' This ownership applies as well to 'the possession of know-how, technology and skill'.[43] There is therefore a strict relationship between man's labour and its intellectual fruit which has particular qualities of ownership.

Although people have the right to ownership, 'Christian tradition has never upheld this right as absolute and untouchable', but has understood it 'within the

[39] *Compendium of the Social Doctrine of the Church*, Pontifical Council for Justice and Peace, 2004, § 164, http://www.vatican.va/roman_curia/pontifical_councils/justpeace/documents/rc_pc_justpeace_doc_20060526_compendio-dott-soc_en.html.

[40] Ibid., § 165.

[41] Ibid., §§ 164–6.

[42] Ibid., § 192.

[43] Pope John Paull II, Encyclical Letter, *Centesimus annus* (1991), 31–2, http://w2.vatican.va/content/john-paul-ii/en/encyclicals/documents/hf_jp-ii_enc_01051991_centesimus-annus.html.

broader context of the right common to all to use the goods of the whole of creation'.[44] 'God gave the earth to the whole human race for the sustenance of all its members, without excluding or favouring anyone. This is the foundation of the universal destination of the earth's goods.'[45] This is not to deny legitimate possession, but to express the social duty essentially inherent in the right, which can never be absolute and unconditioned. Catholic teaching has emphasised that intellectual property, like all property, must 'contribute to putting into effect the principle of the universal destination of goods ... [Knowledge resources] too must be placed in a context of legal norms and social rules that guarantee that they will be used according to the criteria of justice, equity and respect of human rights.'[46]

Another principle of pertinence is the preferential option for the poor, which has always been central to the Church's social mission. 'The principle of the universal destination of goods requires that the poor, the marginalized and in all cases those whose living conditions interfere with their proper growth should be the focus of particular concern.'[47]

In this view, therefore, 'The intellectual property rights system exists not just to protect creative and innovative impetus but also and primarily to serve the common good of the human family. As a universal common good, intellectual property demands that control mechanisms should accompany the logic of the market.'[48]

The principle of solidarity also requires that inventors and IP owners be aware that they too are debtors to society and thus have social moral responsibilities with regard to the use and exercise of their rights, especially in an international development context. Owners 'are debtors because of those conditions that make human existence liveable, and because of the indivisible and indispensable legacy constituted by culture, scientific and technical knowledge, material and immaterial goods and by all that the human condition has produced'.[49]

We may also note in passing that this social function of property is very much part of the Jewish tradition too.[50] Some academic property legal theorists, in addition,

44 Pope John Paul II, Encyclical Letter, *Laborem exercens*, 14 (1981), http://w2.vatican.va/content/john-paul-ii/en/encyclicals/documents/hf_jp-ii_enc_14091981_laborem-exercens.html.

45 *Centesimus annus*, *supra* note 42, 31.

46 Ibid., 32; *Compendium*, *supra* note 39, § 283.

47 *Compendium*, § 182. See also Thomas C. Berg, 'Intellectual Property and the Preferential Option for the Poor', *Journal of Catholic Social Thought*, vol. 5 (2008), pp. 193–233, http://papers.ssrn.com/abstract=966681.

48 Intervention by the Holy See at the Fifth Ministerial Conference of the World Trade Organization (WTO), Address of Msgr. Frank J. Dewane, Cancun, Mexico, 13 September 2003, http://www.vatican.va/roman_curia/secretariat_state/2003/documents/rc_seg-st_20030913_wto-cancun_en.html.

49 *Compendium*, § 195.

50 Gabriel J. Michael, 'Catholic Thought and Intellectual Property: Learning from the Ethics of Obligation', *Journal of Law and Religion*, vol. 25, no. 2 (2009–10), pp. 415–51, pp. 422–3, available at https://ssrn.com/abstract=1349769.

today emphasise the social aspects of property as a counterbalance to the traditional emphasis on individual proprietary rights.[51]

### *Stewardship*

A final relevant principle is that of stewardship. Embedded within the Bible is the notion that we are stewards not only of God's creation in terms of nature (Gen 1: 27–28)[52] and the material goods with which we have been gifted (the parable of the talents: Mt 25: 14–30), but also of our own intellectual talents. Moreover, nations themselves are stewards of the resources within their boundaries – as a power and duty accompanying sovereignty – and should steward those resources to serve the common good. They should aim to benefit all people (giving reasonable access to genetic resources), but particularly those who are most vulnerable (ensuring participation and benefit sharing for vulnerable groups).

### *CST and the Social Function of IP*

From the provided religious ethical principles, a fundamental conclusion follows – namely that there is always a social aspect or dimension to how we use our knowledge and how we relate to our 'knowledge goods'. This should also be an almost intuitive conclusion when one reflects that human knowing and creativity are essentially communal – that is, one's own inventiveness is usually the product of many influences from others, especially educators. Patents are therefore the result of a cumulative inventive process.[53] Accordingly, the private IP rights of an owner allowing him to control access to an inventive process or product ought always to be exercised within a context of social responsibility.[54]

## IV. OVERALL CONCLUSIONS: GR, PATENTS AND SOCIAL RESPONSIBILITY

Developing countries are holders of vast reserves of biologically diverse resources, and yet they have received comparatively little if any benefit from the exploitation of

51 Eric T. Freyfogle, 'Private Ownership and Human Flourishing: An Exploratory Overview', *Stellenbosch Law Review* 24:3 (2013), pp. 430–54, at https://ssrn.com/abstract=2293211.

52 'So God created mankind in his own image, in the image of God he created them; male and female he created them. God blessed them and said to them, "Be fruitful and increase in number; fill the earth and subdue it. Rule over the fish in the sea and the birds in the sky and over every living creature that moves on the ground".'

53 Michael, *supra* note 50, pp. 436–8.

54 The reported examples of social irresponsibility led Pope Benedict XVI to state in his *Encyclical Letter on Integral Human Development in Charity and Truth (Caritas in Veritate)*, 29 June 2009: 'On the part of rich countries there is excessive zeal for protecting knowledge through an unduly rigid assertion of the right to intellectual property, especially in the field of health care'. Ibid., 22, http://w2.vatican.va/content/benedict-xvi/en/encyclicals/documents/hf_ben-xvi_enc_20090629_caritas-in-veritate.html.

these resources by parties from the developed world. Within the developing countries, it is the indigenous peoples, with their communal store of traditional knowledge relating to the beneficial uses of genetic resources, who have lost out most – these being the communities who are the poorest and whose living habitats and way of life are under threat.

It is the view of the developing countries – a view shared by the Holy See (the Vatican) – that introducing a mechanism within the patent regime at an international level requiring mandatory disclosure of the source of the genetic resource with associated traditional knowledge would ultimately benefit the poor and poorest of this world. Such a mechanism would reveal whether or not proper prior consent had been obtained from the relevant authority, together with a benefit-sharing arrangement.

The continued resistance by developed countries (especially the United States) to introducing such changes to the patent regime would appear to reflect a view of intellectual property ownership that does not sufficiently take account of its social dimensions and the overall purpose of intellectual property to promote the common good and not just the individual good and interest of industry and commerce. An understanding of the power politics at the time of the TRIPS negotiations also helps one appreciate the reasons for some of the resistance to making changes to the patent system today.

Catholic Social Teaching principles invite a view of the world that is one of cooperation and mutual support. International legal instruments where appropriate, and where dealing with parties of unequal stature, should also reflect these principles, the international patent regime being no exception. IP rights should benefit and empower the poor in general, and where their own contributions, through traditional knowledge or being careful custodians of biological resources, have led to benefiting others and to ownership claims of IP, they in return deserve in justice to share in the fruits of exploitation and to be empowered, through technology transfer and building of infrastructure, to exercise their own creativity and productivity to better their own futures and to contribute to the cultural, economic, political and social life of the civil community to which they belong.[55]

For the reasons discussed in this chapter, the Holy See, perhaps representing religious observers in general, takes the view that there are compelling reason for embracing policy changes in the IP system. An immediate candidate for such change is the adaptation of the patent system, currently being discussed at the IGC at WIPO, to instigate a formal disclosure requirement for genetic resources and associated traditional knowledge in patent applications at an international level. The Holy See respects the need for experts in the field to work out the technicalities of the patent regime for the subject matter of GR and associated TK (such as the manner by which non-compliance is addressed, the particular trigger points for

[55] Cf. *Compendium*, § 189; Berg, 'Intellectual Property', *supra* note 47, pp. 10–12.

disclosure, the mechanism by which information is shared that can lead to confirmation of prior informed consent, etc.). However, its prudential judgement is that of the need for structural change, especially given the unanimity of views of the poorest countries of the world.

This position of the Holy See also reflects its view that what is at stake is the common good of societies, especially the poorest who stand to lose most if this opportunity for their betterment is passed over. If this view can be shared by all IGC members, the current global IP discord in relation to genetic resources and patents can be overcome. What is needed is the concerted effort of the international community that is at least in part motivated by concern for and solidarity with its poorest members, a concern that is ultimately rooted in justice.

# 13

# Patents for Life: Toward an Ethical Use of Patents on Plant Innovations

*Michael A. Kock*

## INTRODUCTION

Providing sufficient and healthy food to a rapidly growing world population requires maximizing agricultural innovation and its use. Intellectual property (IP) can play an important enabling role, but is also seen by some as a potential obstacle. This paper discusses the root causes of the negative perception of IP and potential solutions to foster a constructive, ethical IP use.

## 1. DEMAND FOR AGRICULTURAL INNOVATION

Humanity is facing the daunting challenge of feeding a rapidly growing world population poised to increase from seven billion today to almost nine billion by 2050. Climate change and decreasing availability of water and farmland adds further complexity to the situation. We need to meet this challenge by producing "more with less" – less land, less water, and fewer inputs like fertilizer.

The "green revolution" of the twentieth century boosted crop yields in many parts of the world. More recently, modern plant breeding and biotechnology further helped boost agricultural productivity.[1] However, the speed of innovation and breeding progress of the past might be insufficient for the future: the demand forecast requires the annual growth rate in agricultural production to increase to 2.5 percent per year, from a rate of 1.5 percent per year over the last twenty years. Unless we are willing to make big sacrifices and to enable a transfer of wealth from the developed to the developing world, this higher calorie demand will require agricultural production to increase by 70 percent by 2050.[2]

It seems, however, that the challenge is less in the quantity of agricultural production than in the quality. Since 2012, the commodity prices for major crops like maize and soya have dropped to almost half and have so far not

[1] Since 1997, the increase in US corn yield moved above the previous linear trend and correlates the commercial biotechnology maize area. Insect resistant (Bt) cotton in India has seen yield increases between 122 and 259 percent across three years.

[2] "OECD-FAO Agricultural Outlook 2010–2019, Highlights," http://www.agri-outlook.org/dataoecd/13/13/45438527.pdf.

recovered.[3] Based on improved varieties but also agricultural practices, global productivity is growing faster than demand. In addition, renewable energy is shifting from biofuels to photovoltaic and wind energy, which is releasing additional calorie resources.[4] On the other hand, the quality of agricultural production, storage, and distribution is still lacking.[5]

Further, "food security" does not only mean having enough calories. The 1996 World Food Summit defined food security as the situation "when all people at all times have access to sufficient, safe, nutritious food to maintain a healthy and active life." Thus, plants with improved nutritional value are also in demand. All this will only be possible if we maximize agricultural innovation from many areas: seeds, biotechnology, modern crop protection, resource-saving agricultural practices, storage and transportation, etc.[6]

Innovation and technology can be a blessing as much as a curse. Pope Francis in his recent encyclical letter *Laudato Si* recognizes that we are the beneficiaries of technological innovation including biotechnology: "'[S]cience and technology are wonderful products of a God-given human creativity,'" which have "remedied countless evils which used to harm and limit human beings."[7] With respect to genetic modification of plants, Pope Francis considers that "[t]he risks involved are not always due to the techniques used, but rather to their improper or excessive application. Genetic mutations, in fact, have often been, and continue to be, caused by nature itself."[8] While "no conclusive proof exists that GM cereals may be harmful to human beings, and in some regions their use ... has helped to resolve problems, there remain a number of significant difficulties. ... In many places, following the introduction of these crops, productive land is concentrated in the hands of a few owners."[9] While technology is certainly not "neutral," in its purest sense it is primarily its purpose and use which make it beneficial or problematic. Not using technology in view of the current challenges is ethically and morally questionable. As Bonhoeffer explains in his *Ethics*, responsible

3 "Commodity Prices (corn), Index Mundi," https://www.indexmundi.com/commodities/?commodity=corn&months=120.

4 This trend could accelerate drastically if, for example, the United States would stop the mandatory maize ethanol blending of fuel for automobiles. Almost 30 percent of the entire US maize production is converted into ethanol. A collapse of this artificial demand would put global maize commodity prices under further pressure.

5 Current industrial agricultural production is far from being sustainable, especially when it comes to the use of water, fertilizer, pesticides, and the lack of crop rotation. The negative impact on soil, water, and biodiversity could jeopardize future agricultural productivity. A shift of the innovation focus from quantity to quality is necessary.

6 Likewise, solutions are necessary to enable the integration of innovations, which are often owned by various parties, into fully integrated solutions.

7 Pope Francis, *Laudato Si (Encyclical Letter on Care for Our Common Home)*, May 24, 2015, ¶102, http://w2.vatican.va/content/dam/francesco/pdf/encyclicals/documents/papa-francesco_20150524_enciclica-laudato-si_en.pdf (quoting John Paul II, *Address to Scientists and Representatives of the United Nations University*, Hiroshima [25.02.1981], 3: AAS 73 [1981], 422).

8 *Laudato Si*, ¶133.

9 *Id.*, ¶134.

action is a highly risky venture that cannot be justified in advance. The demand for responsible action is a demand no one can ignore. We are facing a dilemma: when assaulted by evil, we must oppose it directly. Failure to act is simply to condone evil.[10]

It is easy for people in the developed world, where healthy and nutritious food is plentiful and affordable, to reject new technologies. Certainly, the moratorium for the cultivation of genetically modified crops in Europe has a negative impact on the competitiveness of EU agriculture. The now necessary importation of millions of tons of genetically modified soybeans every year for the EU meat production industry make the position rather hypocritical and shifts the issue from Europe to other places in the world. Such attitudes become worse and ethically questionable if imposed with missionary zeal on the poor in developing countries. For example, NGO activism blocks the utilization of Golden Rice[11] to reduce blindness and other vitamin A deficiency–induced diseases, which are prevalent among the poor whose diets are based mainly on rice.[12] Patrick Moore, a former Greenpeace director, today calls the NGO campaign "a crime against humanity" due to which twenty million people, mostly children, have died.[13] The innovation behind Golden Rice has recently been recognized by the Patents for Humanity Award of the White House Office of Science and Technology Policy and the US Patent and Trademark Office.[14]

In conclusion, the use of technology and innovation *can* be a positive contribution to counter today's challenges, as long as this use is ethical and sustainable.

## II. INCENTIVES FOR PLANT-RELATED INNOVATIONS

Plant breeding has undergone a rapid technification stimulated by the demand for increasing productivity. While a few decades ago breeding was largely empirical, today plant innovations are based on cell biology, genome and proteome research, gene mapping, marker-assisted breeding, and hybrid technology, and are further extending to "new breeding technologies," which enable the creation of new plant varieties in a predictable way.[15]

[10] Dietrich Bonhoeffer, *Ethics*, trans. Neville Horton Smith (New York: Simon & Schuster, 1995), p. 249.

[11] Golden Rice has been genetically modified to produce beta-carotene, the precursor to vitamin A. It is developed by a non-profit organization solely for humanitarian and non-commercial purposes. "Golden Rice Project," http://www.goldenrice.org/.

[12] In 2012, the World Health Organization reported that about 250 million preschool children are affected by VAD, and that providing those children with vitamin A could prevent about a third of all under-five deaths, which amounts to up to 2.7 million children that could be saved from dying unnecessarily. "Allow Golden Rice Now! Preventing It is a Crime against Humanity," http://www.allowgoldenricenow.org/.

[13] *Id.*

[14] United States Patent and Trademark Office (USPTO), "Patents for Humanity" (November 17, 2017), http://www.uspto.gov/patent/initiatives/patents-humanity.

[15] New breeding technologies include, for example, zinc finger nuclease technology, oligonucleotide directed mutagenesis, cisgenesis and intragenesis, RNA-dependent DNA methylation, grafting, reverse breeding, agro-infiltration, and synthetic genomics. See, for example, Maria Lusser et al.,

The resource requirements, high costs, and complexity of this technification – especially for genetically modified plants – opens an opportunity but also a need for patent protection.[16] Developing new crop varieties is a lengthy, risky, and costly task. To bring a new biotechnology trait to the market requires up to US$136 million and, on average, thirteen years.[17] This period is likely to expand due to higher regulatory scrutiny but also the issue of "asynchronous deregulation."[18] On the other hand, the commercial half-life of GM technologies has been demonstrated to be shorter than expected due to resistance issues.[19] As a consequence, new emerging technologies like genome editing[20] could render GM the "nuclear power" of the seed sector: a highly expensive, highly regulated technology, with decreasing profitability but likely long-lasting liability.[21]

Also, non-GM varieties and so-called native traits can take eight to ten years and many millions of euros to bring to market, especially if so-called pre-breeding activities are involved, which require screening of hundreds of plants to find specific

*New Plant Breeding Techniques: State-of-the-Art and Prospects for Commercial Development* (2011), http://www.ab.gov.tr/files/ardb/evt/1_avrupa_birligi/1_9_politikalar/1_9_10_tarim_balikcilik/new_plant_breeding_techniques.pdf; Editorial, "Seeds of Change," *Nature* 520 (April 9, 2015): 131–32, http://www.nature.com/news/seeds-of-change-1.17267.

16 This trend is further intensified by more sophisticated, marker-assisted breeding technologies that allow for an easy circumvention of the traditional protection regimes based on plant variety protection (PVP) and trade secrets. While only a decade ago a normal breeding cycle would be expected to take between five and twelve years, depending on the plant species, nowadays, with new breeding and nursing techniques, this period is reduced by 50 percent or more.

17 Phillips McDougall, "The Cost and Time Involved in the Discovery, Development and Authorisation of a New Plant Biotechnology Derived Trait: A Consultancy Study for Crop Life International" (September 2011), https://croplife.org/wp-content/uploads/2014/04/Getting-a-Biotech-Crop-to-Market-Phillips-McDougall-Study.pdf.

18 A new GM plant requires marketing approval ("deregulation") not only in the cultivation countries but also in the importation countries. Here, most countries follow their own way at their own pace. Launching a new GM variety in the US, for example, could cause trade disruption issues in import countries that have not yet granted marketing approval. Often the EU and China are bottlenecks in this process. Ignoring this challenge could cause substantial liability; see, for example, "Syngenta Agrees to $1.4 Billion Settlement," *CropLife* (September 28, 2017), http://www.croplife.com/crop-inputs/seed-biotech/syngenta-agrees-1-4-billion-settlement/.

19 Both herbicide-resistant weeds and insecticide-resistant pests have emerged quite rapidly. In view of the expanding development times and costs, this could cause a dilemma of increasing investment needs and decreasing profitability, which could make GM technology unattractive.

20 In many countries the majority of characteristics developed with genome editing will be considered "non-GM". In consequence, precision breeding based on genome editing has the potential to develop new plant characteristics in a fraction of the time and costs required for GM traits. This will substantially shorten the product development lifecycle. It will also make "exclusivity" based on a strong intellectual property protection less essential.

21 After a GM technology has been stopped in marketing, it will still exist in trade for many decades as "low level presence." During this time, the owner of the technology will be required to keep at least the dossiers for importation and use in place, which requires – depending on the country – a renewal every five to ten years. This equals a "nuclear waste" scenario with indefinite responsibility, costs, and liability.

properties.[22] New breeding techniques, like genome editing,[23] could speed up breeding drastically and could enable the combination of beneficial alleles from different varieties without the need of crossing and back-crossing.[24] As the adaptation of the new breeding technologies is relatively easy, they have the potential to re-democratize plant biotechnology, which could especially benefit developing countries. The regulatory situation of the new breeding technologies is not yet clear,[25] and the technological process seems to be progressing far faster than the regulatory system or politicians can react.

Nevertheless, public acceptance might not be easy to obtain, and initiatives are emerging to "ban" plants obtained by these technologies from the organic and bio market. The public acceptance will depend, in part, on clear consumer value: potatoes with resistance against potato blight, which do not require the usual sixteen fungicide sprays per season; a fungal-resistant wheat with low mycotoxin content; or a non-browning apple or salad. As the development cost will be low, the innovation space will expand from herbicide and insect resistance in corn, soy, and cotton to almost all crops and all value traits.

The outcome of innovative plant research, literally and figuratively, is a simple yet high-tech product: seed. Since seed can be easily reproduced by farmers and competitors, some form of incentive is required to make such investments. With respect to private sector investment, so far only IP has a proven positive track record to provide such an incentive. While most agree with the need for innovation, there is a debate about the role of IP – especially patents – in this context.[26]

The use of plant variety protection (PVP) as an "open source" IP system is largely undisputed, and until twenty-five years ago, plant innovations were almost exclusively protected by that system. PVP protects a specific variety characterized by all its essential, mostly phenotypical characteristics. Only varieties resembling *all* those characteristics are protected, which makes PVP a type of "copyright" for plant

[22] Formerly pre-breeding was predominantly conducted by public research institutes. As a consequence of budget constraints, these activities have decreased substantially and moved predominantly to the private sector. They are crucial for solving specific problems such as disease resistance and so forth.

[23] Genome editing can involve targeted mutagenesis of single nucleotides (SDN1), the insertion of smaller fragments of (often non-coding regulatory) DNA (SDN2), or insertion of larger sequences (SDN3). See, for example, "Targeted Genome Editing," Belgian Biosafety Server, https://www.biosafety.be/content/targeted-genome-editing.

[24] While genome editing is often looked on as "GMO 2.0," its true potential is in "breeding 2.0": combining favorable genetics and alleles by parallel genome editing instead of conventional crossing and selection, which often requires multiple cycles of back-crossing. Michael W. Bevan et al., "Genomic Innovation for Crop Improvement," *Nature* 543 (March 16, 2017): 346–54.

[25] However, there seems to be a growing consensus that plants that could have been made by conventional breeding, where the genetics has an established history of safe use, should not require a specific approval process. As these plants would not be distinguishable from conventional plants, enforcement of a regulation would be rather difficult.

[26] Conference on Intellectual Property and Public Policy Issues, World Intellectual Property Organization (July 13–14, 2009), http://www.wipo.int/meetings/en/2009/ip_gc_ge/program.html; Michael Kock, "Patents for Life: The Role of Intellectual Property Rights on Plant Innovations," *Bio-Science Law Review* 10(5; 2009): 167–76.

varieties that prevents its unauthorized copying (propagation). PVP laws contain a breeders' exemption that allows the use of a protected variety for further breeding. The exemption thereby enables competitors to use individual characteristics or genes, without the need to start a research program, by simply using the commercial variety as a source. While PVP is sufficient to protect most achievements in plant breeding, it is not suitable – nor intended – to protect specific genes or improved methods of breeding.[27] This is an issue for developments that require a high investment, such as looking for new plant properties in the context of pre-breeding.[28] Here, patents provide for a complementary incentive.

For more than a century, the patent system has proven tremendously successful in providing the "invisible infrastructure" that enables innovation and knowledge sharing in all technology areas, including agriculture and plant breeding. In comparison to PVP, the patent system has higher requirements for protection, such as absolute novelty, inventiveness, and an "enabling disclosure." This disclosure requirement is an important benefit of the patent system that enables others to work with and further improve upon prior inventions that otherwise would be kept as trade secrets.

While patents for genetically modified plants follow the established rules for biotech innovations, or breeding technologies and resulting plants, the patent system is a rather new IP tool, and both patent offices and applicants are still "calibrating" the right standards, especially for inventiveness, enabling disclosure, and the interpretation of exemptions from patentability. This calibration is not unique for the plant area but happens in all emerging areas of technology.

And again, new breeding technologies could trigger a paradigm change: here plants can be developed in a relatively fast and inexpensive way and may be on the market years before a patent is granted.[29] As not only the development time but also the product and technology life cycle will likely shorten, patents may well play a far smaller role than they did for GM technologies. It will probably be handled quite selectively and could also have negative effects on their regulatory and/or public acceptance.[30] The copyright provided by PVP protection in combination with trade secrets could be an attractive alternative. However, from the perspective of knowledge-sharing and technology-dissemination, the patent system could provide advantages, especially in fostering the re-democratization process of new technologies. It is likely that new breeding technologies will trigger a substantial increase in plant related patents, which will have an impact on markets like the EU, where these patents – irrespective of the quite visible political debate – played economically only a very minor role.

27 While having clear advantages, marker-assisted breeding also enables rapid circumvention of PVP. As noted previously, normal breeding cycles have generally been reduced by 50 percent or more.

28 As noted previously, in the text at nn. 22–24.

29 Biotechnology-related patents require at least five to seven years to grant. For controversial cases, such as native traits in Europe, most patents are already pending in examination for more than ten years.

30 One requirement to avoid regulatory constraints is a solid argument for a "history of safe use." Such an argument could contradict an argument for patentability, which often requires arguing "new and surprising effects."

The increasing use of patents triggers a change for breeders. While the PVP system in general doesn't know dependency,[31] patents require breeders to adapt their breeding practice when using competitor's varieties in their breeding programs: it becomes necessary to monitor and avoid the presence of patented genes and traits, or to negotiate licenses. These changes require new capabilities and can complicate business practices. This is seen by some as limiting "freedom to breed" and access to innovations and genetic resources, delaying improvement and slowing down technology cycles. These concerns are expressed especially by EU breeders associations, which argue that patents should not interfere with conventional breeding and should only have a role for plants made biotechnologically.[32] Other associations ask essentially to deny effective patent rights for all plant inventions.[33]

The controversy around "patents on seeds" fits into a general trend of "IP bashing": many NGOs and politicians rarely miss an opportunity to criticize IP as an obstacle to innovation rather than an enabler. While just a few decades ago IP was a subject for technical experts and specialized legislators, it has become a hot topic in social, political, and ethical debate. "Open source," "knowledge sharing," "freedom of information," and so on are buzzwords used to argue for changing the IP framework. And indeed, at times when society becomes increasingly open, the lack of openness of the "old" IP paradigm is sometimes perceived as antiquated, especially for industries that touch on elementary human needs like food, health, and education.

"Patents on life" are often raised as an ethical issue, especially when they relate to higher life forms like plants and animals. Such views are based on the myth that patents grant rights or ownership to what is claimed. However, patents are merely a *right* (i.e., an "option") to exclude a third party from making or using the patented invention. As a consequence, patents confer far less rights to a living organism than does normal property law, and certainly do not provide any ownership on the claimed object. Ethical problems can arise if the patent right is used in an unethical way (i.e., requesting destruction of harvests during times of food scarcity). Thus, it is again their use to exclude that likely shapes the perception of them being "unethical." Increasingly, and perhaps paradoxically, the legitimate use of IP as a "right to exclude" is perceived as inappropriate and triggers reactions beyond reputational

31 Other than for so-called essentially derived varieties (EDV), which resemble a genetically very similar plagiat of the original variety, there is no dependency under the PVP law.

32 The European Seed Association (ESA) and the German Seed Association (BDP) favor an exception from patentability for plants obtained by breeding. ESA, "On Intellectual Property Protection for Plant-Related Inventions in Europe," http://www.euroseeds.eu/system/files/publications/files/esa_12.0100_0.pdf; BDP, "BDP-Position zur Ausgestaltung des Patentschutzes in der Pflanzenzüchtung," http://www.bdp-online.de/de/Ueber_uns/Our_positions/2015-05-20_Positionspapier_Patent-Sortenschutz.pdf.

33 The Dutch breeders' association, Plantum, argued for a full breeders exemption under patents, including commercializing of the new variety, even if it still contains a patented trait, irrespective whether this trait is a native trait or a genetically modified trait. Plantum, "Standpunt Plantum inzake octrooi en kwekesrecht," https://assets.plantum.nl/p/229377/files/Dossiers/Standpunt%20inzake%20octrooi-%20en%20kwekersrecht.pdf.

damage. Courts and legislators, consciously or unconsciously, diminish patentee's rights by excluding critical innovation areas from patentability, from diagnostic markers[34] to imported soya meal[35] to breeding processes.[36] Also, Pope Francis in his recent encyclical letter remarked with respect to technology ownership that "it is extremely risky for a small part of humanity to have it."[37] Use is associated with the rights resulting from a patent, and indeed most of the "side effects" of patents on breeders and farmers relate to the rights resulting from a patent and how these rights are used by the patentee:

- Lack of transparency: whether a plant is covered by a patent cannot be easily determined.[38] In contrast to the PVP regime, patent infringement can occur "accidentally" and unintentionally.
- Uncertain breeding rights: a breeders' exemption exists only in a few patent laws.[39] This results in uncertainty as to whether a patented gene could deprive the entire plant from further breeding.
- High licensing costs and uncertainty around obtaining a license.

While currently the scale of the related "problem" for breeders and farmers can hardly be measured,[40] the fact that part of the seed sector perceives a problem makes it a real problem for the sector. In addition, the progress of science in breeding will enable an increasing number of patents and related products on the market. Ignoring this trend and the potential impact on the sector could lead to issues in the future.

## III. FOSTERING AN ETHICAL USE OF PATENTS ON SEED

It is clear that patents *as such* don't feed the world and are *as such* not a solution for food security. The on-first-sight appealing "solution" to solve the patent problem by simply abandoning them could be short-sighted and, in the long term, diminish the

34 Association for Molecular Pathology v. Myriad Genetics, Inc., 569 U.S. 576 (2013); In re BRCA1- and BRCA2-Based Heredity Cancer Test Patent Litigation, 774 F.3d 755 (Fed. Cir. 2014).

35 Case C-428/08, Monsanto Technology LLC vs. Cefetra BV et al. [2010], 2010 ECR 7, at http://curia.europa.eu/juris/liste.jsf?language=en&num=C-428/08.

36 EPO, Decision of the Enlarged Board of Appeal dated 9 December 2010 G 2/07, OJ EPO 2012, 130 (method for producing broccoli with elevated glucosinolates).

37 *Laudato Si*, ¶104.

38 Patent labeling for conventional seed is not mandatory and is logistically almost impossible to establish.

39 Limited breeders' exemptions are established in the patent laws of Germany, France, the Netherlands, and Switzerland, and for the EU Unitary Patent.

40 In the EU, farmers have the same privileges under patents as under plant breeders' rights (see "Biotech Directive," Directive 98/44/EC of the European Parliament and of the Council of 6 July 1998 on the Legal Protection of biotechnological inventions, Article 11(1), http://www.wipo.int/wipolex/en/details.jsp?id=1440). And there is only one patent litigation relating to plants obtained by breeding. Taste of Nature Holding B.V. v. Cresco Handels-B.V., District Court The Hague, The Netherlands, 8 May 2013, Case No. 416501 / KG ZA 12–452. The patent has been revoked due to lack of novelty.

incentive for urgently needed innovations: Current technology may become freely available, but future innovation would lack an incentive.

It must be remembered that patents are simply an instrument, a tool, provided by society to foster innovation. A tool as such is neither good nor evil. It is the use of the tool which could be bad or good, destructive or constructive. The fact a hammer is sometimes used for destructive purposes like murder would not lead to a conclusion to "ban" hammers, but rather to regulate their negative use.[41] The current negative perception of patents is primarily based on their use to exclude. However, patents can also be used in a positive, constructive way – for example, to foster licensing and technology exchange, and to encourage the disclosure of knowledge and innovation. However, the real question is how to recalibrate the use of patents to maximize the beneficial effects (i.e. preserving the incentives for innovation and knowledge-sharing) while minimizing the effects limiting access. Can patent use evolve from exclusion to inclusion?

## IV. POTENTIAL SOLUTIONS

A solution to a complex problem needs more than one element.[42] Some prototypes that could lead to solutions have already been successfully implemented. For example, a limited breeders' exemption is available in the patent laws of Germany, France, the Netherlands, and Switzerland, and for the EU Unitary Patent.[43] A substantial step toward patent transparency has been achieved by the database PINTO of the European Seed Association, which provides patent information for plant varieties registered in the EU.[44]

Access to patented traits for commercial purposes is likely the most difficult but also the most important part of a solution. Here some recent, promising prototypes exist: some seed companies have made their patents available through e-licensing tools.[45] As a result of a broader industry initiative, the International Licensing Platform (ILP) – Vegetable was recently launched, providing a FRAND (fair, reasonable, and nondiscriminatory) licensing system following the principle of "free access but not access for free."[46] The current ILP members represent more than 50 percent of the global

[41] In view of the fact that the total number of active patents is in the range of ten million, the small number of patents used in aggressive "troll-like" actions per year is likely not higher than the percent of hammers used for murder.

[42] See Kock, "Patents for Life," 1–10, discussing the patent commons project.

[43] This exemption enables breeders to utilize biological material covered by a patent for the purpose of breeding. A license is only required if the resulting new variety still comprises the patented gene and is commercialized. German Patent Act, §11 (2a); Article L613-5-3, French Intellectual Property Code; Dutch Patent Act, Art. 53b(2); Swiss Patent Act, §9e; Agreement on the Unified Patent Court, Article 27(c).

[44] PINTO Database, EuroSeeds, http://pinto.euroseeds.eu/.

[45] Syngenta Global, "TraitAbility," http://www.traitability.com; Seminis, "E-Licensing Contact Form," http://www.seminis-us.com/e-license.

[46] Michael A. Kock and Floris ten Have, "The 'International Licensing Platform – Vegetables': A Prototype of a Patent Clearing House in the Life Science Industry," *Journal of Intellectual*

vegetable seed market, which reflects high acceptance. The ILP provides access to patents on vegetable traits such as disease resistance, improved shelf-life, and nutritional value ("trait patents"), but also to so-called variety patents, which in certain legislations – like the US – protect specific plant varieties. For variety patents, the ILP establishes a cost-free mutual non-assert for breeding of new varieties. This comes as close as possible to a full breeders' exemption under PVP. With respect to trait patents, the ILP enables licenses for breeding and commercializing new varieties. The licensee has to pay a royalty for the commercialization of the new variety, if still covered by the patent, in countries where a patent right exists. Research and breeding is free. Every party can join the ILP irrespective of whether it owns patents or not. The ILP is based on two unique elements:

1. **Pull-in effect:** A party wanting to take a license through the ILP has to become a member and is obligated to make all its own patents relating to vegetables (if any) available to other members. This creates a pull-in effect, and in consequence the ILP already makes more than 220[47] trait patent families and numerous variety patents available. This represents more than 50 percent of the relevant patent families in this field.
2. **Independent FRAND arbitration:** One challenge in FRAND licensing is the definition of "fair." To overcome posturing and negotiation tactics, the ILP utilizes "baseball arbitration." The arbitration process is based on a standard license agreement where the only negotiable term is the royalty percentage on net sales. The parties have to submit to an independent expert committee their written, binding proposal for a fair royalty together with supporting evidence and reasoning.[48] As the unique feature of baseball arbitration, the experts can only pick the one of the two submissions they believe is fairer. Thus, if a submission is unreasonably high or low, automatically the other is adopted.[49] The established royalty is final, binding, and under a most-favored-nation clause, available to all other members.

Despite these first successes, questions remain. What about non-vegetable crops? What about parties that do not join the ILP? In the aftermath of the "Broccoli &

*Property Law & Practice* 11(7) (2016): 496–515; Michael A. Kock, "Plant Breeding Innovations: Free Access but not Access for Free – A New Approach to Facilitate FRAND Licenses for Plant Related Patents," *Bio-Science Law Review* 14(4; 2015): 123–29; Paul van der Kooij and Marian Suelmann, "Het International Licensing Platform: een mijlpaal," *Berichten Industriële Eigendom (BIE)* 12 (2014): 260, http://european-seed.com/a-full-count-for-vegetables/. Detailed information about the statutes, members, and current patent estate can be derived from the ILP web page: http://www.ILP-vegetable.org.

47 For details, see International Licensing Platform Vegetable, "Patent Register Version 04072019", https://www.ilp-vegetable.org/uploads/Bestanden/Patent%20Register/ILP%20Patent%20Register%20-%20total%2004072019%20changes%20marked%20in%20yellow.pdf.

48 Subsequently these proposals are exchanged and the parties are given another 3 weeks to find a bilateral resolution. This has a high likelihood of success if the two submissions are close. However, the parties cannot change their submissions any further.

49 In addition, the "losing" party has to pay the costs for the arbitration proceedings. The costs are currently set at €30,000.

Tomato II" decision of the Enlarged Board of Appeal of the EPO,[50] which confirmed the patentability of plants obtained by conventional breeding, these questions have become pressing. This led the EU Commission, and in consequence also the Administrative Council of the European Patent Office, to declare plants obtained by "essentially biological processes" as non-patentable subject matter. The validity of this new rule, however, is questionable. It was recently revoked by a decision of the Boards of Appeal (T 1063/18) and is now again the subject of a referral (G 3/19) to the Enlarged Board of Appeal by the President of the European Patent Office.[51]

The probably lengthy revision process would have a long, chilling effect on the EU life-science industry,[52] would not be global, and ultimately would not solve the issue: new breeding technologies and the resulting plants will likely remain patentable and gain importance and market share in the EU. If so, the problem of patents for breeders will not go away but increase.

One alternative to a revision of the EU Biotech Directive, and a potential solution to the open questions, could be a combination of voluntary, industry-led collaborative solutions – like ILP or PINTO – supported by regional, legislative measures to encourage the participation of all parties (i.e., a combination of "carrots" and "sticks").

The ILP is a voluntary solution, and there will be patentees who will not join for different reasons.[53] However, the EU patent system provides an adequate "stick" to deal with an unreasonable denial of licenses. The Biotech Directive and the patent laws of all EU member states provide a compulsory-license provision for dependent PVP rights, which requires a "significant technical progress of considerable economic interest."[54] Presumably the ambiguity of this term explains why the provision has so far not been used. Switzerland reduces the ambiguity by linking the term to seed marketing authorization.[55] In consequence, a variety would meet the

50 Reviewed in Michael Kock, "Broccoli and Tomato: Free or Not Free? Decisions 2/12 and G2/13 of the Enlarged Board of Appeal," *Bio-Science Law Review* 14(4) (2015): 167–76.

51 Michael A. Kock and Gareth Morgan, "Broccoli and Tomato: A Never-Ending Story?," *Bio-Science Law Review* 16(3; 2017): 123–28; Michael A. Kock, "New Rules for Plants from Essentially Biological Processes," *Life Science Recht* 1 (2018): 54–58; Michael A. Kock and Herbert Zech, "Pflanzenbezogene Erfindungen in der EU – aktueller Stand," *GRUR* 10 (2017): 1004–16.

52 The development of the current Biotech Directive took ten years from the first draft proposal in 1988 to the final directive in 1998, and it was negotiated with fifteen members only. Today the EU has twenty-eight members, and the EPC has thirty-eight. In addition, it will be difficult to limit a re-opening of the directive to the issue of plants only. Many other topics could likely trigger a lengthy debate: stem cells, diagnostic markers, isolated DNA sequences, and so on. Legal certainty, especially around intellectual property, is of fundamental importance for investors and shareholders.

53 Non-practicing entities (so-called patent trolls) have no need for licenses and certainly no interest in giving away control over their patents' assets. For other parties, strategic or corporate policy reasons may advocate against an "open IP" approach.

54 Biotechnology Directive, Article 12(1), 12(3)(b).

55 Art. 36a-1 of the Swiss Patent Act incorporates Art.12(1) Biotech Directive with respect to compulsory cross-licensing. However, it further clarifies that "[f]or varieties for agriculture and food, the criteria under the Seed Ordinance of 7 December 1982 serve as a reference point." https://www.admin.ch/opc/en/classified-compilation/19540108/index.html. The "value for cultivation and use" (VCU) for seed market authorization requires "*a* clear improvement either for cultivation in general or for the

requirements for a compulsory license once it obtains seed marketing authorization. This "interpretation" can be established without any change of the Biotech Directive – for example, by guidance of the commission – and would be seamlessly compatible with voluntary licensing under the ILP. Likewise, creating transparency through PINTO could be encouraged by a mandatory requirement for patentees to declare the patent-status for their commercialized varieties. A lack of declaration could make the patent unenforceable against a later alleged infringement against an innocent infringer.

These limited changes are compatible with the ILP and other collaborative elements and essentially preserve the incentive character of the patent system. An approach to abandon patents on plants in general is incompatible with the ILP and undermines its sustainability.[56]

## V. THE NEED FOR A CULTURAL TRANSFORMATION

"Culture eats strategy for breakfast." Initiatives like the ILP and PINTO demonstrate that cooperative solutions are possible. They can be prototypes for a new use of IP, which sustains the incentive for future innovation on an open-access basis. However, they may also remain "interesting" singularities, unless the industry undergoes a cultural transformation with respect to the use of IP – on the side of patentees but also the side of technology users.

Today's world is marked by an unprecedented push toward openness. Collaboration and integration are viewed as fundamental drivers of innovation. While the seed industry is somewhat limping behind others, the need for open innovation is high.[57] A new seed variety requires integration of many traits – resistance against biotic and abiotic stresses, higher yield and nutritional value, and so forth – and no single player has all the elements necessary to do this effectively. Just as a changing natural environment requires evolution and adaptation of its inhabitants, a new innovation environment requires IP stakeholders to adapt. IP holders will have to give up control to exclude others, limit access to plant-related invention, and unilaterally determine prices. This would result in a "free access but not access for free" scenario, which could provide a sustainable balance

specific uses which can be made of the crops or the products derived therefrom." "Proposal for a Regulation . . . on the production and making available on the market of plant reproductive material," COM/2013/0262 final – 2013/0137 (COD), Article 58, http://eur-lex.europa.eu/legal-content/EN/TXT/?uri=celex:52013PC0262 (emphasis added). This would be in line with the requirement for "an important technical advance of considerable economic significance" as a prerequisite for the compulsory cross-license.

56 For the large seed companies, the ILP is seen as an acceptable price to sustain patents on plants in Europe and other countries by balancing access and protection. If, however, patents on plants are effectively denied in Europe, this quid pro quo would fall apart. The ILP would de facto become a solution only for the United States, where a restriction of patentable subject matter is unlikely and would not require an ILP compromise.

57 Modern agricultural solutions require integration of multiple forms of innovation: seed, biotechnology, crop protection, grain storage and transport, and so forth.

between the currently diverting interests of different stakeholders. Instead of losing the incentive of patents for plant-related inventions in its entirety, such an approach could actually sustain the incentive and, due to the broader access, actually empower more innovation. In this sense, IP holders would "give up control to gain power." For technology users too, a change of culture will be necessary, both for farmers and for breeders. Instead of challenging IP rights or "free-riding," they will need a culture of "benefit-sharing" and recognizing the contribution of innovators. As in any environmental-change scenario, some players will be able to evolve, while others will not.

Finally and perhaps most importantly, the current confrontational debate needs to evolve to a collaborative dialogue based on trust. A key is to recognize that small and large players, seed companies, farmers, and politicians have a role to play and can only answer the huge challenges in a collaborative approach. Or, to cite Pope Francis once again: "A broad, responsible scientific and social debate needs to take place, one capable of considering all the available information and of calling things by their name. . . . Discussions are needed in which all those directly or indirectly affected (farmers, consumers, civil authorities, scientists, seed producers, people living near fumigated fields, and others) can make known their problems and concerns."[58]

We are at a pivotal moment. Will we continue debating or start working on collaborative solutions? The growing social and economic imbalances and the increasing fragility of our ecosystem require collective action, leaving no solution unexplored. Only through such an alliance can the current perception of "No Patents on Life"[59] evolve into a positive use of "Patents for Life," to support new innovations for the benefit of all.

[58] *Laudato Si*, ¶135.

[59] The "no patents on life" slogan alleges that a patent grants ownership to life, a myth based on a fundamental misunderstanding of the patent system. First, the concept of life is unpatentable. At a maximum, certain genes, traits, and elements isolated from a living organism can be patented. Even if a patent claim "covers" a living organism, it grants no ownership rights, but merely the right to exclude others from practicing the invention. Civil-law ownership rights to animal and plants are much more "rights on life" and are essentially undisputed.

# 14

# Germline Modification of Human Embryos, Patents and the Limits of Markets: Rethinking Equality, Human Diversity and the Question of Innovation Funding

*Katerina Sideri*

## INTRODUCTION

New research tools such as CRISPR-Cas9 have made it possible to modify the human germline, and the ethics of such technological possibilities have stirred heated controversy and hype. The chapter will discuss an aspect of the debate that is rarely brought to the spotlight: the ethics of using a market tool such as patents to bring clinical applications to the market. The problem is twofold. First, the result of patents may be that germline modifications are too costly for public health systems to afford. Germline modifications should be available to all, including the worst off: otherwise, existing inequalities will deepen to the point of balkanizing the social arena. This is the egalitarian objection. Second, the ethics of patenting interweaves with the ethics of information. Genetic information may be used in ways threatening diversity; hence, the proper uses of information should be the object of democratic debate rather than merely an object of choice for consumers. The chapter concludes with thoughts on alternative ways to finance innovation in this area if and when technologies are safe and receive social acceptance. Alternative ways to finance innovation ought to ensure universal access for therapies addressing unmet medical needs.

## 1. CRISPR AND GERMLINE INTERVENTIONS

The discovery of a new research tool called CRISPR-Cas9 has appeared in the news throughout the world several times in recent months. It has been heralded as the medical advance that can prove to be as important in the twenty-first century as vaccines were to the last.[1] The excitement lies in its potential to engineer genomes with precision and speed by inserting beneficial genes and cutting away genes thought to be maleficent. CRISPR-Cas9 is a naturally occurring bacterial defence system against viruses. In essence, bacteria, following invasion by a virus, carry a portion of the viral DNA

[1] Antonio Regalado, 'Engineering the Perfect Baby', *MIT Technology Review*, 5 March 2015, updated 23 April 2015, https://www.technologyreview.com/s/535661/engineering-the-perfect-baby/.

providing a form of 'memory', which is passed on when the cell divides in daughter cells.[2] Following subsequent invasion, the CRISPR system recognizes the virus and produces an enzyme, Cas9, to cut up the viral genetic material.

This is an astonishing discovery in many respects. For example, the CRISPR adaptive immune system works in a way that seems superior to ours (humans do not pass immunity 'memory' to their offspring). What is more, this discovery, often referred to as a 'genetic scalpel', has been modified by human intervention in such a way that it can be purposefully deployed to alter plant and animal DNA. In the past two years, it was used to turn off fruit-*ripening* genes in *tomato* plants, create malaria-free mosquitoes and muscled dogs, alter pig embryos to make their organs more compatible to transplant into humans, create autistic monkeys to serve as human disease models, repair blindness in mice, extract HIV from living organisms, target the command centre of cancer in mice, edit Huntington's disease in mice, and make superbugs that kill themselves.[3] In 2016, a Chinese group was the first to inject a human (cancer patient) with cells edited using CRISPR–Cas9 technique,[4] and in 2017 the US Food and Drug Administration (FDA) approved a clinical trial for gene therapy for a rare form of inherited blindness which partially restored the vision of several patients. In 2018, researchers in China began experimental treatments for terminally ill cancer patients, and various clinical trials were approved in Europe and the United States for diseases such as beta-thalassemia, inherited blindness and sickle-cell anaemia.

CRISPR can be deployed to tinker with cells that play a central role in human's immunity system. These cells could be edited and then infused back into a patient's body.[5] This procedure is called somatic cell editing and builds on an earlier technique known as gene therapy. Controversially, CRISPR can be used to modify a human germline: the egg and sperm, which combine to form an embryo. By editing the DNA of either these cells or the embryo itself, it could be possible to correct genes responsible for increasing the risk of developing a certain disease and pass those genetic alterations on to future generations.

In humans, future scenarios include using this technology to cut genes associated with diseases such as cystic fibrosis and Duchenne muscular dystrophy,[6] and

[2] Colin Barras, 'How Bacteria Invented Gene Editing', *BBC Earth*, 3 February 2016, http://www.bbc.co.uk/earth/story/20160202-how-bacteria-invented-gene-editing.

[3] Tanya Lewis, 'These Are Some of the Crazy Things Scientists Did Using Gene Editing in 2015', *Business Insider*, 28 December 2015, http://www.businessinsider.com/gene-editing-achievements-2015-12; Brad Bergan, '11 Incredible Things CRISPR Has Helped Us Achieve in 2017', *Futurism*, 7 August 2017, https://futurism.com/11-incredible-things-crispr-has-helped-us-achieve-in-2017/.

[4] Jamie Condliffe, 'CRISPR-Modified Cells Have Been Used in a Person for the First Time', *MIT Technology Review*, 16 November 2016, https://www.technologyreview.com/s/602880/crispr-modified-cells-have-been-used-in-a-person-for-the-first-time/; David Cyranoski, 'CRISPR Gene-Editing Tested in a Person for the First Time', *Nature* 539 (2016), 479.

[5] Daesik Kim et al., 'Digenome-seq: genome-wide profiling of CRISPR-Cas9 off-target effects in human cells', *Nature Methods* 12 (2015), 237–43.

[6] For more information, see IAMP, 'Human Gene Editing – Meetings', http://www.iamp-online.org/content/human-gene-editing-meetings; National Academies of Science, Engineering, and Medicine,

according to futuristic scenarios, add genes that offer protection against Alzheimer's and even ageing. Yet, germline engineering is controversial for many reasons. In April 2015, the publication of a paper that described the use of CRISPR technology to modify the germline of human embryos by a Chinese research team sent shockwaves around the scientific world. The embryos in question were obtained from an in vitro fertilization clinic and were stated to be not viable because they had an extra set of male chromosomes.[7] The scientific team tried to correct the gene defect that causes beta-thalassemia, a devastating blood disease, by editing the DNA of fertilized eggs. They reported that the technique is not yet stable and needs further improvement before producing children with repaired genes. Prominent scientists have since then published their views in journals such as *Nature* and *Science*, sparking a renewed debate about the potential use of CRISPR technology to modify the human germ line; and some called for a research moratorium on germline modifications.

National regulatory frameworks on research on human embryos and germline modification vary. In China, regulations clearly stipulate that gene manipulation on the human gamete, zygote[8] and embryo for the purposes of reproduction is banned. In the United States there is no law banning making embryos for research purposes and using embryos in research, but the federal government will not fund research that puts the embryo at risk. This means that research can be funded by individual state governments or private companies. However, there is legislation prohibiting the FDA from reviewing a request for clinical trial of a manipulated embryo with heritable changes.[9] The UK is different, as all aspects of embryo research are regulated by the Human Fertilisation and Embryology Authority (HFEA), and licenses are granted after case-by-case examination.

In fact, in January 2016 Dr Niakan applied for permission to the HFEA to use CRISPR techniques to conduct research on human embryos donated by IVF clinics. The HFEA licenses research on human embryos up to fourteen days after fertilization,[10] as long as the modified embryos are not implanted. The focus of Niakan's research is to understand the role of various genes in this early developmental stage, and their implications for infertility, miscarriages, developmental disorders and the therapeutic use of stem cells; the embryos will be destroyed

'Human Genome Editing Initiative', http://www.nationalacademies.org/gene-editing/consensus-study/meetings/index.htm.

7 Puping Liang et al., 'CRISPR/Cas9-Mediated Gene Editing in Human Tripronuclear Zygotes', *Protein & Cell* 6 (2015), 363–72.

8 Defined as a male or female germ cell which is able to unite with another of the opposite sex in sexual reproduction to form a zygote.

9 National Academies of Sciences, Engineering, and Medicine, *Mitochondrial Replacement Techniques: Ethical, Social, and Policy Considerations* (Washington, DC: National Academies Press, 2016).

10 Experiments are now approved in Sweden, China and the United Kingdom ( Ewen Callaway, 'Gene-Editing Research in Human Embryos Gains Momentum', *Nature* 532 (2016), 289–96).

when they are seven days old. Hugh Whittall, director of the Nuffield Council on Bioethics in the UK, said:

> The changes to DNA made for the purposes of this research could not themselves be used as part of a treatment procedure. There are, however, possible future scenarios in which a modification made in a research context – for example to correct a disease-causing genetic mutation – could, if this were to become permissible, be used in a treatment that would result in the birth of a child. Such research, which could also be licensed under current legislation, would raise a number of significant questions that should be addressed before any such work is undertaken, including about whether, and under what circumstances, a move into treatment (which would require new legislation to be permissible in the UK) could be desirable.[11]

This is an important statement, and its importance becomes more apparent in light of recent technological developments on the stem-cell frontier, which will clearly affect the way technologies enabling the selection and modification of embryos will develop in the future. The new approach currently explored by scientists combines CRISPR with stem-cell science. Scientists try to manipulate stem cells to produce egg and sperm in the laboratory. It may be possible in the future to edit the genes of the stem cells, turn them into eggs or sperm, and finally create human embryos. This may present a new way to edit the human germline genetic identity with CRISPR. As genetic modification targets the genes of stem cells prior to turning them into germ cells, there will be no direct modification of embryos.

OvaScience, a start-up created by Harvard University and Northeastern University, raised $132 million in capital in January 2015 funding this research. The idea is that infertile women will be able to produce hundreds of eggs and hundreds of embryos. Then, with the help of whole-genome sequencing, they could pick the healthiest ones. Modifying the germline identity of embryos may also be possible, for example by means of removing the gene responsible for Huntington's disease.[12]

Germline modifications in the future could potentially go beyond deleting genes. A research team at Imperial College London recently announced that they have found the two networks of genes determining whether people are intelligent or not. According to this research, intelligence is the product of collaborative work of many genes, and mutations in some of the genes lead to dullness of thinking or serious cognitive impairments.[13] Importantly, in news reports the scientists said that they

11 'HFEA Approves Licence Application to Use Gene Editing in Research', *Nuffield Council on Bioethics*, 1 February 2016, http://nuffieldbioethics.org/news/2016/council-response-to-hfea-licence-decision/.

12 Regalado, 'Engineering the Perfect Baby.'

13 Michael R. Johnson et al., 'Systems Genetics Identifies a Convergent Gene Network for Cognition and Neurodevelopmental Disease', *Nature Neuroscience* 19 (2016), 223–32.

believe there must be a 'master switch' regulating the networks and if scientists find it, they could potentially 'switch on' intelligence for everyone.[14]

Indeed, it is possible we will debate in the future whether certain clinical applications are beneficial and should be allowed. While it is true that UNESCO's Universal Declaration on the Human Genome and Human Rights (1997) adopts liberal language stating that germline interventions 'could be contrary to human dignity', the Council of Europe's Convention on Human Rights and Biomedicine (Oviedo Convention, 1997), Article 13, states that '[a]n intervention seeking to modify the human genome may only be undertaken for preventive, diagnostic or therapeutic purposes and only if its aim is not to introduce any modification in the genome of any descendants'. Currently, the Council of Europe is resolutely on its way to reaffirming the twenty-year-old ban on all inheritable interventions involving the human genome, urging members to sign and ratify the Oviedo Convention that bans all inheritable modifications of the human germline. The UK has not signed the Oviedo Convention, and a specific form of human germline modification (mitochondria replacement) has already been permitted.[15] In February 2015, the Houses of Commons and Lords in the UK approved regulations that carve out an exception to the prohibition on human inheritable genetic modification. This will allow performing what has been dubbed as 'three-person IVF' techniques into fertility clinics without human clinical trials, and with no required follow-up of any resulting children. The same year in the United States, a newly appointed committee of the Institute of Medicine (IOM) was set up by the FDA to consider the ethical and social policy issues raised by 'genetic modification of eggs and zygotes to prevent transmission of mitochondrial disease'. The committee concluded that it would be ethically permissible to conduct clinical investigations of mitochondria replacement techniques, as long as they were subject to the strict conditions and principles. The technique is not yet legal in the United States.

Circumstances and public perception of the benefits of germline modification may change in the future. This was clearly stated in the joint statement on genome editing, issued by the National Academy of Sciences (US), Chinese Academy of Sciences and the Royal Society of the UK at the International Summit on Human Gene Editing that took place in Washington, DC, in December 2015.[16] The summit addressed recent scientific developments in human gene editing and the ethical and governance issues associated with these advances. In its concluding statement, the participants of the summit said that clinical applications might be justifiable if the technology proves to be safe, if there is broad societal consensus about

[14] Sarah Knapton, 'Intelligence Genes Discovered by Scientists', *The Telegraph*, 21 December 2015, http://www.telegraph.co.uk/news/science/science-news/12061787/Intelligence-genes-discovered-by-scientists.html.

[15] Marcy Darnovsky, 'A Slippery Slope to Human Germline Modification', *Nature* 499 (2013), 127.

[16] National Academies of Sciences, Engineering, and Medicine, 'On Human Gene Editing: International Summit Statement', 3 December 2015, http://www8.nationalacademies.org/onpinews/newsitem.aspx?RecordID=12032015a.

the appropriateness of the proposed application and if important ethical and regulatory questions have first been explored in an inclusive manner. The joint statement recognized that the cases of most compelling benefit are limited and that many nations have legislative or regulatory bans on germline modification. However, as scientific knowledge advances and societal views evolve, the clinical use of germline editing should be revisited on a regular basis.

These ideas are further reinforced in the consensus report published in February 2017 by an international committee convened by the US National Academy of Sciences and the National Academy of Medicine in Washington, DC. The report concludes that such clinical trials might be permitted, but only following much more research on risks and benefits, and 'only for compelling reasons and under strict oversight'.[17] As Baylis notes:

> In the 2017 report, the 'who' is the public (which is not defined anywhere in the report) and the 'what' is no longer whether (and if so when) human germline gene editing should be permitted, but whether germline gene editing, which is (apparently) permissible for therapeutic purposes, should also be permissible for enhancement purposes.[18]

## II. WHAT ABOUT PATENTS?

Patents are granted by administrative offices which examine whether applications fulfil a set of formal and substantive requirements. A successful application for a patent fully discloses an invention that is new and inventive and can be applied in the industry (or has utility). Patent owners control the uses of the invention for a limited duration of time.

As already mentioned, the United States does *not* prohibit human germline modification. Nuclear genome transfer (to prevent the inheritance of mitochondrial disorders in children) is a form of germline modification, and there is a patent covering this technique. The technical field is in vitro fertilization, the claimed technology is for the production of oocytes for fertilization, and the disclosed methods also provide a prenatal treatment method for mitochondrial disorders.[19] Although there is a patent, the FDA has not approved clinical trials, and performing the technique is not legal in the United States. Patent applications at the EPO for the same invention have been withdrawn.[20] This is no coincidence. In Europe,

[17] National Academies of Science, Engineering, and Medicine, *Human Genome Editing: Science, Ethics, and Governance* (Washington, DC: National Academies Press, 2017).

[18] Françoise Baylis, 'Human Germline Genome Editing and Broad Societal Consensus', *Nature Human Behaviour* 1 (2017), 0103.

[19] Shoukhrat M. Mitalipov, *Methods for Mitochondrial DNA Replacement in Oocytes* (2009), US20120036591. See also CA2759608A1, EP2421958A1 (withdrawn), EP2421958A4 (withdrawn), US9434921.

[20] Ibid.

Directive 98/44/EC on the legal protection of biotechnological inventions (the Biotechnology Directive) introduced harmonized standards on the patentability of biotechnological inventions. Article 6(1) of the Directive states that inventions are not patentable 'where their commercial exploitation would be contrary to *ordre public* or morality'. The meaning of '*ordre public* or morality' is open to interpretation at the level of individual member states, but the directive provides a list with examples of inventions excluded on these grounds. Article 6(2)(b) excludes from patentability '[p]rocesses for modifying the germline genetic identity of human beings', and article 6(2)(c) bans patenting of '[u]ses of human embryos for industrial or commercial purposes'.

The European Patent Convention is an intergovernmental treaty signed by all EU Member States and other countries (such as Turkey), and the patent grant is a bundle of legal rights enforceable in designated member states. Although the Convention is not an EU institution, the two legal orders have tried to coordinate their practices. For example, the essence of the Biotechnology Directive was incorporated into the Implementing Regulations to the European Patent Convention as Rules 23b–e. What is more, in stark contrast to the United States,[21] various civil-society actors in Europe demanded complex ethical reasoning to be incorporated in what has initially been perceived by regulators as a technocratic exercise. The intense debate on the provisions of the Biotechnology Directive and the use of the opposition procedure before the EPO by civil-society actors challenging the patenting of animals, diagnostic tests and stem cells are illustrative of these tendencies.

In the well-publicized *Brüstle* case, the Court of Justice of the European Union (CJEU) gave an interpretation of the meaning of Article 6(2)(c) of the Biotechnology Directive, which rules out patentability for inventions involving uses of human embryos for industrial or commercial purposes, which proved to be highly contested. The CJEU ruled that the patenting of human embryonic stem cells violates human dignity. It said that the concept of 'human embryo' has to be understood broadly (and should include the 14 day blastocysts), because

> the European Union legislature intended to exclude any possibility of patentability where respect for human dignity could be thereby affected. ... Although the EU seeks to promote investment in the field of biotechnology, use of biological material originating from humans must be consistent with regard for fundamental rights and, in particular, the dignity of the person.[22]

However, the use of human embryos for therapeutic or diagnostic purposes which are applied to the human embryo and are useful to it is patentable.[23]

21 Shobita Parthasarathy, 'Whose Knowledge? What Values? The Comparative Politics Of Patenting Life Forms in the United States and Europe', *Policy Sciences* 44 (2011), 267–88.

22 Case C-34/10, *Oliver Brüstle v. Greenpeace* [2011], para 32.

23 Biotechnology Directive, Recital 42.

The problem with this approach is that the definition of 'human embryo' resists uniformity in Europe. In *Brüstle*, the CJEU ruled that the term is to be understood 'in a wide sense', and that any human ovum after fertilization should be regarded as a 'human embryo' within the meaning and for the purposes of the application of article 6(2)(c) of the Directive, since that fertilization commences the process of development of a human being. Nevertheless, in a later judgment in *International Stem Cell Corporation*, the CJEU decided in favour of patent eligibility of non-fertilized human ova stimulated by parthenogenesis.[24] The Court found that in order to be considered a human embryo (and be excluded from patentability), the stimulated ovum must have the 'inherent capacity to develop into a human being'. This permits the patenting of innovative pluripotent parthenotes and their applications.[25] Notably, the *International Stem Cell* decision also leaves considerable discretion to national courts. The Advocate General pointed out in paragraph 43 of his Opinion that Member States are still able to prohibit patentability of parthenotes under Article 6(1) of the Directive on the grounds of public order or morality. This will certainly lead to greater divergence in the treatment of patentability of stem cells across Member States. For example, Poland intervened in the case to argue for excluding parthenotes from patentability in the interests of safeguarding human dignity and is expected to ban such patents on the grounds of Article 6(1).

Moreover, the *International Stem Cell decision* only applies to very specific human embryonic stem cells (parthenotes) and lacks further clarification concerning other non-totipotent human embryonic stem cells, such as stem cells created through somatic cell nuclear transfer. There is still legal uncertainty on the question of patentability, but commentators note that *International Stem Cell* leaves open the possibility of patent eligibility of a wider class of stem cells than parthenotes, such as those with no inherent capacity to develop into a full human being.[26]

What about germline modifications performed in germ cells and early embryos? Remember that Liang et al. modified embryos discarded by IVF clinics that lacked the capacity to develop into a full human being. In this case, it could be argued in the future, article 6(2)(b) of the Directive should not apply to processes for modifying human zygotes, on the basis that they are incapable of developing into full human beings and transmitting modified DNA to offspring. This is important: as already mentioned, scientists are researching ways to use induced pluripotent stem cells ('reprogrammed' cells that can differentiate into almost any cell type) to create sperm or eggs that, in the future if the technology is available and safe, could be

24 Case C-364/13, *International Stem Cell Corporation v. Comptroller General of Patents, Designs and Trade Marks* [2014].

25 The difference between totipotent and pluripotent cells is that only totipotent cells can give rise to a full human body.

26 Maeve Lynch and Duncan Ribbons, 'CRISPR: the shifting sands of patentability', *Life Science IP Review*, 12 November 2015, https://www.lifesciencesipreview.com/article/crispr-the-shifting-sands-of-patentability.

manipulated and implanted using in vitro fertilization techniques.[27] In other words, in the future germline modification of induced pluripotent stem cells may be acceptable, and could be used to produce eggs and sperm for use in IVF. Patents may be allowed, as it could be argued that EPC Rule 23d(b) (implementing article 6(2)(b) of the Directive) does not apply to genetically modified pluripotent cells and the processes for their preparation. However, genetic modification of totipotent stem cells would be prohibited.[28]

I have argued elsewhere that the CJEU made a mistake when it focused on defining the meaning of the embryo, given the divergence of views on the issue.[29] In both decisions discussed previously, the court did not consider the broader effects of patenting stem cells. Patenting has an impact on the conduct of physicians, patients and researchers, and I will argue that the question of stem-cell patenting and germline modification needs to include a consideration of broader consequences such as access to therapies. Although heated discussion on the ethics of research followed the announcement of the publication of Liang et al.,[30] the ethics of patenting presents us with a set of separate questions that have not been adequately addressed. The following analysis will explain this idea.

## III. PATENTING AND THE QUESTION OF ACCESS

### A. *Justifying Patents*

The Biotechnology Directive in Europe explicitly addressed the need to promote the growth of the European biotechnology sector by harmonizing European patent laws.[31] The Directive aimed at market harmonization in a field of economic activity that held the promise of unprecedented growth.[32] When the European Commission proposed the draft legislation, it pointed to legal uncertainty due to the lack of clarity of the provisions of the European Patent Convention 1973 (EPC); the uncertainty resulted in underinvestment in important technologies, while the United States was winning the global competitive race bringing important commercial applications to the market. European companies could move their biotechnology research from the

27 David Cyranoski, 'Rudimentary Egg and Sperm Cells Made from Stem Cells', *Nature News*, 24 December 2014, https://www.nature.com/news/rudimentary-egg-and-sperm-cells-made-from-stem-cells-1.16636.

28 Heike Vogelsang-Wenke, 'Patenting of Stem Cells and Processes Involving Stem Cells According to the Rules of the European Patent Convention', *Biotechnology Law Report* 23 (2004), 155–67.

29 Katerina Sideri, *Bioproperty, Biomedicine and Deliberative Governance: Patents as Discourse on Life* (London: Routledge, 2014).

30 Sheila Jasanoff, J. Benjamin Hurlbut, and Krishanu Saha, 'CRISPR Democracy: Gene Editing and the Need for Inclusive Deliberation', *Issues in Science and Technology* 32 (2015), 25–32.

31 Aurora Plomer and Paul Torremans (eds.), *Embryonic Stem Cell Patents: European Law and Ethics* (New York: Oxford University Press, 2009).

32 Richard Gold and Alain Gallochat, *The European Directive on the Legal Protection of Biotechnological Inventions: History, Implementation and Lessons for Canada* (Ottawa: CBAC, 2001).

European Union to the United States, where patent laws were more permissive, and the European Commission urged the Council to adopt the Directive so as to provide the market incentives and legal certainty required for the biotechnology industry in Europe to flourish.[33] In short, the idea that knowledge is a valuable commodity drove legislative initiative, and the Biotechnology Directive clearly sought to structure demand and supply and consumer needs. This is the prospecting theory of patents, a species of incentives theory, which is thought to be the most prominent theory justifying intellectual property (IP) rights.

Let me pause here to give a brief account of the incentives justification. Ex ante justifications stress the importance of incentivizing the disclosure of information, or they take the form of a Lockean social contract: if individuals could agree not to imitate, they would be better off. The social contract implemented through IP law grants an individual a property right in her innovation, which allows punishing imitation unless the owner gives her permission. If we view knowledge as a public good potentially depleted by free-riding, then IP promotes the production of an important social good. The social contract contributes to the public good by granting an individual a right in the fruits of her labour against misappropriation, and society avoids the inferior outcome in which everyone imitates.[34]

Patents as a tool for commercialization[35] function as incentives in the following way: once an inventor obtains a patent on an invention, the exclusive rights allow the inventor to commercialize the invention and thereby disseminate it to the public. Referred to as a prospecting theory of patents, this ex post justification emphasizes not only the reward that comes to the patent owner in the form of profits from commercialization, but also the benefit to the public that arises from commercial dissemination.[36] Sichelman has even proposed creating an additional commercialization patent right that serves the purpose of bringing a product or process to the market, in addition to the invention patent that serves the purpose of advancing invention.[37] In more practical terms, the working requirement in some patent acts and the acceptance of evidence with regard to commercial success of an invention by courts when examining the requirement of non-obviousness reflect commercialization concerns.

Indeed, various legislative reforms were animated by this rationale. The Biotech Directive discussed above is one example. A second example is the Bayh-Dole Act, adopted by the US government in 1980. The act allowed small businesses,

33 Gerard Porter, 'The Drafting History of the European Biotechnology Directive', in *Embryonic Stem Cell Patents*, ed. Plomer and Torremans, 3–26.

34 Shubha Ghosh, 'Patents and the Regulatory State: Rethinking the Patent Bargain Metaphor after *Eldred*', *Berkeley Technology Law Journal* 19 (2004), 1315.

35 Edmund W. Kitch, 'The Nature and Function of the Patent System', *Journal of Law and Economics* 20 (1977), 267–71.

36 Shubha Ghosh, *Identity and Invention: The Culture and Ethics of Personalized Medicine Patenting* (Cambridge: Cambridge University Press, 2012).

37 Ted Sichelman, 'Commercializing Patents', *Stanford Law Review* 62 (2010),341–413.

universities, and other non-profit organizations funded with public funds to patent inventions. The idea was that if scientists could license a patented invention to a private entity, scientific breakthroughs could more easily be translated into real-world commercial applications. The explicit mandate behind this legislative reform was to bring products to the marketplace.[38]

Similar reforms with similar rationales were passed in European countries. For instance, fostering collaborations between academic and industrial research is a central element in the current policy urge in the UK and elsewhere to shorten the innovation chain by directing universities (traditionally involved in basic 'blue skies' research or 'curiosity driven' science) to seek to engage in research that can be translated in marketable biomedical products (applied science). This is happening not by the traditional means of informal communication ties and in-house development by big companies, but rather by academics and universities becoming directly involved in commercial considerations, developing an entrepreneurial spirit and thinking of research projects that will satisfy or create markets. In short, allowing universities to commercialize their research was underpinned by the following assumptions: publicly funded discoveries with economic potential should be commercialized, science best serves the public good by bringing goods to market and IP is an essential incentive mechanism and a market tool. CRISPR is no exception to this, as I will discuss shortly.

The incentives theory assumes an airtight distinction between public and private purposes, and many have challenged this assumption. An important line of criticism comes from those theorists who view IP as a tool to regulate the marketplace.[39] This view goes beyond the incentives theory and the narrow view of the patent bargain (i.e., rewards and protection of rights). It comes to grips with IP as regulating the production and dissemination of culture and as creating markets for music and books, scientific knowledge and commercial speech. But the marketplace need not be the unregulated marketplace. As Ghosh explains: 'Once patent law is viewed as a means to regulate certain behaviour and to establish market integrity, then patent law works like antitrust law, securities law, consumer protection law, and a host of other regulations that democratic countries enact to bring market values and democratic values in line.'[40] From the vantage point of view of the regulatory theory, patents are not mere private rights, but have a public dimension.

Moreover, the regulatory view of patents resolves the tension between patent law and democracy. The exceptions and limitations of the subject matter of copyright evidently serve democratic values by securing access to educational material and

[38] For criticism, see Bhaven N. Sampat, 'Patenting and US Academic Research in the 20th Century: The World Before and After Bayh-Dole', *Research Policy* 35 (2006), 772–89; Gary Rhoades Gary and Sheila Slaughter, *Academic Capitalism and the New Economy: Markets, State, and Higher Education* (Baltimore: Johns Hopkins University Press, 2004).

[39] Ghosh, 'Patents and the Regulatory State.'

[40] Ibid., 1352.

permitting parody. Ghosh also links patent law and democracy, as it not only creates a property right in science that permits commercialization but also presents a form of science policy and a form of market regulation. In a similar spirit, I have examined the ways in which civil-society movements and IP laws and policies run parallel to each other in knowledge-based economies.[41] The point relevant to the present discussion is that the regulatory view of IP broadens its prescriptive framework to include equality and democracy as animating legislative reforms.

Following this line of thinking, patents raise a different set of questions than the ones commonly raised in debates over the ethics of research: Even if we allow germline modification for certain conditions, how will this technology be developed, by whom, who will benefit and who will have access to it? The public benefit theory provides a useful lens to approach these questions. According to the public benefit theory, patents are meant to bring to the market products (clinical applications) that will benefit society. But what is 'society'? The argument may take several forms: clinical applications can benefit the embryo, parents, patient groups or society at large by eradicating disease and reducing costs. The Biotechnology Directive clearly adopts a deontic approach, emphasizing benefit to the embryo and protection of human dignity and human rights. So how does the analysis proceed from here? Either we view patents as a private right, walled off by a narrow list of exemptions and limitations – with concerns of human rights and dignity lying outside the realm of patent law – or we see patents as promoting a variety of public and private purposes.[42] The distinction has practical importance: In the former scenario we adopt a utilitarian calculus, balancing pre-emption of innovation and fostering of invention, which may be limited if there is a conflict with higher-order principles found in laws and constitutions. This approach commonly results in a very narrow interpretation of exemptions by courts and administrative agencies. In the latter scenario, interpretation and legislative reforms need to reconcile equality, dignity and private property from the start.

### B. *Patents and Public Benefit*

After this brief exposition of justifications, let us now return to the question of germline engineering. If we adopt the regulatory view of patents, the notion of public benefit invites an array of different questions. Is germline engineering medically necessary? It is already possible to test the DNA of IVF embryos and pick healthy ones through a procedure called Preimplantation Genetic Diagnosis (PGD). For example, a man with Huntington's could use IVF to have his sperm fertilize his partner's eggs and pick the ones that do not carry the mutation for

[41] Sideri, *Bioproperty, Biomedicine and Deliberate Governance.*

[42] Ibid.

implantation. If PGD can do the work we want, why go down the route of germline intervention? Should we do it simply to satisfy scientific curiosity?

In the meeting organized in Paris by the Human Gene Editing Initiative,[43] Dr Voit discussed the recent research and clinical trials regarding Duchenne Muscular Dystrophy. He said that research has shown proof of the concept of germline editing of dystrophin DNA mutations: efficacy is limited and improvements are transient, but in principle, full dystrophin correction may be possible for many partial gene duplications. However, Dr Voit said that there is 'no high unmet medical need' for germline editing for the dystrophin gene, as PGD provides a solution to almost all cases.[44]

What about modifying intelligence? Intelligence and educational achievement are highly heritable, and some suggest that children should be genetically screened at the age of four so that an individualized curriculum can be tailored to their needs. 'Understanding the specific genetic and environmental factors influencing individual differences in educational achievement – and the complex interplay between them – could help educationalists develop effective personalised learning programmes, to help every child reach their potential by the end of compulsory education.'[45]

However, other genetics experts have warned that even having intelligence gene networks does not guarantee that children will actually perform well:

> It comes as no surprise that there should be a heritable component to exam success. The approach used in this study is a standard one demonstrating that monozygotic (identical) twins' GCSE marks correlate significantly more closely than dizygotic (fraternal) twins. What we need to be careful of however is leaping to the assumption that this means that people are predisposed to do more or less well in their school exams. Genetics is the science of inheritance, not pre-determinism, and there is no substitute for hard work and application.[46]

How are we going to use this information about the mechanisms of genetics in the future? One way is to say that we will correct serious cognitive impairments. Other options are to say that we will design public policies to tailor curricula to pupils' genetic profile, or that we could intervene to improve everybody's intelligence and use such interventions as tools to alleviate existing socioeconomic disparities. Or we

43 'Workshop on Human Genome Editing in the EU', Paris, 28–29 April 2016, webcast available at http://nationalacademies.org/gene-editing/consensus-study/index.htm. See also Antonio Regalado, 'Can CRISPR Save Ben Dupree?', *MIT Technology Review*, 17 October 2016, https://www.technologyreview.com/s/602491/can-crispr-save-ben-dupree/.

44 'Workshop on Human Genome Editing in the EU.'

45 Kalli Rimfeld et al., 'Pleiotropy across Academic Subjects at the End of Compulsory Education', *Scientific Reports* 5 (2015), 8.

46 Prof. Darren Griffin, University of Kent, quoted in 'Expert Reaction to Genetics and GCSE Achievement', *Science Media Centre*, 23 July 2015, http://www.sciencemediacentre.org/expert-reaction-to-genetics-and-gcse-achievement/.

could decide that social interventions achieve more cost-effective and socially beneficial results.

Let us assume we agree that certain enhancements bring benefit to society or individuals. For example, a discussion paper by the British Medical Association suggests that 'universal access to enhancing interventions would bring up the baseline level of cognitive ability, which is generally seen to be a good thing and is one of the key aims of our ongoing attempts to improve and develop the educational system'.[47] The paper further suggests that socioeconomic inequalities could be addressed by technological means.

The provided reflections are meant to draw attention to the fact that the commercialization of CRISPR introduces a set of separate questions with respect to the role of markets and profits. It is useful to draw an analogy with the commercialization of rDNA science, which attracted multibillion-dollar investment and was built on individual entrepreneurship, start-ups and other university–industry collaborations. In this scenario, knowledge is a very valuable commodity to be traded in markets, as is proven by the patent battle between the University of California, Berkeley, and the Broad Institute over who owns the CRISPR patents. Billions of dollars in royalties are at stake, and freedom of research also means freedom of trade in knowledge goods. But these trends bring additional concerns.

### C. *Patents and the Egalitarian Concern*

Swift commercialization raises concerns about the distribution of medical benefits and whether the research will address unmet medical needs. For an egalitarian, medical interventions (germline interventions) that we consider important should be made available to all.

Let us now look more closely into the egalitarian concern. One particular aspect of the debate is that patents make the cost of therapies unacceptably high by stifling competition, a concern expressed perhaps most famously in the worldwide uproar over patenting the BRCA1 and BRCA2 genes associated with cancer. Access to important genetic testing was hindered due to exclusive licensing strategies of patent owners.[48] A reduction in competition resulted in standard setting that limited the kind of tests to which patients had access.[49] Gene therapies raise the same concerns with respect to access, as viral vectors and gene transfer therapies, cells, cell lines, and higher-order animal models in which a gene that has been inserted or knocked out may attract patent protection as inventions. A 2018 study by the technology

47 British Medical Association, *Boosting Your Brainpower: Ethical Aspects of Cognitive Enhancements* (London: BMA, 2007), 18.

48 Shobita Parthasarathy, 'Reconceptualizing Technology Transfer: The Challenge of Building an International System of Genetic Testing for Breast Cancer', in *Shaping Science and Technology Policy: The Next Generation of Research*, ed. David H. Guston and Daniel Sarewitz (Madison: University of Wisconsin Press, 2006), 333–57.

49 Eileen M. Kane, 'Patent-Mediated Standards in Genetic Testing', *Utah Law Review* (2008), 835.

consulting firm iRunway provides a useful analysis of the patents filed for CRISPR.[50] The study identifies 4,336 filings in various parts of the world, including 410 granted CRISPR patents and 3,926 pending or lapsed applications. Broad Institute is associated with 297 assets, Harvard University and the Massachusetts Institute of Technology have 343 and 316 each, and the University of California has 131 patents. Surprisingly, DowDuPont, the world's largest chemical company, is now the biggest owner of CRISPR patents and applications globally with 514, or 12 per cent of the total. The recent proposal by *MPEG* LA, an IP rights management company, to create a CRISPR licensing facility echoes the concern that a patent thicket in this field may slow down innovation. A different concern is raised by the few university spinoffs holding essential and very broad patents in this area. The patent wars between the team led by the Broad Institute and the team led by the Universities of California and Vienna over who owns the CRISPR-Cas9 technology are well documented. For now, Broad Institute seems to have won the battle in the United States, but nothing is certain, as the decision is under appeal. On 17 January 2018, the EPO revoked one of the Broad Institute's patents relating to a fundamental aspect of CRISPR technology. The opposition board upheld the EPO's preliminary opinion that the claims were invalid (invalid priority claim).[51] The financial importance of the battle is enormous, as the development of relevant technologies in the lucrative market for human therapeutics and treatments is a multibillion-dollar enterprise fuelled by entrepreneurship, venture capital and university spin-off companies. Editas Medicine, Intellia and Crispr Therapeutics are university spinoffs that sprung from the principal researchers involved in developing the CRISPR technology. However, as Contreras and Sherkow explain, each of the key CRISPR patent holders has granted exclusive rights for the development of all therapeutic applications associated with their patents to the spinoff or 'surrogate' company formed by their institutions.[52] Although this presents a common way of thinking about university technology transfer (a university outsources the licensing and commercialization of a valuable patent portfolio to a private company), the authors suggest that it could bottleneck the use of CRISPR technology to discover and develop useful human therapeutics.[53]

[50] iRunway, *CRISPR: Global Patent Landscape* (Bangalore: iRunway, 2018), http://insights.i-runway.com/acton/fs/blocks/showLandingPage/a/4977/p/p-008c/t/page/fm/5.

[51] Rose Hughes, 'EPO Revokes CRISPR Patent – A Clear-Cut Case of Invalid Priority?', *IP Kat*, 22 January 2018, http://ipkitten.blogspot.com/2018/01/epo-revokes-crispr-patent-clear-cut.html.

[52] Jorge L. Contreras and Jacob S. Sherkow, 'CRISPR, Surrogate Licensing, and Scientific Discovery', *Science* 355 (2017), 698–700.

[53] Note that CRISPR patents from both UC and the Broad Institute have been widely licensed to academic and commercial entities. UC, the Broad Institute, and more than seven hundred other institutions have all agreed to make their CRISPR IP available through AddGene, a non-profit repository and patent licensor of CRISPR technologies for academic organizations. Moreover, academic institutions have engaged in broad cross-licensing agreements. Jacob S. Sherkow, 'The CRISPR Patent Landscape: Past, Present, and Future', *The CRISPR Journal* 1 (2018), 5–9.

These are disheartening developments. And there are more problems created by patents in this area. As Sherkow explains, the logic of drug pricing is carried over to the field of gene therapies.[54] The debate with regard to drug pricing is fierce, and the main reason is the opacity of the system. Yet patents are one reason behind the high prices. As explained, the logic of patents is that those developing a useful technology will be able to recoup investment. Pharmaceutical research and development is notoriously expensive and uncertain. Similarly, gene therapies require cutting-edge research and expensive clinical trials, serve limited populations and many times offer one-dose treatments.

Sherkow further offers an illuminating example illustrating these difficulties: Juno Therapeutics (a joint *spinoff* from Fred Hutchinson Cancer Research Center in Seattle, Memorial Sloan-Kettering Cancer Center in New York, and Seattle Children's Research Institute) spent around $300 million on researching CRISPR CAR-T therapy.[55] This procedure involves modifying a patient's T-cells to express an antibody so that antigens on leukemic B-cells are produced, and reinjecting these modified T-cells back into the patient. In Juno's clinical trials, five out of thirty-eight trial participants died. In the future, if Juno develops a safe therapy, the price tag will make up for the losses and the costs of further research. Patents function as an asset by making it possible to demand such prices for an exclusive period of time. What is more, patents allow the therapy's developer to set prices above manufacturing costs; as a result, the ease of using CRISPR therapies does not mean that these will be cheaper than the alternatives.

A regulatory view of patents zooms onto the broader consequences. Bearing in mind that other technologies such as PGD give future parents the possibility to choose healthy embryos. If germline modification techniques are too costly to be afforded by public health systems, then it could be argued that it is unacceptable to allow patents if the inventions will only benefit the well-to-do classes – who will have access to costly therapies – and create a system of competition between private clinics offering relevant services and potentially off-label use of relevant technologies. A related argument is that since markets cannot distinguish between urgent needs and desires, R&D will be directed to create markets for expensive therapies at the expense of the needy and neglected segments of society.[56] This is the egalitarian argument: justice requires equal access to germline modifications, and private appropriation will result in unequal access. This proposition assumes the importance of equal access for all – if this cannot be assured, it is better that no one has access.[57]

54 Jacob S. Sherkow, 'Genome Editing. CRISPR, Patents, and the Public Health', *Yale Journal of Biological Medicine* 90 (2017), 667–672.

55 Ibid.

56 Ghosh, *Identity and Invention*, discusses the argument in the context of race-specific patents.

57 Oliver Feeney, 'Incentives, Genetics and the Egalitarian Ethos', *Ethical Perspectives* 19 (2012), 83–102; Colin Farrelly, *Biologically Modified Justice* (Cambridge: Cambridge University Press, 2016).

This argument has been made by commentators such as Peter Singer, who suggested that the most significant objection to creating a genetic supermarket (the idea that within a free society and a free market, these techniques should be available for purchase) is the threat to the ideal of equality of opportunity.[58] The same concern appears in the aforementioned British Medical Association discussion paper on the ethics of cognitive enhancements and the 2002 Nuffield report on Genetics and Human Behaviour.[59] The Nuffield report also points to the difficulty in drawing a distinction between therapy and enhancement. It will not be possible to draw a line between (a) techniques that we ought to make available to everybody on the basis that they are therapeutic and satisfy an urgent medical need, and (b) techniques that are merely enhancements and hence do not invite support through the public health system. Singer also observes that there is

> no bright line between selection against disabilities and selection for positive characteristics. From selecting against Huntington's Disease it is no great step to selecting against genes that carry a significantly elevated risk of breast or colon cancer, and from there it is easy to move to giving one's child a better than average genetic health profile. Similarly, if almost all of us are willing to abort a fetus that has Down's syndrome, most of us will also be willing to abort one with genes that indicate other intellectual limitations, for example genes that correlate with IQ scores below 80. But why stop at 80? Why not select for at least average IQ? Or a bit above average?[60]

The Nuffield report endorses the egalitarian position. If the well-off have access to germline interventions enhancing traits, the result will be greater social inequalities and polarization. Enhancing traits such as intelligence poses a serious threat to equality of opportunity; therefore such enhancements should be available to everybody, including the worse off. In a similar fashion, Mehlman warns that the rise of a 'genobility' will destroy the foundations of modern democracy; he argues in favour of prohibiting germline interventions, limiting government funding and restricting patenting.[61]

Can patent law take into account these broader social-justice considerations? Ghosh argues that if one accepts that IP law has multiple consequences beyond the effects on creators and innovators, the question becomes: what duties do owners have for the management of new works?[62] One way to answer this is by allowing limitations on exclusive control of owners over the invention. Ghosh continues:

58 Peter Singer, 'Foreword: Shopping at the Genetic Supermarket', in *The Ethics of Inheritable Genetic Modification: A Dividing Line*, ed. John Rasko, Gabrielle O'Sullivan, Rachel Ankeny (Cambridge: Cambridge University Press, 2010), xiii–xxxii.

59 Nuffield Council on Bioethics, *Genetics and Human Behaviour: The Ethical Context* (London: NCOB, 2002).

60 Singer, 'Foreword', xxiv.

61 Maxwel Mehlman, *Wondergenes: Genetic Enhancement and the Future of Society* (Bloomington: Indiana University Press, 2003), 184.

62 Shubha Ghosh, 'Duty, Consequences, and Intellectual Property', *University of St Thomas Law Journal* 10 (2013), 801–19.

> However, deriving duties (whether as affirmative obligations or limitations) from consequences can be complicated and futile, and tracing the effects of intellectual property laws can be indeterminate and speculative. The language of consequence readily gives way to debates over rights and duties based on deontic concepts, such as the right to the fruits of one's labor or the right to health and safety. While such rights talk can be engaging, it suffers from some of the same problems as open-ended consequentialism. How are we to balance when rights are in conflict? Is that balance itself based on consequentialism?[63]

His solution is to use the following guiding principle: effects (consequences) of rights (property) on the duties of others limit the subject matter of patent law. Liddell also argues that the EPO has so far adopted a very narrow interpretation of the morality provision of the EPC by focusing exclusively on objections to the technology and its use/exploitation in the sense of utilization, while completely ignoring broader consequences of patenting.[64] She builds on Burk and Lemley's work to argue that we should understand the morality exclusion in patent law as a 'policy lever' and tailor its interpretation to the utilitarian objective of promoting socially beneficial inventions in a manner compatible with a fair and just social organization.[65] This approach would invite policy analysis and robust normative input by judges and patent examiners, with the goal to develop the law in ways that promote desirable goals and behaviour. In the same spirit, I have argued elsewhere that if patent offices adopt Henry Richardson's method of specificationism, a form of Aristotelian practical syllogism, they could go beyond shallow cost–benefit analysis and a legalistic application of the appropriate norm by administrative agencies.[66] Richardson's model of specifying norms in the light of ideas about good social organization allows both for resolving complex conflicts between norms to come up with a new norm (the product of creative synthesis of the content of conflicting norms), and for continuously revising the ends of policy (broad ideas about what we want as a society) in the light of new information.

To return to the discussion of germline interventions: banning patents is susceptible to the objection that it 'levels down'. The argument is as follows: preserving incentives for investment into research for gene therapy is of crucial importance. Biotechnologies emanate from university research and their spinoffs. These spinoffs need patent protection to attract investors; otherwise important university research will not find its way to the marketplace. By banning patents on germline interventions, we preclude the less well off from access to important therapies at a later point in time; and for this reason it is wrong to 'equalize downwards'. Even if for at least

63 Ibid., 802.

64 Kathleen Liddell, 'Immorality and Patents: The Exclusion of Inventions Contrary to Ordre Public and Morality', in *New Frontiers in the Philosophy of Intellectual Property*, ed. Annabelle Lever (Cambridge: Cambridge University Press, 2012), 140–71, esp. 149.

65 Ibid.; for a summary, see Liddell and Ravenscroft, this volume.

66 Katerina Sideri, 'Practical Reasoning, Impartiality and the European Patent Office: The Legal Regulation of Biotechnology', *European Law Journal* 18 (2012), 821–43.

a limited period of time there will be some who do not have access, we will ultimately reach the goal of universal access. Ronald Dworkin argues: 'The remedy for injustice is redistribution, not denial of benefits to some with no corresponding benefits to others.'[67] For example, an influential libertarian defence of enhancements points out that private schooling is already available to the children of the well-to-do.[68] If these practices already increase chances of success and cultivate intelligence, why ban access to enhancements? No clear distinction can be drawn between genetic interventions and social interventions, and their effects on perpetuating inequalities are the same.

Patents can be justified as promoting the public good, and even if the wealthy gain early access, these technologies will become available to all in the future. Farrelly, for example, argues that a policy of equal access could disincentivize private forces from entering the market and result in poor investment in important health technologies.[69] He further argues that an ideal egalitarian public-policy scenario, which opposes IP rights in an effort to make socially beneficial technologies available to all, is unrealistic.[70] Under the 'market stimulus' view, allowing firms to restrict access based on the ability to pay will incentivize further cost-cutting innovation and in the end will benefit the worse-off as well.[71]

These arguments are familiar. From the vantage point of IP theory, the argument endorses the incentives theory and involves a calculus of incentives to innovation versus pre-emption of innovation. But how essential are patents really? Leaving aside the empirical debate on whether it can be substantiated that patents are important in the process of translating university research into commercial applications, there remains the theoretical debate over patents' purposes. If one adopts the laissez-faire free-market model as an essential incentive mechanism for the production of the public good of knowledge, the answer is that patents are absolutely essential. However, if we adopt the view that markets need to be appropriately regulated and patents present one such form of regulation, then we come up with more sophisticated answers. And for good public policy reasons, we must balance self-interested motivations (from both supply and demand) with other goals such as access to the benefits of important therapies for all, including the worst-off – and we must address the effects of medicalization. The following section will consider the latter idea.

67 Ronald Dworkin, *Sovereign Virtue: The Theory and Practice of Equality* (Cambridge, MA: Harvard University Press, 2000), 440.

68 Nuffield, *Genetics and Human Behaviour.*

69 Colin Farrelly, 'Genes and Equality', *Journal of Medical Ethics* 30 (2004), 587–92.

70 Colin Farrelly, 'Gene Patents and Justice', *The Journal of Value Inquiry* 41 (2007), 147–63.

71 Gillian K. Crozier and Christopher Hajzler,'Market Stimulus and Genomic Justice: Evaluating the Effects of Market Access to Human Germ-Line Enhancement', *Kennedy Institute of Ethics Journal* 20 (2010), 161–79.

### D. *Patents and Democracy*

For those concerned with the effects of medicalization, the creation of markets for germline modifications to satisfy people's desires is met with deep scepticism. The problem is the medicalization of normal characteristics, resulting in loss of this diversity and the loss of 'the merely unusual, or eccentric'.[72] Singer further comments on the work of Antony Rao, a specialist in behavioural therapy in children:

> [M]any middle and upper class parents come to him when their children behave in unusual ways, wanting them to be medicated, because 'they fear that any deviation from the norm may cripple their child's future.' If this is true of behavioural abnormalities that for many children are merely a passing phase, it is likely to be even truer of genetic abnormalities.[73]

It will be easy to confuse abnormal genes with unusual genes, as genetics is not a science of predeterminism but will deliver solutions that fit into existing social and scientific patterns of thinking.

This latter concern should not be underestimated. If many parents decide to terminate the pregnancy, there will be a loss of diversity and perhaps reduction in species' capacity to adapt to changing circumstances.[74] Mukherjee further explains that from research on the structure of DNA to the human genome project, there was a linear pattern of research questions and discoveries – but as genetics shifted its focus from pathology to normalcy, the scientific inquiry became thematic, organizing itself around the genetics of race, gender, sexuality, temperament and intelligence – in one word, personality. One of the most penetrating powers of genomics is to organize classes of similar genomes in groups, but the choice of what groups can be often driven by cultural standards. For example, the genetics of addiction has attracted considerable attention lately. Commentators warn that although behavioural traits such as addiction have a strong genetic influence,[75] the same mutation is present in high risk-takers and people seeking adventure.[76] It is just a question of learning to direct one's talents to the right endeavours.[77] We should not mistake behaviours for traits, and if we do, we underestimate how genes work together.

Relying on market forces and satisfying consumer need for any kind of intervention is fraught with difficulties. Incentives for more innovation targeting consumers who choose what therapy to consume or not is deeply problematic, as it implies the idea of passive consumption of culture. According to this, people will shop for the improvements that fit best their private plans. These plans may vary, as parents may

72 Singer, 'Foreword', xxvi.
73 Ibid.
74 Siddhartha Mukherjee, *The Gene: An Intimate History* (New York: Simon & Schuster, 2016).
75 Robert Plomin et al., 'Top 10 Replicated Findings from Behavioral Genetics', *Perspectives on Psychological Science* 11 (2016), 3–23.
76 Mukherjee, *The Gene*.
77 Marc Lewis, *The Biology of Desire: Why Addiction Is Not a Disease* (New York: PublicAffairs, 2015).

try to increase children's chances of success, or produce the best child there is or children who have a good enough life or a life worth living. But it is naïve to believe that all this genetic information will only have an impact on how decisions are made in the private realm. Public life will be reorganized too. From school curricula to job-candidate selection procedures, decisions will be informed by choices to enhance (or not) and what traits are enhanced. These will be private choices with public repercussions, and too much deference to market interests in defining the meaning of health and disease and who has access to therapies will conflict with democracy. In other words, it is not only inequality but also deference to the market that should concern us: passive consumption of culture (and science and technology is part of culture) is not enough. The future of technological societies should be the proper object of democratic debate, and what the public thinks should be taken into account.

We need to collectively take a stance on the proper uses of genetic information. Anderson wrote that democratic tradition and the notion of collective self-rule (citizens participating in political decision making as equals) requires citizens to debate in politics what goods should be provided by the community as a whole to its members.[78] In certain situations, having a voice in politics cannot be replaced by giving consumers the power to choose to accept or deny a given arrangement. There are cases where we need to shape the background conditions of interaction and the content of goods we need to provide in common. Certain goods need to be distributed in accordance with shared principles, not just to satisfy people's wants. And we need to agree as a society what goods should be made available to all, or should satisfy urgent needs and not merely intense desires.

The argument that we should let people purchase enhancements is too shallow. Markets can give people what they want, but markets are short-term maximizers and people most of the time satisfy parochial desires.[79] Should these forces drive permanent modifications in the human genome? Should we let these forces uncritically drive the reorganization of social life? Or should there be public debate to protect important social goods such as human diversity and deliberate on the socially just distribution of germline therapies? I think the answer to the last questions is 'yes'. This chapter further argues that the decision concerning how innovation is funded is a crucial component in this debate.

## IV. ALTERNATIVE WAYS TO FUND INNOVATION

Patents are not the only way to incentivize research. Recently, the UN High-Level Panel on Access to Medicines proposed that in some instances market tools such as

78 Elizabeth Anderson, 'The Ethical Limitations of the Market', *Economics and Philosophy* 6 (1990), 179–205.

79 Miriam Galston, 'Taking Aristotle Seriously: Republican-Oriented Legal Theory and the Moral Foundation of Deliberative Democracy', *California Law Review* 82 (1994), 329–99.

patents fail to deliver their promise.[80] Rare diseases and diseases of the poor fail to attract investment because they are not profitable markets. Similar problems are presented in the market for antibiotics, as we want new products that will be used sparingly for public health reasons (antimicrobial resistance).

A UNITAID discussion paper explains that delinkage mechanisms present alternative ways to finance innovation to ensure both research and development for new therapies and broad access to new medicines.[81] Moreover, a recent paper addressing health-security issues stemming from antimicrobial resistance proposes, first, to increase public funding of antibiotic research at the preclinical stage (to generate scientific advances in university laboratories and other research institutions); second, to give public-sector support in the form of tax credits, prizes, grants, and enhanced support of clinical trials; and third, to form public–private partnerships (PPPs).[82] After marketing approval and registration, companies producing a truly novel antibiotic should receive rewards according to R&D costs invested or potential therapeutic value. Subsequent annual payments may increase depending on data regarding clinical effectiveness and other data indicating the social value of the medicine.

I propose to think in similar terms for germline interventions. We need a funding mechanism for treatments that will address unmet medical needs and produce therapies that can be provided to all present patients of national health systems. I come to the same conclusion as Singer: 'Shopping at the genetic supermarket has taken us to the surprising conclusion that the state should be directly involved in promoting genetic enhancement. The justification for this conclusion is simply that it is preferable to the most probable alternative – leaving genetic enhancement to the marketplace.'[83] Concerns that the proper uses of information be the subject of democratic debate should not be taken lightly. On a different note, as the authors of the antimicrobial-resistance paper acknowledge, global collective action, rather than isolated efforts, is required to implement delinkage proposals. Such proposals need to be implemented across countries and need to be on the global political agenda, so as to find ways to finance new global institutional arrangements that can ensure collective action.

## CONCLUSION

This chapter defended the idea that patents should be conceived as a tool to regulate the marketplace, rather than an absolute precondition of innovation, and asked a set

80 *Report of the United Nations Secretary-General's High-Level Panel on Access to Medicines: Promoting Innovation and Access to Health Technologies*, September 2016, http://www.unsgaccessmeds.org/final-report/.

81 UNITAID, *An Economic Perspective on Delinking the Cost of R&D from the Price of Medicines* (Vernier: WHO, 2016).

82 Ardal C. et al., 'International Cooperation to Improve Access to and Sustain Effectiveness of Antimicrobials', *The Lancet* 387 (2016), 296–307.

83 Singer, 'Foreword', xxix.

of important questions: Even if we allow germline modification for certain conditions, how is this technology going to be developed? By whom? Who will benefit, and who will have access to it? Delinkage mechanisms are a possible solution addressing the problems described in the chapter; but to be successful, we must seriously consider the need for international coordination. The statement generated at the International Summit on Human Gene Editing in 2015 stressed the need for creating an international ongoing forum of discussion over potential clinical uses of gene editing, to help inform decisions by national policymakers and others, and to formulate recommendations and guidelines and promote coordination among nations.[84] These recommendations point in the right direction.

Yet, what has not been included in the discussion is talk about ways to finance innovation in the area. This is a crucial omission. I hope the chapter has succeeded in highlighting the catalytic nature of this aspect of the debate. Even if the US or Europe ban germline interventions and patenting, we would still have to deal with the fact that we live in a global economy. National bans will not be enough. Solutions must have a global character, given the global nature of threats and possibilities stemming from such potent technologies.

[84] National Academies, 'On Human Gene Editing.'

# 15

# Patent Governance, Ethics and Democracy: How Transparency and Accountability Norms Are Challenged by Patents on Stem Cells, Gametes and Genome Editing (CRISPR) in Europe

*Ingrid Schneider*

## 1. INTRODUCTION

Patents in the field of the life sciences and particularly in reproductive medicine and genetic engineering have been controversial for several decades. Contestations of such patents in civil society were responded to by the European Parliament and by some EU member states. As a result, the Directive 98/44/EC of the European Parliament and of the Council of 6 July 1998 on the Legal Protection of Biotechnological Inventions (hereinafter the Biotech Directive) was passed after ten years of difficult negotiations, and followed the European Parliament's rejection of an earlier draft in March 1995. Vocal protest and dissent led to two new articles in the Directive, which were included to preserve European fundamental values and human rights norms in patent law.[1] Articles 5 and 6 can be considered an attempt at 'upstreaming' ethics and anticipatory impact assessment in biotechnology patent governance.

*Article 5* of the Biotechnology Directive 98/44/EC states that

> (1) The human body, at the various stages of its formation and development, and the simple discovery of one of its elements, including the sequence or partial sequence of a gene, cannot constitute patentable inventions. (2) An element isolated from the human body or otherwise produced by means of a technical process, including the sequence or partial sequence of a gene, may constitute a patentable invention, even if the structure of that element is identical to that of

[1] For the legislative history, see Schneider, I. 2009: 'Can Patent Legislation Make a Difference? Bringing Parliaments and Civil Society into Patent Governance', in Haunss, S./Shadlen, K. (eds.), *The Politics of Intellectual Property*. Cheltenham: Edward Elgar, 2009, 129–57; Gold, R./Gallochat, A. 2001: 'The European Biotech Directive: Past as Prologue', *European Law Journal* 7(3): 331–66; Porter, G.: 'The Drafting History of the European Biotechnology Directive', in Plomer, A./Torremans, P. (eds.) *Embryonic Stem Cell Patents*. Oxford: Oxford University Press, 2009, 3–26. The legislative text and further documents are available at http://ec.europa.eu/growth/industry/intellectual-property/patents/biotechnological-inventions/ (all URLs last accessed 10 June 2018).

a natural element. (3) The industrial application of a sequence or a partial sequence of a gene must be disclosed in the patent application.

In *Article 6*, a 'morality clause' was included. Article 6(1) provides that 'inventions shall be considered unpatentable where their commercial exploitation would be contrary to *ordre public* or morality'. Article 6(2), meanwhile, lists a number of specific examples of biotechnological inventions that are excluded from patentability on moral grounds. Among them are (a) 'processes for cloning human beings', (b) 'processes for modifying the germ line genetic identity of human beings', and (c) 'uses of human embryos for industrial or commercial purposes'.

These Articles, complemented by explanatory recitals,[2] were a political compromise in response to the view expressed by the European Parliament that the patenting of biological materials, especially those of human origin, raises important ethical and social concerns. The non-exhaustive list of examples in Article 6(2) was intended to clarify how the general 'morality' exclusion in Article 6(1) should be applied. However, the interpretation of this Article 6 in relation to patent applications for inventions concerning human embryonic stem cells (hESC) has given rise to extensive, contentious, and lasting legal, ethical, and policy debates. These controversies also led to a number of legal cases, at the European Patent Office's (EPO) Boards of Appeal, in some EU member states, and at the Court of Justice of the European Union (CJEU), which will be referred to later in this chapter.

Legal clauses necessarily must be formulated in a somewhat abstract and general manner to encompass a multitude of empirical cases and also to be applicable to future developments, particularly in fast-moving technological fields. Hence, they always require interpretation, inter alia, by courts and administrative bodies. However, sometimes administrative agencies develop an agenda of their own and thus interpret legal norms in ways which are not necessarily shared by a broader public or by the legislative intent of the European legislator.

The European Patent Office (EPO), as an organisation constituted by the 1973 European Patent Convention (EPC), is not subjected to the treaties and legislation of the European Union. However, the EPO's Administrative Council adopted all the Articles of the Biotech Directive in its own legal framework, the implementing rules of the EPC. Accordingly, Articles 5 and 6 were taken up in Rules 28 and 29 of the EPC.

At the EPO, substantive patent law for new technological fields evolves through the interplay between the granting departments and the Boards of Appeal, which are quasi-judicial bodies. Patent examination, grant and decisions made in opposition

[2] Recitals 37–42 refer to Article 6 directly. Recitals 13, 14, 16, 17, 18, 20, 21 and 37 consider the aim of the Directive. In EU law, a Recital sets out reasons and arguments formally given for the new legal act; conventionally, Recitals start with the word 'whereas'. In contrast to the Articles of a Directive, the Recitals in themselves are not binding EU law but should be taken into account by courts, lawyers and the patent offices, when interpreting the meaning of the Directive.

procedures and by the Boards of Appeal are tacit *policy-making* practices, often masked as mere administrative execution of law. The EPO's technocratic self-determination has raised questions about division of power, transparency, accountability and democratic control. It has therefore been accused of pursuing a *pro-patent bias,* and a misunderstanding of its public mission, understanding itself more as a service provider for industry and researchers than as a truly public agency committed to the common good.[3]

Some of these critiques have also been raised with respect to the EPO's application of Rules 28 and 29 of the EPC. While some researchers have complained about an interpretation too strict,[4] others have suggested that the EPO has made considerable efforts to find loopholes and legal circumventions which would allow controversial human embryonic stem cell–related patents to be granted. In the following, some of these controversies will be described and put to critical inquiry. First, the normative presuppositions of this article will be disclosed. Second, some empirical data on the scientific development of (human pluripotent) stem cell research and respective patent applications and grants in Europe are provided. Third, an overview is given of the case law at the CJEU and the EPO's Boards of Appeal. The fourth part describes the EPO's granting practice and its application of the legal rules. It discloses that the EPO has fabricated a way of granting patents related to human embryonic stem cells which bypasses strict legal prohibitions. This practice of the EPO is then criticised in the following parts, first based on a different interpretation of Article 6 (2)(c) and second on a sound scientific evaluation of the evidence referred to by the EPO, and third with respect to the parthenogenetic stem cell source used as evidence for so-called non-destructive uses of embryos. The next part alerts to more recent developments, coined constructive uses of embryos, as regards the artificial creation of gametes and embryos. Finally, it is argued that patents on new genome editing technologies, in particular CRISPR-Cas, might also jeopardise the prohibitions on human germ line modification enshrined in the Biotech Directive. Pointing to these threats, therefore, the last part calls for attention and action by respective bodies to ensure that the intent of the European legislator in passing the Biotech Directive will be kept alive and applied. This also requires that the administrative agencies, in particular the European Patent Office and the European Commission, will act transparently and account for their application of the legal rules set by the European legislator and interpreted by the CJEU.

[3] The EPO is self-funded by the fees of the patentees. Only applicants' interests are systematically represented within its system. Therefore, the EPO is at risk of treating applicants as customers to be served, and it may thus be susceptible to be captured by its clients. Over time, this has arguably resulted in an expansion of patent eligibility, and in patents granted with a broad scope. See Schneider, I. 2009: 'Governing the Patent System in Europe: The EPO's Supranational Autonomy in Need for a Regulatory Perspective', *Science and Public Policy*, Vol. 36 (8): 619–28.

[4] Mahalatchimy, A. et al. 2015: 'The Impact of European Embryonic Stem Cell Patent Decisions on Research Strategies', *Nature Biotechnology* 33(1): 41–43; Wong, A. et al. 2018: 'Human Stem Cells Patents – Emerging Issues and Challenges in Europe, United States, China, and Japan', *The Journal of World Intellectual Property*, March 2018, https://doi.org/10.1111/jwip.12098.

## II. A TELEOLOGICAL AND HISTORICAL INTERPRETATION OF DIRECTIVE 98/44/EC IS NEEDED

This chapter considers it necessary to recall the legislative history to reveal the intent of the legislator and also to apply an interpretation which takes the purpose of the law into account.[5] Therefore, it departs from the normative view that in codifying Articles 5 and 6 of the Biotech Directive, the European legislator has created a genuine European *ordre public* for biotechnological inventions. This *ordre public* norm must be respected by following a historical and teleological interpretation. Correct interpretation of the law should not be undermined by confinement to the literal meaning of the statutory text, or by skilful drafting of patent claims to circumvent the exclusions, or even by twisting the law.[6]

Inclusion of these Articles 5 and 6 were central preconditions for the European Parliament to pass Directive 98/44/EC in 1998, which it had rejected in 1995. These Articles and the respective Recitals preserve the democratic process of interpreting the *ordre public* clause in European patent law as a public policy clause.[7] Such an understanding, as expressed in the Biotech Directive, has incorporated two meanings and rationalities: First, the *ordre public* and public policy objections to patentability testify for a constitutionalisation patent law, in which such clauses introduce constitutional standards for integrating patent law into the general European order of fundamental norms and values.[8] Second, the legal concept of '*ordre public* and morality' was interpreted by the European legislator with respect to consequentialist considerations: patent law should be harnessed as one of several means to provide for innovation policies which

5 Law theory distinguishes different methods of interpretation of EU legislation – namely literal, contextual, comparative, teleological and historical interpretation. Different methods lead to different outcomes.

6 See CJEU, Grand Chamber, Case C-34/10 *Brüstle v Greenpeace eV*, [2011] ECR 1–0982, [2011] 1 CMLR 41; EPO Enlarged Board of Appeal Decision G2/06, *WARF*; and Comments of the President of the EPO on G2/06, 2006.

7 The author's opinion is based on the insights into European patent governance gained in her academic work, in which she has traced in detail more than two decades of the controversial legislative history of Directive 98/44/EC, from its first Agenda Setting in 1988 to its final adoption in 1998, and subsequent contentions on the implementation in all the members states in the EU and the EPC. To this purpose, she has conducted thorough analyses of legal and policy documents and interviews with dozens of MEPs, legal scholars and stakeholders. The analysis was successfully accepted as her habilitation thesis in political science in 2009 and was published as *The European Patent System. Shifts in Governance through Parliaments and Civil Society* in 2010 (*Das Europäische Patentsystem. Wandel von Governance durch Parlamente und Zivilgesellschaft*, 771 pp., Frankfurt/New York: Campus). Parts of the analysis are available in English, e.g., Schneider, I. 2009 (see note 1); Schneider, I. 2012: 'To Be or Not IP? Exploring Limits within Patent Law for the Constitutionalization of Intellectual Property Rights and the Governance of Synthetic Biology in Human Health', *Law and the Human Genome Review* 37: 193–233. The author also was a member of the European Commission's Expert Group on the development and implications of patent law in the field of biotechnology and genetic engineering (2013–16) and has submitted two dissenting opinions to the Expert Group's final report.

8 See Geiger, C. 2006: '"Constitutionalising" Intellectual Property Law? The Influence of Fundamental Rights on Intellectual Property in the European Union', *Int'l Rev. Intell. Prop. & Competition Law* 37: 371–406.

are responsible, sustainable and promote public welfare. The latter entails a modern understanding of patent governance as a mode of proactive, sociopolitical regulation of technological trajectories, in accordance with European social values and norms, as enshrined in the Charter of Fundamental Rights and in the precautionary principle.

Such a shift in the understanding of patent governance has resulted in special specifications and a non-exhaustive list of exclusions. For example, in exempting '(a) processes for cloning human beings' and '(b) processes for modifying the germ line genetic identity' in Articles 6(2), the EU legislator has regulated scientific inventions which at that time were not yet feasible. The EU legislator thus provided patent legislation based on an anticipatory impact assessment, in which certain techniques were excluded from patentability because they were considered socially undesirable and/or violative of human dignity. Pursuing the idea of an ex ante control of the social desirability of an invention, patent non-eligibility serves to express the societal disapproval of an invention. It is assumed that non-patentability induces a disincentive for respective research and development. It must be emphasised, however, that non-patentability is not penal prohibition. It only removes an incentive – the legal exclusivity of said invention and the expected return on investment – and thus it is used as an indirect mode of regulating R&D without forestalling the freedom of research.

Taking this understanding of Directive 98/44/EC into account, it would be inadequate to restrict the substance of these provisions by a too narrow or merely formalistic interpretation.[9] Implementing the law in both letter and spirit must become an integral part of granting practices at the European Patent Office and national patent offices, of rulings in case law and of interpretations issued by the European Commission.

## III. EMPIRICAL FACTS, STATISTICS AND CONTEXTS ON HUMAN PLURIPOTENT STEM CELLS

It is not possible to provide a full account of the scientific evidence and empirical developments in science, law, economy and society in the field of stem cells in this article's context. However, it should be noted that starting from James Thomson's first discovery and successful cultivation of human embryonic stem cells (hESC) in 1998, there was a first wave of international research into hESC, including a lot of funding in R&D.[10] Extracted from the inner cell mass of the embryo in its earliest stages post-fertilisation (blastocyst stage, at around 5 days after fertilisation), hES cells in respective laboratory conditions are capable of self-renewing extensively and are competent to generate cells of all tissues and organs, a property referred to as

[9] 'It must be borne in mind, further, that the meaning and scope of terms for which European Union law provides no definition must be determined by considering, inter alia, the context in which they occur and the purposes of the rules of which they form part.' *Brüstle v Greenpeace eV* (see note 6), para. 31.

[10] Thomson, J.A. et al. 1998: 'Embryonic Stem Cell Lines Derived from Human Blastocysts', *Science* 282: 1145–47.

*pluripotency*. Legal and ethical controversies relate to the fact that in order to derive pluripotent stem cells, the embryo must be destroyed.

Big shifts in the field of stem cell research were caused by the advent of induced pluripotent stem cells (iPS cells) as another type of pluripotent stem cells that can be generated directly by reprogramming from mature adult cells. The iPS technology was pioneered in 2006 by Shinya Yamanaka in Japan, who was awarded the Nobel Prize in 2012.[11] iPS has offered the means of producing patient-tailored cells without recourse to embryo destruction or cloning (SCNT). In contrast to human embryonic stem cells, iPS need not be derived from human embryos, and they can be used to create individual, patient-matched stem cell lines which in the future could be used to generate transplants without the risk of immune rejection. The field of adult stem cells has also developed further, and trans-differentiation of adult cells into other cells also allows for the preparation of patient-specific cells.[12] Thus many technologies are available today which not only allow bypassing the need for human embryos, but also provide unlimited supplies of autologous cells. Therefore, the ethical controversies around human embryonic stem cells can largely be avoided by these new means. The advent of iPS cells has definitely changed the scientific field and mitigated the ethical debate, because other sources of stem cells which do not rely on the use of human embryos have become available.

By and large, the field of pluripotent stem cells is still in the stage of basic research, with very few clinical trials carried out to date. Moreover, many hurdles and risks have to be overcome until therapeutic applications, both for iPS and embryonic stem cells, can become a clinical reality. Therefore, honest scientists have repeatedly warned against promising too much too early. Based on their characteristics of unlimited self-renewal and high proliferation rate, the risks associated with all pluripotent stem cell therapies include tumour formation, unwanted immune responses and the transmission of infectious or other agents.[13]

By now, several of the first broad patents on embryonic stem cell–related inventions have expired or are soon to expire because they were filed twenty years ago, or longer. According to patent information based on the Global Patent Index, from 1986 to 2017, 1,866 patent applications involving human embryonic stem cells were filed as international PCT applications and 1,311 European patent applications were filed at the EPO. According to the same source, by 2017, 328

11 Takahashi, K. et al. 2007: 'Induction of Pluripotent Stem Cells from Adult Human Fibroblasts by Defined Factors', *Cell* 131: 861–72.

12 See for instance: Dave, S. 2014: 'Mesenchymal Stem Cells Derived In Vitro Transdifferentiated Insulin-Producing Cells: A New Approach to Treat Type 1 Diabetes', *Adv Biomed Res.* 31(3): 266; Forcales, S.V. 2015: 'Potential of Adipose-Derived Stem Cells in Muscular Regenerative Therapies', *Front Aging Neurosci* 13(7):123. doi: 10.3389/fnagi.2015.00123. eCollection 2015; Novak D., et al. 'From Skin to Other Cell Types of the Body', *J Dtsch Dermatol Ges.* 2014(9)12: 789–92.

13 See Herberts, C.A., Kwa, M.S.G., Hermsen, H.P.H. 2011: 'Risk Factors in the Development of Stem Cell Therapy', *J. of Translational Medicine* 9:29, https://www.ncbi.nlm.nih.gov/pmc/articles/PMC3070641/pdf/1479-5876-9-29.pdf.

respective European embryonic stem cell patents have been granted (see Table 1 in annex). Table 2 (in annex) lists respective embryonic stem cell patent applications per year from 1986 to 2017. It shows that up until 2008, fewer than 100 embryonic stem cell applications were published per year, with a steady increase up to the peak of 163 European patent applications in 2012 at the EPO.[14] According to another statistic provided by the EPO, and based on manual sorting of cases, applications reached a peak in 2011/2012 at about 105 PCT embryonic stem cell patent applications per year, and 100 respective European applications, and ranged between 20 and 100 EP embryonic stem cell patent applications per year between 2001 and 2013 (see Table 3 in annex).

Another counting method, based on manual examination of European patents granted on methods and products which include claims on human embryonic stem cells or respective methods, has revealed 109 patents granted by the EPO from 1996 to March 2018.[15]

As the EPO's patent granting practice had been protracted for several years due to legal uncertainties, a large number of patent applications involving human embryonic stem cells have been granted in the last couple of years or are still due to be granted. Therefore, the actual granting practice of the EPO is relevant and matters.

## IV. CASE LAW BY THE CJEU AND THE EPO'S BOARDS OF APPEAL

Controversies concerning the EPO granting patents in the field of human embryonic stem cells started in 2000 at the occasion of the *Edinburgh* patent (EP 0695351), which sparked high media attention worldwide and for which the EPO admitted that it had granted the patent erroneously.[16] In the *Edinburgh* opposition proceeding, not only Greenpeace but also several EU member states – namely Germany, Italy and the Netherlands – were among the fourteen opponents. Another important case at the EPO was the *WARF* (*Wisconsin Alumni Research Foundation*) case and its subsequent ruling G2/06 by the EPO's Enlarged Board of Appeal, which held that Rule 28(c) EPC 'forbids the patenting of claims directed to products which – as described in the application – at the filing date could be prepared exclusively by a method which necessarily involved the destruction of the human embryos from which the said products are derived, even if the said method is not part of the claims'. The decision G2/06 also held that 'it is not of relevance that after the filing date the

[14] Table 1 and 2 by courtesy of Dr Ruth Tippe of 'No Patents on Life' and the genetic-ethics foundation GeS (www.gen-ethische-stiftung.de/). The author wishes to thank Dr Tippe for her work and her decade-long efforts of monitoring the EPO's applications and granting practice. She has also established a respective patent database, available at http://www.kein-patent-auf-leben.de/patentdatenbank/.

[15] Information meticulously collected and kindly provided by Dr Ruth Tippe, 28 March 2018.

[16] See Schneider 2010 (note 7): 403ff.

same products could be obtained without having to recur to a method necessarily involving the destruction of human embryos'.[17]

The first CJEU decision, in Case C-34/10 *Brüstle v Greenpeace*, was fully in line with the former decisions of the EPO's judiciary and provided broad definitions for the term 'human embryo', which hitherto had not been legally defined in EU law, as well as for the term 'uses' of human embryos. The CJEU held that Article 6(2)(c) of the Biotech Directive 98/44/EC must be interpreted as meaning that 'any human ovum after fertilisation, any non-fertilised human ovum into which the cell nucleus from a mature human cell has been transplanted, and any non-fertilised human ovum whose division and further development have been stimulated by parthenogenesis constitute a "human embryo"'. The Court also stated that the exclusion from patentability concerning the use of human embryos for industrial or commercial purposes set out in Article 6(2)(c) 'also covers the use of human embryos for purposes of scientific research, only use for therapeutic or diagnostic purposes which is applied to the human embryo and is useful to it being patentable'. Furthermore, the CJEU decided that Article 6(2)(c) 'excludes an invention from patentability where the technical teaching which is the subject-matter of the patent application requires the prior destruction of human embryos or their use as base material, whatever the stage at which that takes place and even if the description of the technical teaching claimed does not refer to the use of human embryos'.[18]

The Court held that the definition of the term 'human embryo' must be uniformly understood throughout the territory of the European Union in a wide sense, to ensure necessary respect for human dignity. Accordingly, any human egg cell (oocyte) must, as soon as fertilised, be regarded as a 'human embryo', since fertilisation commences the process of development of a human being.[19] This judgment is legally binding to all EU member states. Both the CJEU as well as national practice consider the fact as irrelevant that a destruction of a human embryo may have occurred at a stage long before the invention, which is what happens when stable established hESC lines are used as the starting material of the patented procedure. This view is also in accordance with the standing practice of the EPO.

The second CJEU decision in Case C-364/13 reversed C-34/10 with regard to parthenotes. It clarified 'that an unfertilised human ovum whose division and further development have been stimulated by parthenogenesis does not constitute a "human embryo" ... if, in the light of current scientific knowledge, it does not, in itself, have the inherent capacity of developing into a human being'.[20]

17 EPO Enlarged Board of Appeal, *WARF*, G2/06 (November 25, 2008), 5 OJ EPO 306. See also: Treichel, P. 2009: 'Case Comment: G2/06 and the Verdict of Immorality', *Int'l Rev. Intell. Prop. & Competition Law* 40: 450ff.

18 *Brüstle v. Greenpeace eV* (see note 6), para. 24 *et seq.*; para. 32 *et seq.*

19 *Id.*

20 CJEU, Grand Chamber, Case C-364/13, *International Stem Cell Corporation v Comptroller General of Patents, Design & Trademarks* (18 December 2014), OJ C 65. Human oocytes can be induced to 'commence' embryonic development without fertilisation, using electric and chemical activation,

## V. THE EPO'S GRANTING PRACTICE AND APPLICATION OF THE LEGAL RULES

Even though judgments by the CJEU on the interpretation of Directive 98/44/EG are not binding for the EPO, they are considered to be persuasive.[21] On 4 February 2014 in the *Technion Research and Development Foundation Case*, the European Patent Office followed the CJEU *Brüstle v. Greenpeace* decision, excluding from patentability inventions using hESCs obtained by destruction of human embryos, whenever such destruction takes place.[22]

However, the EPO, following G2/06 and CJEU C-34/10, has henceforth developed a practice which distinguishes different grades and forms of uses of a human embryo. It distinguishes among the following:

a) a *direct use* of a human embryo 'wherein the embryo is destroyed', and where the patenting prohibition extends to cells derived 'directly and exclusively from such a destructive use';
b) products derived *indirectly* from such destructive use, which are also excluded from patentability, and where the point in time of the embryo destruction is irrelevant; and
c) a *direct use* of a human embryo wherein the embryo is *not destroyed*, which is also excluded from patentability under Article 53(a) and Rule 28(c) EPC, as Rule 28(c) EPC excludes from patentability *any use* of human embryos and not only destructive uses. The patenting prohibition extends to cells derived *directly* from such a non-destructive use.
d) However, the EPO considers products derivable *indirectly from a non-destructive use of a human embryo* as in principle patentable under Article 53(a) EPC. As examples of such products of a non-destructive use of a human embryo, the EPO refers to the human ES cell lines obtained through the single blastomere biopsy methodology disclosed by Chung et al. on 10 January 2008.[23]

a process referred to as parthenogenesis. Parthenotes undergo certain stages of cell division but later usually arrest in their development. Very few studies have shown that this route can be used to generate blastocysts from human oocytes, which can serve as source for the derivation of parthenogenetic ES cells. See Mai, Q. et al. (2007): 'Derivation of Human Embryonic Stem Cell Lines from Parthenogenetic Blastocysts', *Cell Res.* 17:1008–19; Lin, G. et al. 2007: 'A Highly Homozygous and Parthenogenetic Human Embryonic Stem Cell Line Derived from a One-Pronuclear Oocyte following In Vitro Fertilization Procedure', *Cell Res.* 17: 999–1007; Revazova, ES et al. 2007: 'Patient Specific Stem Cell Lines Derived from Human Parthenogenetic Blastocysts', *Cloning Stem Cells* 9: 432–49. However, these studies are merely a proof of principle, because parthenogenetic stem cells are not used in the vast majority of research on hESCs.

21 See EPO Technical Board of Appeal decisions T 2221/10 and T 441/13.

22 EPO Technical Board of Appeal decision T2221/10.

23 These distinctions are laid out and cited according to the 'Final Report of the Expert Group on the Development and Implications of Patent Law in the Field of Biotechnology and Genetic Engineering', 17 May 2016, p. 150, available for download at http://ec.europa.eu/growth/industry/intellectual-property/patents/biotechnological-inventions/. The publication referred to is Chung, Y. et al. 2008: 'Human Embryonic Stem Cells Lines Generated without Embryo Destruction', *Cell Stem*

The latter category (d) has turned into the largest one for recent EPO granting decisions. That means that despite the clear rules set in both the EPC's and the EU's jurisdictions, the EPO's granting practice has managed to actually *patent around the prohibitions* set under Article 53(a) and Rule 28(c) EPC. In 2013, the EPO took the single scientific publication by Chung et al. (2008) as justification to allow patents on products and processes 'indirectly' derived from hESC, and took 10 January 2008 as a cut-off date. This practice was not revealed in the EPO's Guidelines for Examination, but was circulated in presentations of EPO examiners at patent law conferences, and was later included in T1441/13. Accordingly, the EPO accepts disclaimers containing language that the stem cells derived can be obtained by means of a process in which human embryos are not destroyed.

Furthermore, in October 2015, the EPO in its granting practice even shifted and thus preponed this cut-off date to 5 June 2003, based on a protocol to derive human parthenogenetic embryonic stem (hpES) cells from activated oocytes (parthenotes), which was disclosed in the PCT application WO 03/046141. This international patent application was filed as European patent EP1456374. It contained only one claim, was only briefly examined and was silently withdrawn on 25 November 2009 after fees were not paid several times. Nonetheless, according to the latest revision of its granting practice, and in interpreting the CJEU's ruling of C-364/13, the EPO now considers inventions relating to human pluripotent stem cells, including hES cells, to their uses and to products derived from them as patentable from 5 June 2003 on. Respectively, the EPO revised its Guidelines for Examination.[24] Hence, the current EPO practice does allow for granting patents with claims involving human embryonic stem cells if such patent applications are filed after 5 June 2003 and provided that other EPC conditions (such as novelty, inventive step and industrial application) are met. In practice, this means that the bulk of patent applications filed after 2002 (compare Tables 2 and 3) can and will be granted, except those directly derived from destructive embryo research or direct use of embryos.

## VI. ARTICLE 6(2)(C) PROHIBITS PATENTABILITY OF *ALL* THIRD USES OF HUMAN EMBRYOS

In the following, it is argued that the current EPO practice outlined previously does not comply with the provisions set in Directive 98/44/EC. To this purpose, and to provide the reasons, a different interpretation of several legal provisions in the light of scientific developments is articulated, together with recommendations for

*Cell* 2(2): 113–17. The article by Chung et al. 2008 is first cited in the EPO's Technical Board of Appeal decision T1441/13. The Final Report also cites several European applications that have been granted or proposed for grant: e.g., granted patents EP 2424977, EP 2424976, EP 2139989 and EP 08839475 claiming human stem cells per se.

24 EPO, Guidelines for Examination, valid from 1 November 2015, Part G Chapter 5.3. (iii).

clarification and specification by either the Commission, the European Parliament and the Council, the CJEU or the EPO itself.

Article 6(2)(c) of the Directive considers unpatentable the 'uses of human embryos for industrial or commercial purposes'. The Directive does not distinguish between 'destructive' and 'non-destructive' uses of human embryos. Therefore, the Directive can only mean both possibilities of embryo use. If the legislator had intended to restrict the use to 'non-destructive uses', it would have provided respective clarification in the Article and/or in the Recitals. It is arbitrary to exclude 'destructive uses' and to allow 'non-destructive' uses of human embryos. Furthermore, the term 'use' in patent law is normally understood in broad terms, for instance in the rules on second or further medical use of known pharmaceutical products.[25]

Moreover, the CJEU in Case C-34/10 also states that Article 6(2)(c) 'includes any invention of patentability wherein the technical teaching which is the subject matter of the patent application requires the prior destruction of human embryos *or their use as a base material*, whatever the stage at which that takes place and even if the description of the technical teaching claimed does not refer to the use of human embryos'. The CJEU has not suggested that it wanted to allow non-destructive uses of human embryos, and there is no CJEU decision that has explicitly allowed it.[26]

Therefore, the term '*or* their use as base material' must be interpreted as excluding *any* use of a human embryo from patentability, unless the invention is 'for therapeutic and diagnostic purposes which are applied to the human embryo and are useful to it' (Recital 42). In conclusion, only a broad interpretation of the term 'uses' of human embryos in Article 6(2)(b) is in compliance with the Directive.

This interpretation has already been backed by the ruling of the EPO's opposition division in the *Edinburgh* case. Here, the board stated that 'only a broad ruling of Rule 23d(c) can have been intended' by the European legislator:[27]

> If the legislator had intended a narrow interpretation of Rule 23d(c) . . . he would not have introduced both Rule 23d(c) and Rule 23e(1) into the EPC or correspondingly, Article 5(1) and Article 6(2)(c) in the Directive. Rule 23e(1) EPC excludes from patentability the human body, at the various stages of its formation and development, and recital 16 even states that the human body at any stage in its formation and development cannot be patented. The human embryo being an early stage in the development of the human body is thus already included in the scope of Rule 23e(1) EPC, and therefore, a narrow interpretation of Rule 23d(c) EPC would result in a 'redundancy' over Rule 23e(1) EPC. The fact that Rule 23d(c) EPC refers to 'uses' for 'industrial or commercial purposes' is not of relevance in the given context .... If the patenting of a product is ethically unacceptable, it is hardly conceivable that the patenting of 'uses' of this product can be judged differently.

25 Second medical indication, EPO Guidelines for Examination; G. VI, 7.1.

26 *Brüstle v. Greenpeace* (see note 6) (emphasis added).

27 The former Rule 23d (c) EPC is now Rule 28(c); the former Rule 23e(1) is now Rule 29(1) EPC.

> Thus it is considered that the exclusion of human embryos from patentability under Rule 23e(1) also pertains to the 'uses' of human embryos for whatever purpose.[28]

The broad exclusion of *any* use of human embryos for third-party interests is also backed by the legislative history of Article 6(2)(c) of Directive 98/44.[29] And even if the CJEU in Case C-34/10 *Brüstle v. Greenpeace* had been completely silent about 'prior destruction of human embryos', it would be a logical fallacy to deduce that this would then allow patentability of 'non-destructive' uses of human embryos. The court was just not asked to judge this question. In addition, it must be emphasised that the list of inventions in Art 6(2) of Directive 98/44/EC is non-exhaustive and therefore does not enumerate all possible uses intended by the legislator. Several Recitals (38 to 40) of said Directive already provide further examples – for instance, Recital 38, which states that 'processes, the use of which offend against human dignity, such as processes to produce chimeras from germ cells or totipotent cells of humans and animals, are obviously also excluded from patentability'.

The patent exclusion in Article 6(2)(c) of Directive 98/44/EC is based on the Kantian rationale for human dignity, which states that humans must always be treated as an end in themselves and not as a mere means to an end.[30]

In conclusion, considering the term 'uses' of human embryos for industrial or commercial purposes' in Article 6(2)(b), only a broad and comprehensive interpretation of what constitutes 'use' is in accordance both with the legal provisions and with the spirit, intentions and will-formation of the European legislator. Reducing the term 'uses' of human embryos to 'non-destructive' uses, if cells are 'indirectly' obtained, as suggested and practised by the EPO, is not a justified means of interpretation.

## VII. SCIENTIFIC VALIDATION: CHUNG 2008 DOES NOT PROVIDE SOUND EVIDENCE FOR 'NON-DESTRUCTIVE' USE OF HUMAN EMBRYOS

However, it is not adequate to assume that the method disclosed by Chung et al. (2008) would be on an equal footing with 'classical' (hESC) or 'reprogramming' (iPS) methods. In particular, it is inappropriate to use the method published by

28 Cited from the EPO, Edinburgh Case, Grounds for the decision, Application No. 94913174.2, 21.07.2003, p. 22.

29 Article 6(2)(c) was introduced as amendment 55 by the European Parliament and held as unpatentable 'methods in which human embryos are used'. After acceptance by the Commission, this wording was changed in the Common Position adopted by the Council to 'uses of human embryos for industrial and commercial purposes'. For references, see Comments of the President of the EPO on G2/06, 2006: 38–9.

30 Reference to human dignity as a fundamental norm for the interpretation of the Directive is also made in Recital 16 of the Biotech Directive, reasserted in CJEU Case C-377/98 *Netherlands v. Parliament and Council*, [2011] ECR I-7079, [2001] 3 CMLR 1173, paras. 71 and 76; and reiterated by the CJEU in *Brüstle v. Greenpeace* (see note 6), paras. 33–34.

Chung et al. in 2008 as a loophole to allow patentability of methods and products 'indirectly' derived from human embryonic stem cell lines for patents which were filed after 10 January 2008, as applied by the European Patent Office (EPO). Chung's publication claims to have enabled the provision of hES lines without destroying a human embryo in any production step for the first time.[31] It is based on one single cell line (No. 5) said to be produced from a single blastomere, without co-culture with hESC.

However, the belief that Chung's method indeed changed the field of human embryonic stem cell research must be contested. The most relevant reasons for contestation are as follows:

1) The existence of the cell line in question (No. 5 in Chung et al. 2008) is questionable, as it is not available to international researchers and no published studies are listed that utilise this cell line.
2) The method has no real-world applicability with respect to the transfer of the remaining embryo (from which a single blastomere was extracted) to a woman's uterus. All biopsied embryos used by Chung et al. 2008 were frozen (cryopreserved) but never implanted.
3) The method applied in practice – taking a blastomere from a human embryo without destroying it – could cause harm to the embryo, without any indication and medical benefit for the embryo itself. (Hence, Recital 42 of Dir. 98/44 does not apply.)
4) The blastomere taken from an embryo at the eight-cell stage may still be considered totipotent; therefore, the totipotent cell must be attributed the same legal status as a human embryo.[32]

31 Chung Y. et al. 2008 (see note 23).

32 *Totipotent cells* are mentioned in the exclusions from patentability in Recital 38 of Directive 98/44/EC. In some national legislations (e.g., Germany), *embryo* is defined as any human totipotent cell which, exposed to the necessary conditions, has the potential to divide and develop into an individual (2002 Stem Cell Act, §3(4)). There are different scientific definitions concepts of totipotency. It is agreed that blastomeres from the 2 to 4 cell stages are totipotent (see Geens et al. 2009: 'Human Embryonic Stem Cell Lines Derived from Single Blastomeres of Two 4-Cell Stage Embryos', *Human Reproduction* 24(11), 2709–17). According to scientific reviews of the state of the knowledge to date, totipotency can also not be excluded for blastomeres from the 6- or 8-cell stage because cleavage does not happen for all blastomeres simultaneously. For legal and ethical reasons, single blastomeres or manipulated human embryos are not transferred into a uterus to test their potency, and thus totipotency will never scientifically be proved in the human. De Paepe C. et al. 2014: 'Totipotency and Lineage Segregation in the Human Embryo', *Mol Hum Reprod.* 20(7): 599–618, p. 602.

Therefore, it is necessary to draw the conclusion that a single blastomere in the 8-cell stage, as used by Chung et al. 2008, has to be legally treated as totipotent. As it is about human embryos, it is prudent to err on the side of caution. Such a position is also backed by the view of experimental animal studies in mice and rabbits, which have provided strong indications for totipotency of at least some of the blastomeres, even in the 16-cell stage embryo (Ziomek et al. 1982, Chisholm/Fleming et al. 1984 and further references cited in Littwin, T./Denker, H.W. 2011: 'Segregation during Cleavage in the Mammalian Embryo?', *Histochem Cell Biol* 135: 553–70).

5) On the basis of information currently available, to date no stem cell lines produced by this method are used in any human embryonic stem cell research for therapeutic purposes.

In conclusion, the EPO appears to be using this method as *a legal artefact* without sufficient technical basis. One single cell line in a single publication does not provide sufficient and robust scientific evidence to legitimise such a far-reaching shift in patentability which has effects on several hundred human embryonic stem cell applications filed after 10 January 2008 (compare Table 1). Moreover, patents granted on the basis of this method give adverse incentives to the field of stem cell research, as it would encourage such methods fraught with medical and ethical difficulties. Alternative strategies which do not require human embryos at all, such as iPS cells, and further strategies which are based on 'direct reprogramming' by 'direct conversion' of human cells, and thus are 'bypassing pluripotency', will be more conducive for stem cell research, especially as recent research indicates that human pluripotent stem cells '*in vitro* can form structures that have some resemblance to embryos' post-implantation.[33]

In conclusion, it amounts to a violation of the will of the European legislator to bypass non-patentability of 'uses of human embryos' (Art.6(2)(c)) by theoretical and technical means which are not based on sound, evidence-based science and real-world applicability. Therefore, the Chung et al. (2008) technology is not an adequate means for allowing the granting of patents on inventions derived from human embryonic stem cells.

## VIII. PARTHENOTES AS STEM-CELL SOURCE ARE NOT IDENTICAL TO EMBRYONIC STEM CELLS

As explained previously, in the EPO's recent revision of the Guidelines for Examination, valid from 1 November 2015, the EPO refers to the 'state of the art at the date of filing'; and it considers inventions relating to human pluripotent stem cells (including hES cells), to their uses and to products derived from them as patentable from 5 June 2003 on, based on parthenogenesis, disclosed in the PCT application WO 03/046141.[34] Accordingly, the EPO's new first-instance practice, adopted around October 2015, takes 5 June 2003 as the new cut-off date. Indeed, the PCT application WO 03/046141 (Advanced Cell Technology) discloses in Example 4 a protocol for the 'Production of Autologous Cells by Parthenogenetic Activation of Oocytes' (pp. 45–46). However, the resulting cells are human parthenogenetic embryonic stem cells (hpESC), which are deemed 'ES-like cells' (p. 47). Therefore,

33 See Pera M.F. et al. 2015: 'What If Stem Cells Turn into Embryos in a Dish?', *Nature Methods*, 12(10): 917–19; see also Denker, H.W. 2012: 'Time to Reconsider Stem Cell Induction Strategies. Review', *Cells* 2012(1): 1293–1312.

34 WO 03/046141 (Advanced Cell Technology).

even though these hpES cells are similar to human embryonic stem cells (hESC), they are not identical to hESC. A conclusion of identity is scientifically inadmissible.[35]

Activated human eggs called parthenotes do not involve the union of male and female germ cells, and so genetic material will be derived exclusively from the female oocyte donor.[36] Therefore, they have important biological differences compared to fertilised human embryos.[37] Even though 'pESCs are similar to ESCs', they lack paternal imprinted genes.[38] This is one of the reasons why parthenotes normally do not develop into full human beings.[39]

Moreover, the conclusion drawn by the EPO from the CJEU judgment in C-364/13 is wrong. The CJEU was only asked whether parthenotes were included in the term 'human embryos' in Article 6(2)(c) of Directive 98/44/EC. The CJEU responded that

> Article 6(2)(c) . . . must be interpreted as meaning that an unfertilised human ovum whose division and further development have been stimulated by parthenogenesis does not constitute a 'human embryo', within the meaning of that provision, if, in the light of current scientific knowledge, it does not, in itself, have the inherent capacity of developing into a human being, this being a matter for the national court to determine.

Hence, the court neither made a decision as to whether parthenotes are eligible for a patent, or whether processes to obtain parthenogenetic embryonic stem cells (hpES) derived from parthenotes are patentable. Parthenotes may fall under the exclusions in Art. 5 and Art. 6(1) of Directive 98/44/EC. But even if these aforementioned products and processes were patentable, the conclusion drawn by the EPO that human embryonic stem cells (hESC), their uses and the products derived from them are patentable must be considered as incorrect, because hpES are not identical to hESCs.

Moreover, it is necessary to draw attention to the fact that the literature has described cases which demonstrate the ability to produce the birth of live parthenogenetic mice which were able to produce offspring when the appropriate imprinting of key genes by a maternal genome were expressed.[40] In humans, this concept was

35 See Swelstad B.B., Kerr C.L. 2009: 'Current Protocols in the Generation of Pluripotent Stem Cells: Theoretical, Methodological and Clinical Considerations', *Stem Cells Cloning* 22(3): 13–27.

36 Cibelli J.B., et al. 2006: 'Embryonic Stem Cells from Parthenotes', *Methods Enzymol.* 418: 117–35; Kim K., et al. 2007: 'Recombination Signatures Distinguish Embryonic Stem Cells Derived by Parthenogenesis and Somatic Cell Nuclear Transfer', *Cell Stem Cell* 1: 346–52.

37 Brevini T.A./Gandolfi F. 2008: 'Parthenotes as a Source of Embryonic Stem Cells', *Cell Proliferation* 41 Suppl 1: 20–30.

38 Daughtry B./Mitalipov S. 2014: 'Concise Review: Parthenote Stem Cells for Regenerative Medicine: Genetic, Epigenetic, and Developmental Features', *Stem Cells Translational Medicine* 3(3): 290–8.

39 Wu Q. et al. 2006: 'Regulated Expression of Two Sets of Paternally Imprinted Genes Is Necessary for Mouse Parthenogenetic Development to Term', *Reproduction* 131(3): 481–8.

40 Kono T. et al. 2004: 'Birth of Parthenogenetic Mice That Can Develop to Adulthood', *Nature* 428(6985): 860–4; Loebel, D./Tam, P.L. 2004: 'Genomic Imprinting: Mice without a Father', *Nature* 428: 809–11; Wu Q. et al. 2006(see note 39).

also demonstrated in a single case report of spontaneous parthenogenetic chimerism in which the patient survived with a mixed makeup of normal and parthenogenetic cells.[41] Therefore, the question as to whether human parthenotes in themselves have the inherent capacity of developing into a human being has not yet been answered conclusively in scientific knowledge.

To sum up, the EPO's granting practice regarding inventions relating to human embryonic stem cells has managed largely to escape restrictive limits set by the European legislator and the CJEU. This application of Rule 28 EPC and its inherent redefinition of the morality exclusions on patentability require urgent critical re-assessment.

## IX. THE CREATION OF ARTIFICIAL GAMETES AND EMBRYOS MUST REMAIN EXEMPTED FROM PATENTABILITY

Recent developments in reproductive research suggest that other new methods, like the artificial creation of human germ cells, may also become available that could lead to the artificial creation of human embryo-like entities. Inventions related to such methods or their products would be excluded from patentability, if these entities fulfil the definition used by the CJEU of being 'inherently capable of developing into a human being', as they then have to be considered human embryos. New methods which could potentially lead to the creation of artificial germ cells or germ cell lines are excluded as such from patentability according to Article 5 and Recital 16 of the Biotech Directive. Additionally, inventions involving an embryo produced by means of artificial germ cells should be treated in the same way as any inventions related to a 'natural' embryo produced by the fusion of an oocyte and a sperm cell.

Moreover, it is necessary to make sufficiently clear that the term 'uses of human embryos' in Article 6(2)(c) may also extend to uses of human embryos in which an embryo is being technically created. It is important to take into account recent developments of such possible constructive uses[42] of human embryos. The term *constructive use* is introduced here to comprise the creation of an artificial embryo by means other than normal in vitro fertilisation (IVF). One example for such a new method for the construction of an embryo is 'mitochondria replacement', also termed 'three-parent babies'.[43]

[41] Strain L. et al. 1995: 'A Human Parthenogenetic Chimaera', *Nature Genetics* 11(2): 164–9. Also cited in Swelstad B.B./Kerr C.L. 2009: 'Current Protocols in the Generation of Pluripotent Stem Cells', *Stem Cells Cloning* 22(3): 13–27, on p. 20.

[42] The term *constructive use* is coined here in contrast to the aforementioned term *destructive use* of an embryo (in the context of Article 6(2)(c) of the Biotech Directive). It aims to highlight that in normative and legal terms, it is not only the destruction of embryos (for instance, for hESC) which requires political and moral attention, but also new forms of creating embryos or embryo-like entities from stem cells or other forms and methods of modifying human biological material.

[43] In 'mitochondria replacement', also termed 'three-person in vitro fertilisation', an embryo with genetic material from three different people (two females and one male) is created, which also results in inheritable genetic modification (changes that would be passed on to future generations). Some

Another novel method for the construction of an embryo by new means is to use iPS (induced pluripotent stem cells) or embryonic stem cells to create artificial gametes and embryos. The latter technology might allow same-sex partners (e.g., two males), using their own genetic material, to create an embryo that is genetically related to both partners. It might also allow one individual to reproduce without another genetic parent, by conjoining artificially created male and female germ cells from the single individual.[44]

In light of these recent developments, it would be reasonable that both the European Commission and the EPO specify and clarify that the term 'germ cell' (Directive 98/44/EC, Recital 16) applies not only to naturally created egg and sperm cells, but also to artificially created egg and sperm cells (e.g., from iPS cells).

For the time being, the EPO's interpretation of patent exclusions related to germ cells appears to be as follows: The Guidelines for Examination in the EPO explicitly state that according to Rule 29(1) EPC, 'the human body, at the various stages of its formation and development, ... cannot constitute patentable inventions. Such stages in the formation or development of the human body *include germ cells* (Directive 98/44/EC, Recital 16).'[45] It follows that human germ cells, be they artificially created or naturally occurring, are plainly unpatentable under the EPC. Germ cells are mentioned in Recitals 16 and 38,[46] which have an impact on the interpretation of Articles 5 and 6. Germ cells, in particular oocytes, are relevant inter alia in the previously mentioned use of parthenotes as a source of human embryonic stem cells and in SCNT (somatic cell nuclear transfer) techniques, where the resulting cloned embryo could be used for the derivation of human embryonic stem cells. SCNT, being a process for cloning human beings, is explicitly excluded from patentability under Rule 28(a) EPC. It is worth noting that, according to the Guidelines for Examination in the EPO, for the purpose of this exception, 'a process for the cloning of human beings may be defined as *any process*, ... designed to create a human being with the same nuclear genetic information as another ...

observers have warned that this would open the door for human germline modification. Darnovsky, M. 2013: 'A Slippery Slope to Human Germline Modification. The United Kingdom's Decision to Trial the Technique of Mitochondrial Replacement Is Premature and Ill-Conceived', *Nature* 499: 127.

44 For the scientific background, see the 2014 report of the German Ethics Council (Deutscher Ethikrat), 'Stem Cell Research – New Challenges for the Ban on Cloning and Treatment of Artificially Created Germ Cells?', http://www.ethikrat.org/files/recommendation-stem-cell-research.pdf. A review article in 2000 already noted that '[a]lternative sources of gametes are not merely science fiction, but already are a concrete fact'. Tsai et al. 2000: 'Alternative Sources of Gametes: Reality or Science Fiction?' *Human Reproduction* 15(5): 988–98, on p. 995.

45 EPO Guidelines for Examination, G-II, 5.3 (emphasis added).

46 Recital 38 refers to processes to produce chimeras from germ cells or totipotent cells of humans and animals, which are excluded from patentability under Art. 53(a) EPC. A narrow interpretation of this wording suggests that totipotent cells are only excluded when used for chimeras. A broad interpretation means that totipotent cells are excluded anyway. The author follows the second interpretation, as totipotent cells are equivalent to a human embryo, which is excluded in Article 5(1)/Rule 29(1) EPC.

human being (EU Dir. 98/44/EC, rec. 41)'.[47] However, that conjoining of artificially created male and female germ cells from a single individual is functionally equivalent to a (cloned) human embryo would require scientific and legal clarification.

In any event, such methods for producing artificial embryos and the resulting entity itself should not be eligible for patentability, either under Article 6(2)(a)/Rule 28(a) EPC (cloning), under Article 6(2)(b)/Rule 28(b) EPC (modification of the germ line genetic identity of human beings), under Article 5(1)/Rule 29(1) EPC (the human body at the various stages of its formation and development) or ultimately under Article 6(1)/ Art. 53(a) EPC. It is necessary for both the Commission and the EPO to specify and clarify that such methods and products would fall under the prohibitions of Articles 5 and 6 of Directive 98/44/EC.

Here again, it is necessary to stress that such a provision of non-patent eligibility would not ban respective research, but would remove some commercial incentives for such research and development. As patents are granted according to the 'winner-take-all' principle, only the first inventor who has filed the patent gets the temporary exclusivity. Hence, patents can have an accelerating effect on spurring such new techniques. Democracy, however, needs time to deliberate whether such new forms of procreation should be made available or not.

Proceeding with caution and without patents as an incentive for the construction of artificial gametes and embryos will allow European society sufficient time to debate whether such developments are socially desirable or not and whether national legislation of such developments should be set on the agenda. Granting patents on such inventions for creating artificial gametes and embryos would set both an incentive for pursuing such research and an unfavourable precedent for other areas of law.

Moreover, it is important to state that any use of human oocytes for third parties and/or non-procreative purposes, for instance for parthenotes or for somatic cell nuclear transfer (SCNT) as sources of human embryonic stem cells, also carries the risk for women to be exploited as oocyte providers. The 2005 UN Declaration on Human Cloning (59/280) called upon the nations 'to prohibit all forms of human cloning', both reproductive cloning and research cloning (SCNT). It also called upon member states 'to take measures to prevent the exploitation of women in the application of life sciences'.[48] This call was reiterated in a resolution of the European Parliament.[49]

In conclusion, constructive uses of human embryos and gametes, by the creation of artificial gametes from stem cells, and by the conjoining of such artificial germ

47 EPO Guidelines for Examination, G-II, 5.3 (emphasis added).

48 United Nations Declaration on Human Cloning (59/280). Resolution adopted by the General Assembly at its 82nd meeting, on 8 March 2005, http://legal.un.org/docs/?symbol=A/RES/59/280.

49 European Parliament 2005: Planned egg cell trade. Resolution on the trade in human egg cells. P6_TA(2005)0074; http://www.europarl.europa.eu/sides/getDoc.do?pubRef=-//EP//TEXT+TA+P6-TA-2005-0074+0+DOC+XML+V0//EN. See, in particular, numbers 9–13.

cells to a totipotent cell, must remain exempted from patentability. Such techniques increase the risks of inducing and transmitting serious harms to the embryo and developing human being. Such techniques also presuppose that women provide a vast number of oocytes for enucleation and/or act as surrogates to carry the artificially created embryo to term. Hence, they rely on the instrumentalisation of women and violate their dignity. For cross-compliance and cross-coherence of law, it is important that patent law adhere to higher-ranking legal norms such as the wellbeing of the child-to-be and the dignity of women.

## X. GENOME EDITING (CRISPR-CAS) MUST NOT BECOME PATENTABLE FOR HUMAN GERMLINE MODIFICATION

According to Article 6(2)(a) and (b) of the Biotech Directive, equally excluded from patentability are processes for cloning human beings and claims to the processes for modifying the germline genetic identity of human beings. These clauses will gain more importance in light of recent genome editing technologies, which can also be applied in pluripotent and other stem cells, in gametes and in the early stages of the development of a human embryo. Therefore, it is necessary briefly to elaborate on these new genome editing technologies and the associated scientific and legal debate.

Recent scientific advances in genome editing have received much attention in scientific, policy and public discussions. CRISPR-Cas9 and CRISPR-Cpf1[50] promise precise, fast, inexpensive and easy-to-handle ways of genome editing and can be employed in all kinds of DNA. These and other genome editing techniques, such as Zinc-finger nucleases (ZFNs), transcription activator-like effector nucleases (TALENs) and oligonucleotide-directed mutagenesis (ODM),[51] have been hailed as revolutionary, with significant advantages for research and therapeutics as well as for economics and intellectual property.[52] However, if genome editing techniques such as CRISPR were applied in heritable germline modification, they could have an unpredictable effect on future generations. Hopes and fears are not only present in media and policy debates, but also among researchers themselves. On the cautious end of the spectrum, scientists in Europe and the United States have called for a moratorium and a public debate on CRISPR in the human germline.[53] On the

50 The acronym CRISPR stands for Clustered Regularly Interspaced Short Palindromic Repeats.

51 Church, G., Regis, E. 2012: *Regenesis. How Synthetic Biology Will Reinvent Nature and Ourselves*. New York: Basis Books; Gaj, T. et al. 2013: 'ZFN, TALEN, and CRISPR/Cas-Based Methods for Genome Engineering. Review', *Trends in Biotechnology*, 31(7): 397–405.

52 Ledford, H. 2015: 'CRISPR, the Disruptor', *Nature* 522: 20–4.

53 Baltimore, D. et al. 2015: 'A Prudent Path Forward for Genomic Engineering and Germline Gene Modification', *Science* 348: 36–8; Lanphier, E. et al. 2015: 'Don't Edit the Human Germ Line', *Nature* 519(7544): 410–11; Leopoldina 2015: *The Opportunities and Limits of Genome Editing. Statement*, https://www.leopoldina.org/en/publications/detailview/publication/chancen-und-grenzen-des-genome-editing-2015/; BBAW (Berlin-Brandenburg Academy of Sciences and Humanities) 2015: *Human*

other end, researchers in China have published that they have conducted the first experiments in using CRISPR for editing the genomes of human embryos.[54] As of summer 2018, three other genome editing experiments on human embryos in China,[55] one in the United States[56] and another in the UK[57] have been published.

The British HFEA (Human Fertilisation and Embryology Authority) has signalled support for germline-editing research and has issued a first licence for respective research for the purpose of improving fertility and IVF success rates.[58] Parts of the international scientific community, such as the Hinxton Group, have recommended that editing of the human genome in embryos should be allowed for basic research and suggested that the door should not be totally closed towards allowing genetic engineering of humans in the future.[59] Similarly, the US National Academies of Sciences and National Academy of Medicine (NAS-NAM) stated, 'Heritable germline genome editing trials must be approached with caution, but caution does not mean they must be prohibited'.[60]

These developments are also reflected in patent filings and in patent arms races. Ledford, who mentions around 200 patent applications worldwide, recognises that 'a patent battle' has intensified, with a sharp increase in patent filings in 2014.[61] Kupecz refers to anonymous 'third-party observations' filed at the EPO aimed at preventing the granting of a European patent.[62]

For Europe, tentative research as of March 2018 has revealed more than 700 PCT applications and almost 300 European patent applications at the European Patent

*Genome Surgery – Towards a Responsible Evaluation of a New Technology*, http://www.gentechnologiebericht.de/bilder/BBAW_Human-Genome-Surgery_PDF-A1b-1.pdf.

54 Liang, P. et al. 2015: 'CRISPR/Cas9-Mediated Gene Editing in Human Tripronuclear Zygotes', *Protein and Cell* 6(5): 363–72; Cyranoski, D./Reardon, S. 2015: 'Chinese Scientists Genetically Modify Human Embryos', *Nature News*, 22 April 2015, https://www.nature.com/news/chinese-scientists-genetically-modify-human-embryos-1.17378.

55 Kang, X. et al. 2016: 'Introducing Precise Genetic Modifications into Human 3PN Embryos by CRISPR/Cas-Mediated Genome Editing', *J Assist Reprod Genet*, 33(5): 581–8; Tang et al. 2017: 'CRISPR/Cas9-Mediated Gene Editing in Human Zygotes Using Cas9 Protein', *Mol Genet Genomics* 292(3): 525–33; Liang, P. et al. 2017: 'Correction of β-thalassemia Mutant by Base Editor in Human Embryos', *Protein Cell* 8(11): 811–22.

56 Ma, H. et al. 2017: 'Correction of a Pathogenic Gene Mutation in Human Embryos', *Nature* 548: 413–19.

57 Fogarty, N.M.E. et al. 2017: 'Genome Editing Reveals a Role for OCT4 in Human Embryogenesis', *Nature* 550: 67–73. (For brief discussion of developments post-summer 2018, see the opening and closing chapters.)

58 UK science bodies 2015: 'Genome Editing in Human Cells – Initial Joint Statement', https://wellcome.ac.uk/sites/default/files/wtp059707.pdf; Sample, I. 2015: 'GM Embryos: Time to Decide, Say Scientists', *The Guardian*, 1 September 2015, https://www.theguardian.com/science/2015/sep/01/editing-embryo-dna-genome-major-research-funders-ethics-debate.

59 Hinxton Group 2015: *Statement on Genome Editing Technologies and Human Germline Genetic Modification*, http://www.hinxtongroup.org/hinxton2015_statement.pdf.

60 The National Academies of Sciences, Engineering, and Medicine 2017: *Human Genome Editing*. Washington, DC: National Academies Press, 102.

61 Ledford, H. 2015 (see note 52).

62 Kupecz, A. 2014: 'Who Owns CRISPR-Cas9 in Europe?', *Nature Biotechnology* 32(12): 1194–6.

Office, of which more than 60 have been granted. (Of 500 patent application documented in a database, 188 are on plants and plant methods, 162 on animals or animal cells or methods on animals, and 83 on therapies.) Eighteen patents on CRISPR as such have been granted, most of them on methods. Twelve CRISPR patents were granted to the US Broad Institute of MIT and Harvard and its member Feng Zhang. However, all of these received oppositions.[63] In January 2018, a first patent granted to the Broad Institute was revoked at the EPO.[64] UC Berkeley structural biologist Jennifer Doudna and microbiologist Emmanuelle Charpentier, now at the Max Planck Institute for Infection Biology in Berlin, who collaborated with Doudna on early CRISPR technology, seem to be more successful in receiving key patents in Europe than at the USPTO.[65] As of March 2018, one patent for Charpentier/University of California and two for Doudna/University of California (of which one has been opposed) were published as granted by the EPO.

Some of the patents filed shall or will include disclaimers such as 'non-human', 'with the proviso that the human germline is not modified' or 'wherein the cells are not germ cells'. The EPO also has to take into account the prohibition of patents for 'methods for treatment of the human or animal body by surgery or therapy and diagnostic methods practiced on the human or animal body' (Art. 53(c) EPC). For patent applications on animals, the 'balancing test' between 'suffering' and 'medical benefit' has to be applied by the EPO.

The potential for using this technology in germ cells and embryos, or in stem cells which were later introduced in enucleated oocytes (SCNT), has definitely reignited the debate on human germline interventions. At present, however, on a worldwide scale, human germline editing on human embryos in the clinical context is prohibited in 29 of 39 countries investigated, including fourteen European states, based on either legislation or guidelines.[66] International and European regulations also consider human germline modification as unethical human experimentation or an abuse of human rights.[67]

63 Information kindly provided by Dr Ruth Tippe, 29 March 2018.

64 See Servick, K. 2018: 'Broad Institute Takes a Hit in European CRISPR Patent Struggle', *Science Magazine*, 18 January 2018, http://www.sciencemag.org/news/2018/01/broad-institute-takes-hit-european-crispr-patent-struggle. For the US situation, see 'Berkeley Statement Regarding Patent Board's Decision on CRISPR-Cas9 Gene-Editing Technology', 15 February 2017, http://news.berkeley.edu/2017/02/15/berkeley-statement-regarding-patent-boards-decision-on-crispr-cas9-gene-editing-technology/.

65 See Cohen, J. 2017: 'How the Battle Lines over CRISPR Were Drawn', *Science Magazine*, 15 February 2017, http://www.sciencemag.org/news/2017/02/how-battle-lines-over-crispr-were-drawn.

66 Araki, M./Ishii, T. 2014: 'International Regulatory Landscape and Integration of Corrective Genome Editing into In Vitro Fertilization', *Reproductive Biology and Endocrinology* 12: 108, p. 8, https://rbej.biomedcentral.com/articles/10.1186/1477-7827-12-108.

67 The UNESCO's Universal Declaration on the Human Genome and Human Rights indicates in Article 24 that 'germline interventions' could be 'contrary to human dignity'. The Council of Europe's Convention on Human Rights and Biomedicine enshrined in Article 13 that 'an intervention seeking to modify the human genome may only be undertaken for preventive, diagnostic or therapeutic purposes and only if its aim is not to introduce any modification in the genome of any descendants'.

As already stated, Article 6(2)(b) of Directive 98/44/EC excludes patentability for 'processes for modifying the germ line genetic identity of human beings'. However, at the international level, there is no legal consensus as to whether human embryos are to be defined either as 'human beings' or as mere 'human life', with the latter implying less legal protection. The European Court of Human Rights in *VO v. France*, and referring to the Oviedo Convention on Human Rights and Biomedicine, also observed that at the European level

> there is no consensus on the nature and status of the embryo and/or foetus. ... At best, it may be regarded as common ground between States that the embryo/foetus belongs to the human race. The potentiality of that being and its capacity to become a person – enjoying protection under the civil law ... require protection in the name of human dignity, without making it a 'person' with the 'right to life' for the purposes of Article 2'.[68]

Therefore, it is paramount that both the European Commission and the EPO specify and clarify that Articles 6(2)(b) and 6(2)(c), which exclude patentability of human germline modification and the use of human embryos for commercial and industrial purposes, apply to CRISPR technologies, if practised in human germ cells and human embryos. Only such clarifications will ensure that patents do not serve as incentives to pursue controversial research of a kind which is contrary to European values, *ordre public*, and public policy.

## XI. CONCLUSIONS

More public debate, deliberations in the European Parliament and the Council, and inclusion of civil society organisations are needed in determining the appropriate interpretation of Directive 98/44/EC. At present, the discussion is often confined to patent attorneys, patent lawyers and officials of patent offices, which often restricts reasoning to formalistic legal arguments.[69] Furthermore, the EPO's granting practice and governance structures lack transparency and accountability to the citizens of Europe, an issue which needs to be addressed in current patent reforms. More transparency and an overhaul of the EPO's governance structure would be welcomed both by national and European legislators and by the general public. It would also be in line with the EPO's public mission and with the principle of disclosure,

[68] European Court of Human Rights, *Case of VO v. France*, Application no. 53924/00, 8 July 2004, Point 84. p. 35, http://www.pravo-medicina.sk/___files/Rozne/Vo%20vs.%20France%202004.pdf.

[69] The EU Commission's Expert Group (see the previous discussion in notes 7 and 23) could have been an opportunity for a broad deliberation following Habermasian standards on discourse. Unfortunately, it has not used this opportunity, as arguments and validity claims brought forward in many instances were not refuted by reasoned argumentation but were just cut off by the power of majority votes. See the Final Report with its dissenting opinions (note 23), and Godt 2016: 'Experts and Politics, Patent Policy', *Int'l Rev. Intell. Prop. & Competition Law* 7: 960–80.

which is one of the fundaments of patents' legitimacy. Transparency and accountability of the EPO would require (1) providing data on the stem cell patent applications filed and granted, and (2) for the EPO's granting practices on human embryonic stem cells, revealing the scientific and legal rationales of the EPO's new interpretation of the EPC's Article 53(a) and Rule 28 with reference to Chung et al. (2008) and PCT application WO 03/046141 in the EPO's Guidelines for Examination.

Moreover, the European patent system urgently needs to be democratised to serve the public interest and to comply with the social contract upon which it is constituted.[70] A broader view on both the empirics and impact of patents is needed, one which recognises that patents are embedded in social and economic contexts; therefore, interpretation of legal statutes must not be confined to a literal interpretation, but must also include the history and the purpose of the EU Biotech Directive. Understanding steps leading up to the Directive's implementation must take into account the socio-technical contexts and the normative implications for human dignity and social justice.

There is a strong need for a better balance to secure the proper interpretation of the '*ordre public* and morality' exclusion in European patent law, in accordance with the purposes and intentions of the European legislator and with the EU's Charter of Fundamental Rights. This requires the European Commission to take the initiative in strengthening the patent exclusions in Articles 5 and 6, and the European Parliament and the Council to dedicate themselves to policy reforms. The first step taken by the Commission should be to provide adequate clarification and precise guidance for the correct interpretation of the EU Biotech Directive.[71] These clarifications would establish new, binding rules for the interpretation of current patent law without

70 For elaboration on the social contract theory for patents, see Schneider, I. 2012: 'The Politicisation of Intellectual Property as a Challenge for Patent Theories. The Janus-Face of Patent Rights. A Plea for the Renewal of IP's Social Contract' (Die Politisierung des Patentrechts als Herausforderung für Patentrechtstheorien: Von der Janusköpfigkeit des Patentrechts zum Pladoyer fur eine Erneuerung des ihm zugrundeliegenden Sozialvertrags), *Political Science Quarterly* (*Politische Vierteljahresschrift PVS*) 46: 208–47.

71 Despite two and a half years of intense working of the aforementioned expert group, and its Final Report submitted in May 2016, the European Commission, by April 2019, has not delivered its respective 'Report on the Development and Implications of Patent Law in the Field of Biotechnology and Genetic Engineering', as requested in Article 16(c) of the Biotech Directive. Even though Article 16(c) demands an annual report on the development and implications of patent law in the field of biotechnology and genetic engineering, the Commission has published only two reports – one in 2002 and the second (and by now last) in 2005. In addition, Article 16(a) requires a report every five years on any problems encountered with regard to the relationship between this Directive and international agreements on the protection of human rights to which the member states have acceded. To date, the Commission has never fulfilled this 16(a) reporting requirement. Thus it must be stated that the monitoring effects expected by the European Parliament and the Council, and enshrined in the Biotech Directive, have failed. But nobody in the political arena seems to care.

changing the text of Directive 98/44/EC. As a fallback position, a further step could be a separate legislative action without altering the Directive as such. Another step, and a final fallback position in case the former steps were unsuccessful, would be legislative action that may include a thorough revision of the EU Biotech Directive in order to incorporate robust and legally defined limits of patentability. Taking no political action is not and cannot be an option.

## ANNEX

TABLE 1 *EPO Patent Applications and Grants on Human Embryonic Stem Cells (1986–2017, Cumulative)*

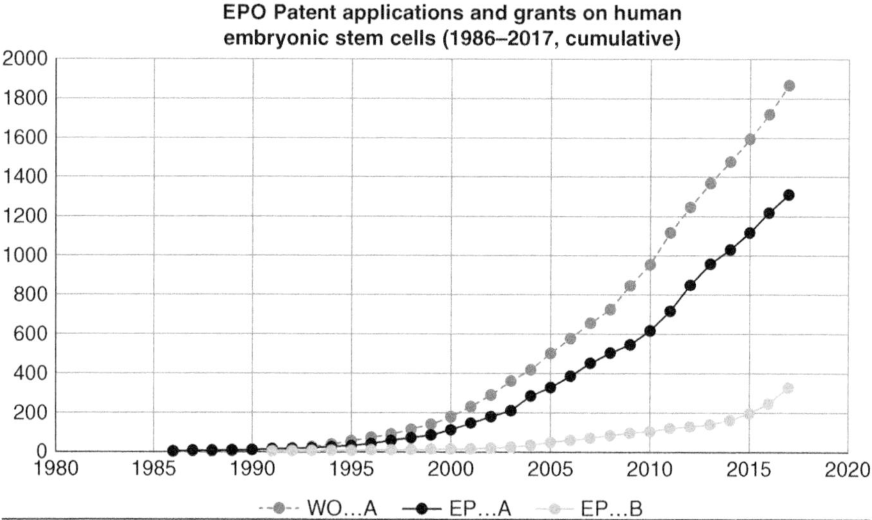


*Source*: Global Patent Index, Courtesy of Dr Ruth Tippe
WO ... A = PCT Applications
EP ... A = European Applications
EP ... B = Granted European Patents

TABLE 2 *EPO Patent Applications and Grants on Human Embryonic Stem Cells (1983–2017, Per Year)*

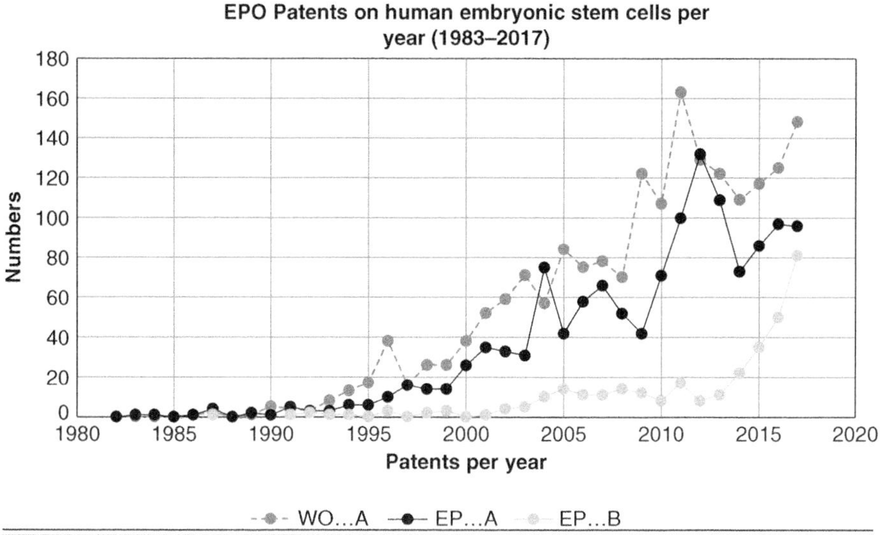


*Source*: Global Patent Index, Courtesy of Dr Ruth Tippe
WO . . . A = PCT Applications
EP . . . A = European Applications
EP . . . B = Granted European Patents

TABLE 3 *Number of Published Embryonic Stem Cell Patent Applications per Year (1995–2013), as Provided by Source EPO*

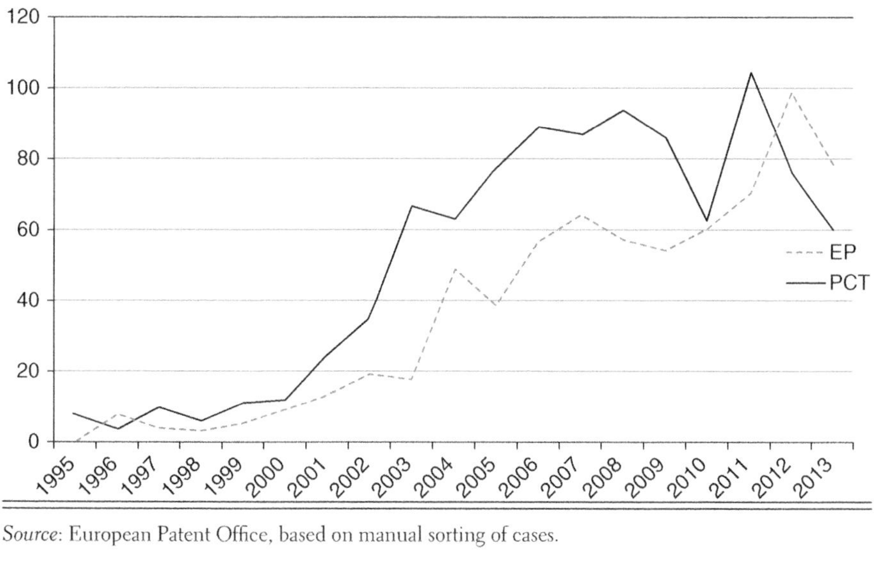


*Source*: European Patent Office, based on manual sorting of cases.

# Closing

# 16

# Life Patents, Religion, and Justice: A Summary of Themes

*Thomas C. Berg*

The chapters in this book bring religious perspectives, among others, to bear on the social and ethical questions raised by patents on life (i.e., on genetic material or processes concerning humans, other animals, or plants). Although a large literature of religious bioethics exists, relatively little of it deals with the role of patents or other intellectual property (IP) in biotech. That is so even though patents in the life and medical sciences have recently caused heated moral and political debates. For example, do pharmaceutical patents improperly render lifesaving medicines unaffordable in developing nations? Do global corporations commit "biopiracy" when they use traditional knowledge or genetic resources (TK/GR) to obtain their own patents?

Religion is absent partly because the IP-oriented world – the broad technology sector – is highly secular, even more so than industry, law, and academia in general. In an episode of the HBO comedy *Silicon Valley*, Richard, head of a tech start-up who is meeting to discuss a potential deal with executives of a larger company, lets it slip that one of his lead software developers is a practicing Christian.[1] The revelation tanks the deal. The humor relies on the widely acknowledged fact of the tech world's secularity.

That secularity extends to high-profile debates about IP in the world of bio- and medical technology. For example, in the debate over enforcing drug patents against developing nations, the anti-patenting activists who have appealed to "access to medicines" as a human right have made little effort to ground their arguments in religious terms, even in part.

This book gathers religious, secular moral, legal, and sociopolitical perspectives in one place. It aims to be a resource so lawyers, policy activists, and policymakers in patent debates might better understand what religious perspectives have to offer, and so religious thinkers and leaders might better understand biotech patents and thus have more to offer. Three themes emerge from the balance of the chapters. First, patents on life call for evaluation under criteria of morality and social

[1] *Tech Evangelist*, Silicon Valley, Seas. 5, Ep. 4 (April 15, 2018).

justice. Second, religious thought can contribute to (although of course it must not dominate) such moral and social evaluation. Finally, however, for religious thought to contribute effectively, it must be more informed and sophisticated than it has been, about both patent law and biotechnology.

## I. PATENTS CALL FOR MORAL AND SOCIAL-JUSTICE EVALUATION

For years, the dominant rationale for patents was economic, focused on optimizing incentives for the production and commercialization of innovation. Under this the sole aim is to strike the best balance between (a) granting rights to exclude, which enable returns that encourage invention and commercialization, and (b) preserving room for downstream innovation through a commons of freely available information. This rationale does not necessarily mandate expansive patent rights: empirically, narrower property rights and a larger commons might optimize innovation in some or many settings. But the incentives rationale tends in an expansive direction. The recent high-profile disputes, however, have made clear that IP is also fraught with moral, cultural, and social implications. Patent holders such as drug companies call infringement of their rights "theft" (see Margo Bagley's analysis here of that claim);[2] patent opponents respond with moral claims of, for example, a human right to access lifesaving medicines or treatments. "[T]he accessibility, affordability, and widespread diffusion of the goods regulated by [IP] have tangible effects upon human beings"; they can be "'matter[s] of life and death.'"[3]

From these moral-social debates, a different view of patents emerges than the incentives approach – what Katerina Sideri's chapter calls the "regulatory view." This view "broadens [patent law's] prescriptive framework to include equality and democracy as animating legislative reforms."[4] For example, as to technology for modifying germlines (egg and sperm, thus transferring the modifications to offspring), the regulatory view asks questions such as "[H]ow is this technology going to be developed? By whom? Who will benefit, and who will have access to it?"[5]

The regulatory view does not ignore incentives for innovation; but by adding considerations of equality (access) and democracy (public regulatory power) to the calculus, it tends to call for narrower patent rights. Several authors here acknowledge or endorse the regulatory view. Bagley, who presents a moral case for certain compulsory licenses on drug patents, emphasizes familiar arguments for considering patent rights not as full-fledged property, but as a "social bargain" between

2 Bagley chapter.

3 Gabriel J. Michael, *Catholic Thought and Intellectual Property: Learning from the Ethics of Obligation*, 25 J. Law & Relig. 415, 418 (2009–10) (quoting Christopher May and Susan K. Sell, Intellectual Property Rights: A Critical History 7 (Lynne Rienner Publishing, 2006)).

4 Sideri chapter, p. 251.

5 *Id.* at p. 262.

government and patentee, under which government can condition rights on preserving poor people's access to lifesaving drugs.[6] Ingrid Schneider likewise argues that European patent laws envision "patent governance as a mode of pro-active, socio-political regulation of technological trajectories, in accordance with European social values and norms, as enshrined in the Charter of Fundamental Rights."[7]

The regulatory view is not the exclusive premise in this book; other authors here see the balance of economic incentives as central.[8] But a common premise here is that one cannot naively assert that incentives are the only consideration. As the chapter by Kathy Liddell and Simon Ravenscroft says, "Broad ethical and moral considerations are here to stay in patent law whether we like it or not."[9]

## A. *Morality Critiques in European and US Law*

European and US law reflect different attitudes toward moral criteria for patentability. Our transatlantic roster of authors captures these differences.

The European Patent Convention (EPC) and the European Union (EU), although separate legal regimes, have matching rules forbidding patents on various biotechnologies, reflecting explicit or implicit judgments that such technologies raise ethical concerns. Both the EPC, in article 53(a), and the EU's 1998 Biotechnology Directive (98/44/EC), in Article 6(1), have a general clause forbidding patents on inventions whose "commercial exploitation would be contrary to *ordre public* or morality." And Directive Article 6(2) prohibits patents on specific technologies: (a) processes for cloning human beings; (b) processes for modifying the germ line genetic identity of human beings; and (c) uses of human embryos for industrial or commercial purposes. In turn, the European Patent Office (EPO), which implements the EPC, has adopted these prohibitions in its own Rules 28 and 29.

These prohibitions have spawned considerable EPO and European-court case law, analyzed at length here in chapters by Christopher Rennie-Smith as well as Liddell/Ravenscroft, Sideri, and Schneider. Some cases interpret the exclusions broadly, such as the European Court of Justice holding in *Brüstle v. Greenpeace* that many forms of embryonic stem-cell research fall within the ban on patenting commercial uses of embryos. Schneider argues, however, that various EPO decisions have allowed applicants to "patent around the prohibitions set" in the EPC and the rules; more generally, she criticizes the EPO for displaying a "technocratic" and "pro-patent bias" and acting "as a service provider for industry and researchers."[10]

[6] Bagley chapter, p. 190 at n.17 (emphasis in original) (noting that ideas are nonrivalrous – that is, can be shared without interfering with anyone's enjoyment – lack clear boundaries giving notice of infringement, and so forth).
[7] Schneider chapter, p. 267.
[8] Broyde/Weiner, Heald, and Kock chapters.
[9] Liddell and Ravenscroft chapter, p. 30.
[10] Schneider chapter, p. 265.

Nevertheless, as Liddell and Ravenscroft emphasize, the European prohibitions preserve some notion that public morality should place limits on patents. Thus Rennie-Smith, former judge on EPO Boards of Appeal, writes from experience not only about the scope of the prohibitions, but also about interventions on the moral issues by "non-commercial" parties – primarily pressure groups opposed to animal experimentation and commercialized research. He concludes they have had some effect – although they could have had even more – and that churches and secular ethical groups could do more to help determine the meaning of "*ordre public*" by presenting "evidence of accepted norms in European culture." "Churches and ethicists," he says, "have a contribution to make to patent oppositions which has so far not been made."[11]

Recognition of morality limits on patentability is the overwhelming majority position in the world. As Liddell and Ravenscroft report, one survey of "seventy-three member countries of the World Trade Organization . . . found that all but six of them had explicit morality exclusions in their patent regimes along similar lines [to Europe]."[12]

The United States, however, is one of those six nations; morality plays almost no direct role in US patentability decisions. Contributor Margo Bagley has elsewhere described the US approach as "patent first, ask questions later": the US Patent and Trademark Office (USPTO) and reviewing courts disregard questions about an invention's morality, leaving it to Congress to act after patents have been issued (which Congress almost never does).[13] A doctrine under which inventions with immoral purposes were deemed to lack "moral utility" has died out in recent years. In its 1980 decision opening the door to patents on genetically engineered organisms, the US Supreme Court said that worries about consequences of genetic manipulations – worries that they might "spread pollution and disease, . . . result in a loss of genetic diversity, [and] depreciate the value of human life" – "should be addressed to the political branches of the Government, the Congress and the Executive, and not to the courts."[14]

As a corollary, the USPTO, even more than the EPO, has managed to retain an aura of "apolitical," "technocratic neutrality" – even when issuing patents for inventions as potentially controversial as abortifacient devices and drugs.[15] Notably, in Europe (as several chapters here discuss[16]) moral debates over

[11] Rennie-Smith, p. 80.

[12] Liddell and Ravenscroft, p. 30.

[13] Margo A. Bagley, *Patent First, Ask Questions Later: Morality and Biotechnology in Patent Law*, 45 Wm. & Mary L. Rev. 469, 470 (2003). There are exceptions: statutory prohibitions on claiming a human organism, America Invents Act, Pub. L. No. 112–29, § 33(a) (2011), and on suing doctors for infringement of medical-process patents in treating patients, 35 U.S.C. § 271.

[14] Diamond v. Chakrabarty, 447 U.S. 303, 317 (1980).

[15] Kara W. Swanson, *Patents, Politics, and Abortion*, Northeastern University School of Law Faculty Publications, Paper 71, at 9, 13, 16 *et seq.* (2016), http://lsr.nellco.org/nusl_faculty/71.

[16] Liddell/Ravenscroft; Rennie-Smith; Sideri; Schneider.

embryonic stem-cell research focus on whether such technology can be patented; but in the US they focus on whether government should fund it with tax dollars.[17] (The difference may reflect a greater American tendency – one emphasized by Paul Heald here – to view patents as a form of private ordering rather than as a form of government regulation or subsidy.[18])

Instead, the chief US limit on patenting biotechnology has been the patent-ineligibility of natural phenomena: laws, processes, and substances. In *Mayo Collaborative Services v. Prometheus Laboratories*,[19] the Supreme Court held that a diagnostic test for appropriate drug dosages was unpatentable when it involved merely administering the drug, measuring the patient's resulting metabolite levels, and describing levels above which harmful side effects would likely occur and below which the dosage would likely be ineffective. The correlations between metabolite levels and toxicity or effectiveness were unpatentable natural phenomena; the administration and measurement steps were merely conventional, adding nothing novel or inventive ("nonobvious") to the natural phenomena. And in *Association for Molecular Pathology v. Myriad Genetics*,[20] the Court held that isolated gene sequences – in that case, sequences coding an increased risk of breast cancer – were unpatentable natural substances, although it simultaneously held that cDNA, the reverse sequence with "non-coding" material removed, was an artificial substance eligible for patent.

But the natural-phenomena exclusions have implicit moral foundations, even if they are not explicitly so described today. Joshua Sarnoff's chapter describes the rich history of the exclusions, showing that they arose from overlapping scientific, philosophical, and religious affirmations that "claiming as one's own origination an invention derived proximately from nature, and seeking property rights in it, was an act of hubris and contrary to morality."[21]

The Supreme Court also offers a public-policy rationale for excluding natural phenomena: because they are "the basic tools of scientific and technological work," conferring exclusivity over them would "tie up their use [and] inhibit future innovation premised upon them."[22] This public-policy concern is central, as Michael Broyde and Steven Weiner emphasize in their chapter evaluating gene patents from a Jewish perspective, even though the Court does not directly attempt to decide whether a given patent would preempt "basic tools" but rather applies the "natural

17 Varnee Murugan, *Embryonic Stem Cell Research: A Decade of Debate from Bush to Obama*, Yale J. Biology & Med. 101 (September 2009), https://www.ncbi.nlm.nih.gov/pmc/articles/PMC2744932/; Sonya Davey et al., *Interfacing of Science, Medicine and Law: The Stem Cell Patent Controversy in the United States and the European Union*, 3 Frontiers Cell. Dev. Biol. 71 (2015), https://www.ncbi.nlm.nih.gov/pmc/articles/PMC4639617/.

18 Heald chapter.

19 566 U.S. 66 (2012).

20 569 U.S. 576, 590–95 (2013).

21 Sarnoff chapter, p. 41.

22 *Mayo*, 566 U.S. at 86 (quotation omitted); *Myriad*, 569 U.S. at 589 (quoting *Mayo*).

vs. artificial" distinction as a "more easily administered proxy" for such preemption.[23]

Although the US court's worry about ownership of "basic tools" lies in the effect on "future innovation premised upon them," such ownership can also have another effect with moral implications: it can affect access to medical technology for persons who lack the resources to pay the higher prices that patents support. The American Civil Liberties Union (ACLU), representing the challengers in *Myriad*, emphasized that the company's patent monopoly had allowed it to price breast-cancer diagnostic tests beyond many women's means and to control "whether the quality of their tests can be assessed, whether testing can be improved and alternative tests developed through research, and whether patients can get a second opinion."[24] Once Myriad lost its patent on isolated gene sequences, a competitor began offering tests at about half of Myriad's $4,100 price.[25]

Such concerns about relative access to, and distribution of, medical-technology benefits are thus only implicit, not explicit, in US case law – and likewise in Europe.[26] But the concerns have been prominent in ethical and socio-political debates over patents.

### B. *Access/Distributional Concerns in International Debate*

Concerns about patents' access and distributive effects first became prominent in the 1990s debate over the use of generic drugs to combat epidemics, like AIDS and malaria, wracking the least-developed countries (LDCs). When drug companies sought to block generics, invoking the strengthened patent norms required by the 1994 global TRIPS agreement (Trade-Related Aspects of Intellectual Property), the resulting controversy led the World Trade Organization (WTO) to clarify that TRIPS gave nations flexibility to address public-health emergencies by limiting patents through measures such as compulsory licenses.[27] In this book, Margo Bagley defends LDCs' use of compulsory licenses to make lifesaving drugs affordable, since among other things, patent holders reap little reward from such markets anyway. Her argument bears on biotechnology patents, because they too may raise concerns about access to medicine – and because the new trend in pharmaceuticals, "biologic" drugs

[23] Broyde-Weiner chapter, p. 118, quoting *Mayo*, 566 U.S. at 88–89.

[24] ACLU, *BRCA FAQs*, https://www.aclu.org/other/brca-faqs?redirect=free-speech/brca-faqs.

[25] Joe Carlson, *"The Cost Will Drop": Myriad Genetics' Shares Drop as Competitors Enter*, Modern Healthcare, June 15, 2013, http://www.modernhealthcare.com/article/20130615/MAGAZINE/306159974.

[26] The EPO Board of Appeal has held that access/distributional harms do not trigger EPC Article 53(a)'s exclusion on "inventions the commercial exploitation of which would [be immoral]," on the grounds that such harms stem from exploitation of the patent, not the invention. *Breast and ovarian cancer/ UNIVERSITY OF UTAH*, T 1213/05; Rennie-Smith, [at n.11].

[27] WTO, Declaration on the TRIPS agreement and public health (November 14, 2001), WT/MIN(01)/DEC/2, https://www.wto.org/english/thewto_e/minist_e/min01_e/mindecl_trips_e.htm.

drawn from genes and other living materials, are highly costly and often "covered by a web of 'life' patents."[28]

It bears mention that the same global trade system that has toughened patent rights is now under attack from the United States, once its strongest supporter. President Donald Trump has criticized international trade agreements, called the WTO a "catastrophe" for Americans,[29] and withdrawn the United States from the proposed Trans-Pacific Partnership (TPP).[30] By withdrawing, the United States foregoes various TRIPS-plus expansions of patent protection in the TPP, notably for drugs, including biologicals.[31] This does not mean that transnational IP protection has ceased to be a priority of US policy. Like its predecessors, the Trump administration rages over Chinese theft of American IP.[32] Moreover, expanded IP tops the US agenda in bilateral trade negotiations, a setting that Trump favors over multilateral negotiations.[33] Moreover, the US will likely return to a posture of seeking multilateral agreements. But for now, the US attack on the trade system may hamper the system's protection even of rights that the US favors, such as patents.

## II. RELIGIOUS VALUES CAN CONTRIBUTE TO THE MORAL EVALUATION

The book's second large theme is that religious perspectives have something valuable to contribute to the moral and social evaluation of biological patents. As already noted, this runs counter to prevailing practice.

Leading IP scholar Mark Lemley recently published an article attacking what he called "faith-based intellectual property."[34] He actually targeted not traditional religions as such, but rather a more generic category of "faith-based" positions – namely justifications of IP rights that rest on deontological moral assertions (such as "IP respects the dignity and autonomy of creators," or "violating a copyright is theft") rather than on the pragmatic consequences of recognizing such rights. Lemley charges that defenders of strong IP rights, unable to justify their position by empirical evidence, have retreated to the argument "'I [the creator] made it and so I own it,'

28 Bagley, pp. 189–90. But see Broyde/Weiner, p. 128 (responding that costs may drop after initial rise).

29 See, e.g., Tom Miles, *Trump's Bonfire of the Treaties Sweeps towards the WTO*, Reuters, May 18, 2018, https://www.reuters.com/article/us-usa-trade-wto-analysis/trumps-bonfire-of-the-treaties-sweeps-towards-the-wto-idUSKCN1IJ1K9/.

30 https://ustr.gov/about-us/policy-offices/press-office/press-releases/2017/january/US-Withdraws-From-TPP.

31 Deborah Gleeson et al., *The Trans Pacific Partnership Agreement, Intellectual Property and Medicines: Differential Outcomes for Developed and Developing Countries*, Global Social Policy (October 13, 2017), http://journals.sagepub.com/doi/10.1177/1468018117734153.

32 Alan Rappaport, *Intellectual Property to Take Center Stage as Trump and Xi Meet*, N.Y. Times, November 28, 2018, https://www.nytimes.com/2018/11/28/us/politics/intellectual-property-trump-xi.html.

33 Remarks by President Trump to the World Economic Forum, January 26, 2018, https://www.whitehouse.gov/briefings-statements/remarks-president-trump-world-economic-forum/.

34 Mark A. Lemley, *Faith-Based Intellectual Property*, 62 UCLA L. Rev. 1328 (2016).

often attributing that sentiment to John Locke, or Hegel, or, more recently, Rawls."[35] In this book at least, authors with distinctively religious perspectives, far from accepting strong patent rights on faith, are more often ambivalent, even skeptical, about them.[36]

Yet Lemley's critique hardly bodes well for religious perspectives, since he claims that the problem with deontological IP is that it "is a form of religious belief" and religious belief "is remarkably resistant to evidentiary challenge."[37] Inserting faith-based arguments into IP policy debates means that "we will make policy based on our instincts without being able to engage in a meaningful conversation about the wisdom of that policy."[38] All this takes for granted, as James Grimmelman observes, that "'faith' is a pejorative in the language of academic discourse," since "[b]y and large, [IP] scholars are heirs to the Enlightenment tradition."[39] What are the answers to such a critique? Put differently: "What's faith got to do with it?"[40]

## A. *Religious Resources*

Kathy Liddell and Simon Ravenscroft's chapter here offers one set of answers to that question. In their view, if moral-ethical criteria may both justify and limit patent rights, religious perspectives may contribute to formulating those criteria. "[I]t is unfair to characterize non-religious, reflective bioethical reasoning as completely different from reflective theological reasoning," they argue, since both reasoning modes make "assumptions about the nature of the world and what it means to be human" and therefore rely "to some extent on a type of faith or faiths."[41]

Indeed, some theorists argue that the deepest grounds of legal principles necessarily involve some reference to convictions or assumptions that are, broadly speaking, religious. An example is Michael Perry's writing on universal human rights, a concept regularly invoked in global policy debates, including as a reason to limit patents when they affect vulnerable people's access to medicines or control over their genetic resources.[42] Perry argues that a "fundamental constituent of" human rights – "the conviction that every human being is sacred, . . . is 'inviolable,' has

35 *Id.* at 1338–39. See *id.* at 1337 n.34 (distinguishing such IP defenders "from those who seek guidance in existing religions to evaluate IP doctrines").

36 Neves; Colecchi; El-Said; Wojda; Bagley; Cholij ch. 12.

37 Lemley, *supra* note 34, at 1337–38.

38 *Id.* at 1346.

39 James Grimmelman, *Faith-Based Intellectual Property: A Response*, The Laboratorium (2d ser.), April 21, 2015, https://2d.laboratorium.net/post/117023858730/faith-based-intellectual-property-a-response.

40 Thomas M. Mengler, *What's Faith Got to Do with It? (With Apologies to Tina Turner)*, 35 U. Toledo L. Rev. 145 (2003–4).

41 Liddell and Ravenscroft, p. 29.

42 See, e.g., Paul L.C. Torremans, ed., Intellectual Property and Human Rights (3d ed., Wolters-Kluwer 2015).

'inherent dignity'" – is "inescapably religious."[43] Only the "existential" religious conviction that "the world . . . hangs together" and "the Other is, finally, one's own sister/brother" can ground "loving concern" for others in cases of deep conflict or great distance.[44] Likewise, in an essay responding to Lemley, IP scholar David Opderbeck explains how pragmatic or utilitarian reasoning cannot capture the sense of human meaningfulness and sacredness:

> [A]ny inquiry constrained by the rules of methodological naturalism cannot tell us what a phenomenon actually *is*. Such an inquiry can only tell us that certain results were obtained using certain assumptions. . . . [T]his compound, administered in this way to people with this form of cancer, kills cancer cells without damaging other cells. From the perspective of bare "nature," there is no significance to this result at all – cells survive, cells die, genes are passed on or not passed on – something has been described, but nothing at all has been *explained*. If the cancer patient is a close friend or relative, however, this result may be cause for joy. This *person*, whom I *love*, may be saved. This treatment matters to *me*, to *my spouse*, to *my children*.[45]

Unlike naturalism, religion can explain why the effect of patents on human beings should "matte[r] to me" – and why the circle of those who matter extends, in some form, to all human beings.

To say that religious affirmations are indispensable to justifying human dignity is a strong, and sharply debated, claim. A weaker version of the claim is that religious affirmations can contribute to justifying such dignity and understanding its implications. The great religious traditions have long, rich histories of reflection on the human condition and on what human beings are due. As Opderbeck elsewhere puts it, even when religiously based proposals parallel their secular counterparts, the religious basis can root them "in a richer social imagination that gives them meaning and weight."[46]

## B. *Religion and Societal Norms*

The relevance of religion is empirical as well as normative. Religion remains a potent source of public culture and societal values worldwide, especially in developing countries, where the justification and scope of patent rights is particularly at issue. Nations have flexibility under TRIPS to decide whether to protect genetic materials and whether to limit patent rights for public-health purposes.[47]

43 Michael J. Perry, The Idea of Human Rights: Four Inquiries 13 (Oxford Univ. Press 1998).

44 *Id.* at 21.

45 David W. Opderbeck, *Lex Machina Non Est: A Response to Mark Lemley's "Faith-Based Intellectual Property,"* Seton Hall Public Research Paper (September 6, 2017), at 12, https://papers.ssrn.com/sol3/papers.cfm?abstract_id=3022065 (emphases in original).

46 David W. Opderbeck, *Beyond Bits, Memes, and Utility Machines: A Theology of Intellectual Property as Social Relations*, 10 U. St. Thomas L.J. 738, 773 (2013).

47 TRIPS Art. 27(3)(b) (members may exclude "plants and animals other than micro-organisms, and essentially biological processes for the production of plants or animals"); Art. 31 (permitting compulsory licenses, without a prior offer of "reasonable commercial" compensation, "in the case of

Exercising those flexibilities might – and under democratic theory, should – reflect the population's moral values, including its religious values. Accordingly, insight into how religious values relate to IP can help a nation's policymakers incorporate public sentiment in their decisions and better explain their decisions to the public. And it can help outsiders better understand a nation's attitude toward controversial IP questions.

Consider Muslim nations. Polls show "that vast majorities in Muslim societies desire to see Islam embedded in their constitutional and legislative framework."[48] These nations vary in how fully they enshrine sharia (Islamic law) in their constitutions or apply it in legislation. Yet sharia certainly plays a role – one that, surprisingly, may increase in importance the more that a Muslim nation becomes committed to democratic implementation of popular values.[49] Accordingly, there is great benefit in Mohammed El-Said's chapter setting forth Islamic understandings of the proper purposes of biotechnology and the justification and limits of patents.[50] El-Said concludes by suggesting how legislation in Muslim nations might reflect the Islamic prioritization of life, for example when patent-based monopolies limit access to medical treatments.

Christian concepts figure even more prominently in this book. But Christianity is likewise vital in developing nations. In Latin America as of 2014, 69 percent of the population was Catholic and 19 percent Protestant (primarily evangelicals).[51] In sub-Saharan Africa, Christians constituted 62.9 percent of the population in 2010, and although their share is declining, Pew Research Center estimates it will still be 58.5 percent in 2050.[52] Moreover, as Christianity's influence declines in Western Europe and even the United States, the African brands of the faith are predicted to fill the gap and tilt Christian social thought toward Global South, developing-world perspectives.[53]

This collection, therefore, includes several chapters exploring Christian themes.[54] Osvaldo Neves, Stephen Colecchi, and Paul Wojda set forth Roman Catholic institutional and theological positions on patents, ownership, access to knowledge,

a national emergency . . . or in cases of public non-commercial use"); both at https://www.wto.org/english/docs_e/legal_e/27-trips_04c_e.htm.

48 Dawood I. Ahmed and Moamen Gouda, *Measuring Constitutional Islamization: The Islamic Constitutions Index*, 38 Hastings Int'l & Comp. L. Rev. 1, 9 (2015) (citing Pew and Gallup polls).

49 *Id.* at 7–9.

50 El-Said chapter.

51 Pew Research Center, *Religion in Latin America: Widespread Change in a Historically Catholic Region* (November 13, 2014), http://www.pewforum.org/2014/11/13/religion-in-latin-america/.

52 Pew Research Center, *The Future of World Religions: Sub-Saharan Africa* (April 2, 2016), http://www.pewforum.org/2015/04/02/sub-saharan-africa/. Pew predicts that increases in Muslims will make up the difference. *Id.*

53 David McClendon, *Sub-Saharan Africa Will Be Home to Growing Shares of the World's Christians and Muslims*, Pew Research Center (April 19, 2017), http://www.pewresearch.org/fact-tank/2017/04/19/sub-saharan-africa-will-be-home-to-growing-shares-of-the-worlds-christians-and-muslims/.

54 Many of which, course, have their roots in Judaism.

human dignity, and development. Margo Bagley argues that the Biblical injunction "Thou shalt not steal" does not condemn compulsory licenses for essential medicines once it is understood in the light of the broader command to love one's neighbor. And Roman Cholij argues, based on the social view of property drawn from Catholic teaching as well as from other sources, that patent systems worldwide should include a requirement of disclosure of the origin of TK/GR, so as to facilitate equitable benefit sharing with communities of vulnerable people.[55]

Even if government does not constrain industry's reliance on patents, industry decision makers might take guidance from the values their religious communities teach. For example, a Vatican publication, *Vocation of the Business Leader*, calls on businesspeople to "produce goods and services that meet genuine human needs and serve the common good, while taking responsibility for the social and environmental costs of production and the supply chain and distribution and watching for opportunities to serve the poor."[56] In this book, agricultural-biotech consultant (and former industry executive) Michael Kock proposes a licensing model designed to foster the "constructive, ethical use" of seed patents so as to "maximize the[ir] beneficial effects (i.e. preserving the incentives for innovation and knowledge-sharing) while minimizing the [bad] effects" (restrictions on access and on farmers' practices).[57]

## C. *The Role of Religious Organizations*

Religion is also relevant because religious institutions are active, concretely, in matters related to patents on life. The Catholic Church is perhaps the most active, as the Neves and Colecchi chapters document: the Vatican, a participant-observer at international organizations like the WTO and the UN, has regularly spoken on issues concerning international IP and human development. Throughout the controversy over patents on essential medicines, the Vatican has emphasized that patents, although needed to encourage innovation, should also be tempered to spread "the benefits of the innovations as widely as possible," since IP law "is there primarily *to serve the common good of the* [entire] *human community*."[58] Moreover, as Colecchi details, US and Latin American bishops' conferences have repeatedly

[55] Cholij, ch. 12.

A full collection of religious reflections on patents would also include Hindu, Confucian, and other contributions. See, e.g., Shubha Ghosh, *Duty, Consequences, and Intellectual Property*, 10 U. St. Thomas L.J. 801 (2013) (applying insights from the *Bhagavad Gita*); Peter K. Yu, Intellectual Property and Confucianism, in Irene Calboli and Srividhya Ragavan, eds., Diversity in Intellectual Property: Identities, Interests and Intersections 247 (Cambridge Univ. Press 2015).

[56] Pontifical Council for Justice and Peace, Vocation of the Business Leader: A Reflection 3 (2014), https://www.stthomas.edu/media/catholicstudies/center/ryan/publications/publicationpdfs/vocationofthebusinessleaderpdf/PontificalCouncil_4.pdf.

[57] Kock, pp. 227, 235.

[58] Intervention by the Holy See at the 6th Ministerial Conference of WTO, Address of H.E. Mons. Silvano Tomasi (December 18, 2005), http://www.vatican.va/roman_curia/secretariat_state/2005/

advocated that trade agreements should share with indigenous peoples the rewards from use of their traditional knowledge and should not uphold patent rights that seriously inhibit farmers' practices such as reuse of seeds.

The Catholic Church, as Colecchi emphasizes, also has extensive "on-the-ground experience" serving people with serious diseases, subsistence farmers, and other persons whom patent policy may affect.[59] If Doctors without Borders deserves a hearing for its views on patenting of essential medicines,[60] Catholic Relief Services deserves a hearing for its views on seed patents, genetic resources, and other matters.

## III. RELIGIOUS-ETHICAL EVALUATIONS MUST BE MORE INFORMED AND SOPHISTICATED

But if religious perspectives can be fruitful in evaluating life-patent issues, they often have failed to do so. As bioethicist Audrey Chapman has described, religious leaders' public statements on gene patenting have been "intermittent" and often merely "prophetic" utterances lacking "careful theological and ethical analysis" or examination of "the specifics of patent law."[61] Chapman criticized religious leaders' 1995 "Joint Appeal against Human and Animal Patenting" – which, in a single short paragraph, broadly opposed patenting on the ground that "humans and animals are creations of God, not [of humans]" – as well as other religious statements that equate gene patents with "playing God" or usurping "divine ownership of life."[62] Critiques by several contributors here follow Chapman's lead.[63]

One problem with the "ownership of life" objection, Paul Heald argues here, is that it "misconceives patent law."[64] A patent gives its holder "no right to use or exploit the invention," only the right to stop others from doing so. And even the broadest gene patents issued give no control "over the normal human use of our own DNA," only control over "the isolated form of the gene as it existed outside the human body."[65] One purpose of this book is to help religious leaders understand patent law better and thereby sharpen their theological evaluation of life patents.

documents/rc_seg-st_20051218_wto_en.html; Intervention by the Holy See at the World Trade Organization, Address of H.E. Msgr. Diarmuid Martin (December 20, 2002), http://www.vatican.va/roman_curia/secretariat_state/documents/rc_seg-st_doc_20021220_martin-wto_en.html (emphasis in original).

59 Colecchi, p. 107.

60 See, e.g., Medecins Sans Frontieres, *A Matter of Life & Death: The Role of Patents in Access to Essential Medicines* (November 2001), https://www.msf.org/sites/msf.org/files/doha_11–2001.pdf.

61 Audrey R. Chapman, *Religious Contributions to the Debate on the Patenting of Human Genes*, 10 U. St. Thomas. L.J. 650, 681–82 (2013).

62 *Id.* at 665–76 (critically analyzing such arguments and generally finding them oversimplified and underdeveloped).

63 Broyde/Weiner; Heald; Wojda; Kock.

64 Heald, p. 154. See also Kock, p. 239 ("The 'no patents on life' slogan alleges that a patent grants ownership to life, a myth based on a fundamental misunderstanding of the patent system.").

65 Heald, p. 154.

Above all, Chapman, like Heald, thinks that religious critics of gene patents have failed to integrate their moral critique with concerns about how patent-supported higher prices may effectively deny the poor access to medical technology.[66] Multiple chapters of this book take up that concern.

The chapters here do not oppose patents altogether. Several, indeed, endorse the role of patents in encouraging invention and commercialization.[67] But the chapters, taken as a whole, do identify three sorts of moral criticism of life patents to which religious perspectives might contribute.

### A. *Patents on Nature*

Some examples of biotechnology may be unpatentable because they are natural phenomena. Admittedly, it is too simple to condemn such patents as "playing God." And religious analyses cannot tell us exactly where the natural/artificial line should be drawn: after all, religious concepts of "stewardship" emphasize the vocation of humans to work upon divine creation "to make this world a better place."[68] Nevertheless, as Josh Sarnoff's chapter emphasizes, the exclusions grew in part out of the religious sense that natural phenomena had "divine origins as commons property" – a gift of God – that "imposed moral duties to freely share knowledge of nature and thus of God."[69] Compared with the utilitarian rationale for the exclusions, this rationale offers the sort of "richer social imagination," in Opderbeck's words, that may sustain the exclusions against pressures to commodify all aspects of life.

### B. *Objectionable Technologies or Uses*

Some examples of biotechnology might be objectionable for other moral reasons, either inherently or because of their use. Examples could include germline modifications – at least those that step over the line into practicing eugenics – some uses of human embryos, or the use of genetic screening to deny jobs, insurance, or other important benefits.[70] European patent law treats some such technologies as unpatentable. And as the chapters by Schneider and Sideri emphasize, if a certain

66 Chapman, *supra* note 61, at 681 ("The religious community did not discuss the detrimental impacts of genetic patents on scientific and medical research [or] the concerns . . . that the willingness to extend life patents restricted the availability of needed genetic testing and obstructed patient care."); Heald, p. 155. See also Keith Douglass Warner, *Are Life Patents Ethical? Conflict between Catholic Social Thought and Agricultural Biotechnology's Patent Regime*, 14 J. Agricult. & Envir'l Eth. 301 (2001).

67 Heald; Kock; Broyde/Weiner.

68 Ted Peters, *Patenting Life: Yes*, First Things, May 1996, https://www.firstthings.com/article/1996/05/004-patenting-life-yes.

69 Sarnoff, p. 42.

70 As this book entered production, in late 2018, a scandal erupted when a Chinese researcher announced that he had altered the genes of two embryos using CRISPR to reduce susceptibility to HIV, had introduced the embryos into the mother's uterus, and had produced twin pregnancies. The

product or process is morally objectionable, the government ought not to support or encourage it with a patent monopoly.[71]

### C. *Access and Distributional Concerns*

Finally, as already discussed, in some situation patents might have negative effects, on balance, on the access of poor people to important technologies. A recurring feature of the chapters here is the integration of this social-justice concern into religious evaluation of patents. Of course, patents may also spur the creation of technologies that otherwise would not exist. Access and distributional concerns require careful examination and balancing. The chapters here examine:

1. The use of compulsory licenses to ensure that patents do not price essential medicines (and by extension other essential technologies) out of the reach of people living in poverty – at the very least when the license does not deny patent holders a significant reward they would otherwise reap in the market in question (Margo Bagley).
2. The development of licensing schemes for seed-patented crops that attempt to balance fairly the interests in encouraging innovation and commercialization with the interest in preserving poor farmers' ability to pay and to maintain some flexibility in their farming practices (Michael Kock).
3. Reliance on non-patent means of encouraging innovation and commercialization, such as "tax credits, prizes, grants, and … public–private partnerships" (Katerina Sideri).[72] Sideri examines such concerns particularly in the context of the CRISPR gene-editing tool, where some broad patents have been granted, and where, "[i]f the well-off have access to germline interventions enhancing traits, the result will be greater social inequalities and polarization."[73]
4. Enforcement and further development of norms against piracy of indigenous peoples' traditional knowledge and genetic resources. Chapters focus on ensuring "fair and equitable sharing" of the benefits of such knowledge and resources under, among other things, the Nagoya Protocol to the Convention on Biological

experiment appeared to manifest multiple ethical and scientific pitfalls: the editing appeared imprecise, may have introduced susceptibilities to other diseases, may have lacked informed consent, and did not respond to any existing disease in the embryos. See Ed Yong, *The CRISPR Baby Scandal Gets Worse by the Day*, THE ATLANTIC, December 3, 2018, https://www.theatlantic.com/science/archive/2018/12/15-worrying-things-about-crispr-babies-scandal/577234/ (describing it as "the worst possible scenario for introducing the first gene-edited babies into the world").

[71] On the other hand, patent defenders argue that the proper response to an objectionable technology is to prohibit it, not to declare it unpatentable (which may actually allow more persons to practice the technology). Indeed, occasionally opponents of a particular technology accept patents on it as a means to limit its spread: anti-abortion activists kept the drug RU-486 out of the United States in part because they could concentrate their pressure on the German company that held the patent. Swanson, *supra* note 15, at 17–18.

[72] Sideri, p. 261.

[73] *Id.*, p. 256.

Diversity (Osvaldo Neves);[74] and mandating disclosure of the origin of TK/GR used in inventions for which national patents are sought (Roman Cholij).

## CONCLUSION

The issues discussed in this book will continue to develop and will remain in flux. With respect to gene-editing technologies like CRISPR, new patents will issue, and conflicts will continue over who owns patent rights to particular processes (or altered genes).[75] The ethical questions will also become increasingly pressing, as more researchers are likely to implant gene-edited embryos (following the Chinese researchers' lead) or pursue other forms of editing the human germline.[76]

In the United States, the Supreme Court's *Myriad* and *Mayo* decisions restating the patent-ineligibility of natural products and processes may be far from the last word. As this book entered final production stages, draft Senate legislation being pushed by a Democrat and a Republican (Sens. Christopher Coons and Thom Tillis) would provide: "No implicit or other judicially created exceptions to subject matter eligibility, including 'abstract ideas,' 'laws of nature,' or 'natural phenomena,' shall be used to determine patent eligibility under section 101, and all cases establishing or interpreting those exceptions to eligibility are hereby abrogated."[77] Whether this proposal succeeds—in whole or in some part (for example, restoring patentability for some diagnostic tests)—the bipartisan support for it dramatizes the influence of the biotechnology and pharmaceutical industries, which argue that expansive patent protection is necessary to encourage innovation.[78] On the other side, opponents of the proposal claim that it is—in the words of Joshua Sarnoff, a contributor to this book—"bad utilitarian innovation policy," "likely unconstitutional," "violative of human rights," "immoral," and (in terms of Christian theology) "religiously sinful" because it would eliminate public access to scientific discoveries

[74] Neves, part III; Nagoya Protocol on Access to Genetic Resources and the Fair and Equitable Sharing of Benefits Arising from Their Utilization (2010), https://www.cbd.int/abs/doc/protocol/nagoya-protocol-en.pdf.

[75] See, e.g., Sharon Begley, *Patent office reopens major CRISPR battle between Broad Institute and Univ. of California*, STAT, June 25, 2019, https://www.statnews.com/2019/06/25/crispr-patents-interference/ (noting USPTO's declaration of interference proceeding to determine priority between CRISPR-process inventions).

[76] See, e.g., David Cyranoski, *Russian biologist plans more CRISPR-edited babies*, NATURE, June 10, 2019, https://www.nature.com/articles/d41586-019-01770-x (noting that such steps have triggered "renewed calls for an international moratorium on germline editing").

[77] Draft of May 22, 2019, https://www.tillis.senate.gov/services/files/E8ED2188-DC15-4876-8F51-A03CF4A63E26.

[78] Paige Winfield Cunningham, *The Supreme Court banned patenting genes. But Congress might change that*, WASH. POST, June 3, 2019, https://www.washingtonpost.com/news/powerpost/paloma/the-health-202/2019/06/03/the-health-202-the-supreme-court-banned-patenting-genes-but-congress-might-change-that/5cf1987f1ad2e52231e8e91b/?utm_term=.1c4cf91f27c1.

and "unjustly enrich patent claimants beyond the scope of human creativity."[79] It is uncertain whether the work of excluding natural products and processes can be done by other patent doctrines, like novelty or the constitutional limitation of patents to "the useful arts."

As the world continues to debate the policy, ethical, and social-justice ramifications of life patents, religious insights hold the potential to enrich that discussion – although they should not and will not dominate it. We hope this collection will deepen lawyers' and policy makers' understanding of these religious insights – and deepen the religious insights as well, by increasing religious leaders' understanding of patent law and policy.

[79] Testimony of Joshua D. Sarnoff, Senate Committee on the Judiciary, Subcommittee on Intellectual Property Hearing, *The State of Patent Eligibility in America: Part I* (June 4, 2019), at 5–7, https://www.judiciary.senate.gov/imo/media/doc/Sarnoff%20Testimony.pdf. The full hearing and witnesses' written submissions are at https://www.judiciary.senate.gov/meetings/the-state-of-patent-eligibility-in-america-part-i.

# Index